A PRACTICAL GUIDE TO GROUNDWATER AND SOLUTE TRANSPORT MODELING

A PRACTICAL GUIDE TO GROUNDWATER AND SOLUTE TRANSPORT MODELING

Karlheinz Spitz
Dames & Moore, Jakarta, Indonesia

Joanna Moreno
Dames & Moore, Denver, Colorado

A WILEY-INTERSCIENCE PUBLICATION

JOHN WILEY & SONS, INC.

New York • Chichester • Brisbane • Toronto • Singapore

This text is printed on acid-free paper.

Library of Congress Cataloging-in-Publication Data:

Spitz, Karlheinz, 1955-
A practical guide to groundwater and solute transport modeling / Karlheinz Spitz, Joanna Moreno.
p. cm.
ISBN 0-471-13687-5 (cloth : alk. paper)
1. Groundwater flow—Mathematical models. I. Moreno, Joanna. II. Title.
TC176.S65 1996
551.49'01'5118—dc20 95-35407

10 9 8 7 6 5 4

To our families
and the memory of Steve Moreno

CONTENTS

SYMBOL REFERENCES

a	[L]	Distance between plates
A	[L^2]	Area
c	[T^{-1}]	Leakance of a leaky layer
c	[M/L^3]	Concentration
c_a	[M/L^3]	Adsorbed concentration
c_B	[M/L^3]	Concentration of a water solute in boundary flux
c_L	[M/L^3]	Concentration of a water solute in leakage flux
c_N	[M/L^3]	Concentration of a water solute in natural groundwater recharge
c_o	[M/L^3]	Initial or source concentration
C_O	[1]	Courant number
c_R	[M/L^3]	Concentration of a water solute in groundwater recharge from surface water
c_s	[M/L^3]	Concentration on soil
c_W	[M/L^3]	Concentration of a water solute in well recharge
c_w	[M/L^3]	Concentration in groundwater
d	[L]	Thickness of surface water bed
d	[L]	Thickness of aquitard
D	[L^2/T]	Dispersion coefficient
D_L	[L^2/T]	Longitudinal dispersion coefficient
D_o	[L^2/T]	Diffusion coefficient
D_T	[L^2/T]	Transverse dispersion coefficient
$erf(x)$	[1]	Error function
$erfc(x)$	[1]	Complimentary error function
E	[volts]	Electrical voltage
f_{oc}	[1]	Organic carbon weight fraction
g	[L/T^2]	Acceleration due to gravity
$\hat{h}$	[L]	Approximate head
h	[L]	Head

h_B	[L]	Elevation of the surface water bed
h_o	[L]	Elevation of the water level prior to pumping
h_R	[L]	Elevation of the surface water
H	[L]	Surface water elevation
i	[1]	Hydraulic gradient
I	[amperes]	Electrical current
k	[L^2]	Permeability of the porous medium
K	[L/T]	Hydraulic conductivity
K	[1]	Modified Bessel function of the second kind and of zero order
K_d	[L^3/M]	Distribution coefficient
K_H	[atm.L^3/mol]	Henry's law constant
K_{oc}	[L^3/M]	Octanol-water partition coefficient
K_{ow}	[L^3/M]	Octanol-water partition coefficient
k_1	[L^3/ohms]	Quantity of water/resistance
k_2	[L/volts]	Head/voltage
k_3	[L/T. amperes]	Flux/current
k_4	[1]	$\text{Time}_n/\text{time}_m$
L	[L]	Length
m	[1]	Subscript for model
m	[L]	Saturated aquifer thickness
M	[M]	Mass
M_s	[M]	Mass of contaminant on soil
M_w	[M]	Mass of contaminant in groundwater
n	[1]	Number of values
n	[1]	Subscript for natural groundwater system
n	[1]	Total porosity
N	[L/T]	Groundwater recharge per unit area
n_e	[1]	Effective porosity
p	[M/LT^2]	Pressure
p	[M/LT^2]	Vapor pressure
Pe	[1]	Peclet number
q	[L/T]	Darcy flux
q	[L/T]	Specific flux
Q	[L^3/T]	Local sources and sinks
Q	[L^3/T]	Well discharge or recharge
Q	[L^3/T]	Total flux
q_B	[L^2/T]	Boundary flux per unit length of boundary

Q_B	[L^3/T]	Subsurface flow over boundary
q_L	[L/T]	Leakage flux per unit area
Q_L	[L^3/T]	Leakage component coming from adjacent aquifers
Q_N	[L^3/T]	Natural groundwater recharge and discharge
q_o	[L/T]	Uniform groundwater flow
q_o	[L/T]	Diffusive flux
q_R	[L]	Surface water infiltration or exfiltration rate per unit area
Q_R	[L^3/T]	Exchange with surface water
Q_S	[L^3/T]	Groundwater in storage
Q_W	[L^3/T]	Local infiltration or exfiltration
r	[1]	Subscript for the natural-model aquifer ratio
r	[L]	Radial distance from the coordinate origin
R	[ohm-L]	Electrical resistance
R	[1]	Retardation factor
s	[L]	Drawdown
s	[L]	Distance
S_s	[L^{-1}]	Specific storage
S	[1]	Storativity
S	[M/L^3]	Solubility
S	[M/L^{3}T]	Source or sink of contaminant
t	[T]	Time
$\bar{t}$	[T]	Average replacement time
T	[L^2/T]	Transmissivity
T_m	[M/L]	Uniform membrane tension
$T_{1/2}$	[T]	Half-life
v	[L/T]	Advective transport velocity
V	[L^3]	Aquifer volume
$v_{x,y}$	[L/T]	Flow velocity in x,y-direction
W_m	[M/L^2]	Weight of the membrane per unit area
$W(u)$	[1]	Well function
$W(y,r/\lambda)$	[1]	Function similar to the well function, but with two parameters in the integral
x	[L]	Space coordinate in x-direction
y	[L]	Space coordinate in y-direction
z	[L]	Space coordinate in z-direction

z	[L]	Elevation of membrane surface
$Z_{1,2}$	[1]	Random variable
α	[L]	Dispersivity
α	[T^{-1}]	Coefficient
α	[degrees]	Angle between base flow direction and x-axis
α_L	[L]	Longitudinal dispersivity
α_T	[L]	Transverse dispersivity
β	[degrees]	Angle between x-axis and line connecting well with location
θ	[deg. Kelvin]	Temperature
θ	[1]	Volumetric moisture content
θ	[1]	Weighting factor
σ	[Ohm/L]	Electrical conductivity
σ	[—]	Standard deviation

ACKNOWLEDGMENTS

This guide is influenced by the ideas, comments, and views of our teachers, colleagues, and students. Many thanks are due to all these professionals. We especially want to thank our former teachers, Professor Helmut Kobus, Professor Emil O. Frind, Dr. Bernard Johns, and Dr. Devraj Sharma for introducing us to the hydraulics of groundwater and to modeling. Special thanks are further due to Professor Wolfgang Kinzelbach for having shared his knowledge on random walk models.

It is with pleasure that we acknowledge the indirect help of our students at various institutions, in particular at the Directorate of Environmental Geology, Bandung, Indonesia, at the Geological Survey of Malaysia, Ipoh, and at Dames & Moore. From our experience of teaching groundwater modeling at these places, we gained many ideas for presenting the concepts of practical groundwater modeling.

We appreciate the support given to us by Peter Sinton and Diane Germain in reviewing our manuscript.

Thanks are extended to Theresa Daus-Weber for editing our manuscript and improving our English, Frank Hanstein for his excellent computer drafting, and Sue Taylor for her excellent typing of the manuscript.

Finally, we recognize a great personal debt to our families, Lucy and Clarissa, and Peter, Ross, Malcolm, Amanda, and Natalie. The time we spent writing this guide was at the expense of time spent with our families.

INTRODUCTION

Numerical models, because they provide the most versatile approach to groundwater problems, have outclassed all other types of groundwater models. The well-earned popularity of numerical models, however, may lead to overestimation of their potential. Groundwater systems are complicated beyond our capability to evaluate them in detail.

To grasp the groundwater problem under investigation, assumptions and simplifications are required. A model, no matter how sophisticated, will never describe the groundwater system under investigation without deviation of model simulations from the actual physical processes. As a consequence, in applying a numerical model to a field study, the model user should always understand the implications of simplifying assumptions.

Numerical models are a tool to solve problems but not the solution itself. Numerical models do not give accurate answers to insufficiently defined problems. Neither do they magically release model users from their duty to carefully study the groundwater system. Hence numerical modeling is not primarily a question of understanding the numerical techniques but of understanding model application.

This book is designed as a nonmathematical approach to numerical groundwater modeling. It is written for model users interested in applying models to groundwater problems but avoids the underlying numerics. Thus, the book provides an understanding of model application, and not of model theory. In applying groundwater models, the model user must have a good understanding of geology, hydrogeology, and groundwater hydraulics.

Although understanding theory is helpful in groundwater modeling, theory does not necessarily help to apply a computer program to a field case. In the application of models, model users usually face problems originating from model use rather than from model theory. They encounter the problem of defining a suitable model area and its boundaries, of compiling model input data, and of calibrating a model. They learn the difficulties of predicting the impact of changes in the groundwater system, even though they may believe that the model works. Model users will struggle through these problems in each model study with little support from the literature. This book is intended to fill this gap. The book represents many years of practical modeling experience, and years of teaching modeling.

The arrangement of the text reflects the fact that an understanding of basic principles of geology, hydrogeology, and groundwater hydraulics is essential in groundwater modeling. This leads to the following structure of the book.

Chapter 1, "Understanding Groundwater Modeling," describes the basic methods and concepts of groundwater modeling, with reference to various types of models. In some cases knowledge of models using physical analogues is useful because the model user can imagine the effect of mathematical modeling in physical terms.

Chapter 2, "Reviewing the Mathematical Model for Flow," deals with the physics of groundwater movement, summarizing the mathematical model describing groundwater flow from the viewpoint of the model user. The focus is on the practical aspects of flow problems, such as how to conceptualize groundwater systems or on how to define suitable boundary conditions.

Chapter 3, "Reviewing the Mathematical Model for Transport," summarizes the dominant mechanisms of solute transport and their mathematical formulation from the viewpoint of the model user. The chapter is presented for the convenience of a model user seeking background information on solute transport issues.

Chapter 4, "Using Analytical Models for Idealized Flow and Transport Problems," takes advantage of the reader's familiarity with analytical solutions as an introduction to the more demanding numerical modeling. This chapter is particularly useful for the inexperienced model user. This chapter uses analogies between analytical and numerical models as an aid to understanding key issues in modeling such as developing the conceptual model or calibrating the numerical model. It also demonstrates that for many practical problems, the model user may find a solution without applying sophisticated numerical models.

Chapter 5, "Understanding Numerical Flow Models," provides introductory material on groundwater flow models and mathematics. Emphasis is on understanding the potential and limitations of numerical flow models. A simple finite difference model is developed to illustrate the differences between applying a numerical model and developing one.

Chapter 6, "Understanding Numerical Transport Models," reviews finite difference, finite element, and random walk models, and models based on the method of characteristics. The aim is to give sufficient insight into the underlying theory of each model type for the model user to appreciate the advantages of each model and to realize its limitations.

Chapter 7, "Dimensionality of Flow and Transport Problems," refers to numerous case studies to demonstrate how and when to reduce the number of dimensions considered in a model study. This is one of the most vital aspects of modeling. Three-dimensional model studies are restricted to a few field investigations, due to the significant modeling effort and the large amount of model data input required.

Chapter 8, "Applying Numerical Models," discusses all aspects of modeling analysis from compiling data to auditing models. Numerous references to selected field studies are made to demonstrate practical model application. Emphasis is on practical, everyday uses of models rather than experimental or research-oriented approaches to modeling analyses.

Chapter 9, "Special Topics," refers to some aspects of modeling that go beyond most model applications but are of immediate concern to the model user.

The chapters and sections are as independent from each other as possible. Thus the reader experienced in hydrogeology and contamination issues could skip Chapters 2 and 3 and understand the discussion in the following chapters. Similarly the experienced model user could focus on Chapters 7 to 9, only briefly reviewing previous chapters.

To ensure undisturbed flow of the text and the presentation of the field studies, a series of representative case studies is included in Appendix A. The case studies reflect a wide range of model applications and illustrate topics to include in a modeling report.

Published data on hydrogeological parameters and on pollutant characteristics are listed in Appendixes B, C, and D. Even though such a list could never be complete, it can serve as a reference if site-specific field data are not immediately available to the model user.

Although it is impossible to present in detail all of the current knowledge and experience in groundwater modeling, this book guides model users in the application of numerical models. Such applications include the following:

- In cases where considerable quantities of geological and hydrological data exist, a numerical model is helpful in summarizing, interpreting, and presenting available data. Unlike other models, it combines all available field data and, assuming careful application, allows the estimation of missing data. In this way models can increase the utility of available data.
- In cases where site-specific data are sparse, a conceptual model may be developed to examine the validity of different working hypotheses on site conditions. Different hypothetical groundwater flow conditions can be assessed and compared.
- In the planning of proposed future activities, numerical models are best suited to compare responses of the flow system to alternate activities. In cases where relative rather than absolute values are investigated, the numerical model is both appropriate and versatile.
- In the estimation of the flow system performance under groundwater recharge and discharge, numerical models can be used to analyze flow systems without too much difficulty. In water management problems, numerical models find wide application in predicting changes to the watertable due to human activity. In solute transport problems, numerical models may be used to estimate the present extent of contamination as well as its historical evolution. They are well suited to evaluate monitoring networks or alternate remediation measures.
- By means of a sensitivity analysis, numerical models may be used to highlight the site parameters that are most influential to the flow or

transport conditions. They also help in the understanding of inadequacies in the existing database.

- Numerical models are widely applied to study various specific groundwater problems such as saltwater intrusion, heat transport, land subsidence, displacement of immiscible fluids, or solute transport in fractured rock.
- Once a working model has been prepared it also can be used to support groundwater management decisions such as apportioning liability for contamination, estimating liability based on cleanup costs, and focusing negotiations with regulatory agencies or the public.

Since model predictions may be used to optimize future investigations or operations, it can be argued that the expense of modeling efforts is actually an overall cost saving.

1 Understanding Groundwater Modeling

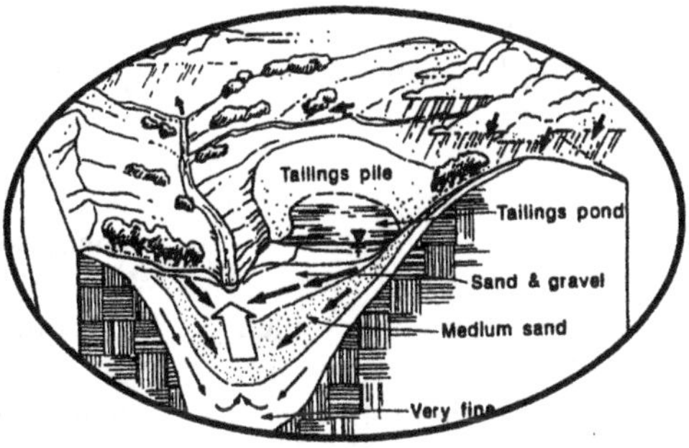

In most groundwater investigations it is not practical to monitor all aspects of the groundwater flow and solute distribution. Information between and beyond monitoring locations and in the future are needed to understand the site and make informed decisions. Groundwater models, which replicate the processes of interest at the site, can be used to complement monitoring and laboratory bench-scale studies in evaluating and forecasting groundwater flow and transport. However, every reliable model is based on accurate field data.

In designing a groundwater model, the model user combines numerous modeling components. These components, listed in Table 1.1, are:

- Natural system for which the model is designed.
- Conceptual model as an idealized representation of the natural system.
- Mathematical model representing controlling mechanisms in mathematical terms.
- Solution of the mathematical model.
- Calibration of the solution by adjusting simulated to observed responses of the natural system.
- Validation of the accuracy of the model predictions.
- Simulations based on the calibrated solution of the conceptual model.

As the table indicates, the term "model" has different meanings. "Groundwater model" usually stands for the combination of all model components, but the term "model" is also used in the context of the various solution methods.

TABLE 1.1 Modeling Components

Component	Key Elements	Examples
Natural System	Geometry	Lateral extent, thickness, source volume
	Dimensionality	One-, two-, three-dimensional
	State	Transient, steady
	Hydrogeology	
	Material properties	Porosity, hydraulic conductivity, dispersivity, storativity, chemical properties
	Observed responses	Water level, concentration
	Groundwater problem	Extraction, contamination
Conceptual Model	Idealized system	
	Relevant units	Aquifer, aquitard, aquiclude
	Boundary and initial conditions	Initial condition
	Controlling processes	Flow, capillarity, gravity, transport, chemical reactions
Mathematical Model	Physical laws	Conservation of mass Conservation of energy Equilibrium of forces Constitutive relationships Material relationships
	Differential equations	Laplace equation
	Boundary conditions	First-kind condition Second-kind condition Third-kind condition
	Initial conditions	Specified head or concentration
Solution	Analytical model	
	Porous media (bench-scale) model	
	Analog model	Viscous fluid model Membrane model Electrical analog model
	Empirical model	
	Mass balance (single-cell) model	
	Numerical model	Finite-difference model Finite-element model Random walk model Method of characteristics Boundary element method
Calibration	Solution versus observation	

TABLE 1.1 ***(Continued)***

Component	Key Elements	Examples
	Adjustment of model input data	
Validation	Testing of model predictions versus observations not used in calibration	
Simulations	Parameter sensitivity Predictive simulations Analysis of uncertainty	

Section 1.1 reviews different solutions, or models. The value of the model review lies in fostering the understanding of numerical modeling. Numerical models, being the most versatile approach to complex groundwater systems, have outclassed all other models. Section 1.2 introduces the main phases in numerical model design stipulated by the modeling components.

1.1 REVIEWING GROUNDWATER MODELS

The role of groundwater models in the study of groundwater flow and transport has long been a topic of interest for earth scientists. Among the numerous models developed, we will consider these eight:

- Analytical models
- Porous media models
- Viscous fluid models
- Membrane models
- Electrical analog models
- Empirical models
- Mass balance models
- Numerical models.

Advantages and disadvantages of each model are discussed briefly in the following subsections.

1.1.1 Using Analytical Models

The use of traditional analytical solutions is restricted, due to the rigorous simplification of the real world that is required. Whenever analytical solutions for the investigated groundwater problem exist, however, they are in general more efficient than other model types. The influence of individual parameters can be studied with little effort, and the method is often simple and time efficient. Traditionally analytical solutions were widely used in the analysis of pump tests (Kruseman and de Ridder 1970). Analytical solutions also find wide application in describing two-dimensional steady-state flow in a homogeneous flow system. Model users benefit from the fact that the Laplace equation, the governing flow equation, is the most studied differential equation in mathematics and physics. Solutions derived for problems in other engineering disciplines that follow the same differential equation can be easily adopted for the investigated groundwater problem. This holds true for the analogy between heat and groundwater flow. Numerous helpful analytical solutions addressing heat transport in solids are combined in Carslaw and Jaeger (1959). For solutions of the groundwater flow equation, refer to Polubarinova-Kochina (1952) or Strack (1989).

In transport problems, analytical solutions often become so complex and unwieldy that the advantages of the analytical approach, over a simple numerical solution, shrink. Examples of relatively simple analytical solutions that enjoy broad application are the one-dimensional solution of longitudinal transport of Ogata and Banks (1961) and the one-dimensional solution for transverse spreading of Harleman and Rumer (1963). Other helpful solutions are combined in Bear (1972) and Freeze and Cherry (1979).

1.1.2 Using Porous Media Models

Porous media or bench-scale models belong to the group of hydraulic models widely used in hydraulic engineering. The groundwater system is represented in an appropriate scale in the laboratory with its boundary included, as shown for one-dimensional flow in Figure 1.1. Figure 1.1 also gives examples of the viscous fluid, the membrane, the electrical analog, the mass balance, and the numerical model discussed in the following subsections. In the porous media model, the hydrogeological properties are distributed in space and magnitude according to the natural system. The porous media model is then manipulated and the flow system responses recorded, yielding insight into the behavior of the real system. When studying unconfined groundwater flow, corrections for the capillary rise are necessary, since it is disproportionately large in the model compared to field conditions. Streamlines can be produced by injecting dye.

Similarities between the natural and model system are defined by

Geometry (linear scale ratio L_r):

$$L_r = \frac{L_n}{L_m} \tag{1.1}$$

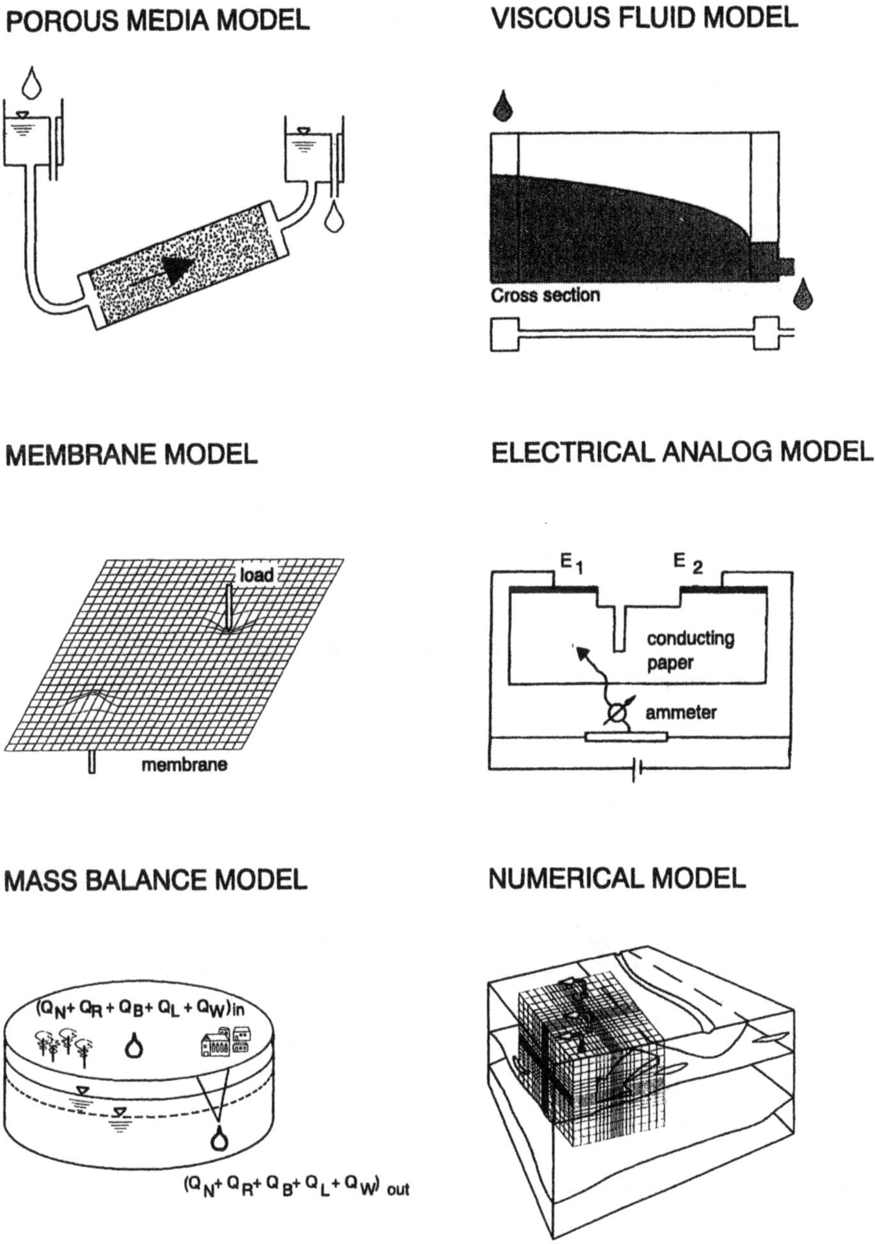

Figure 1.1 Examples of groundwater models.

Area (A_r):

$$A_r = L_r^2 \tag{1.2}$$

Specific discharge (Darcy flux q_r):

$$q_r = K_r i_r \tag{1.3}$$

Total flux (Q_r):

$$Q_r = K_r i_r L_r^2 \tag{1.4}$$

Time (t_r):

$$t_r = \frac{n_e L_r}{K_r i_r} \tag{1.5}$$

where

n = subscript for natural groundwater system
m = subscript for model groundwater system
r = subscript for the nature-model ratio
L = length in [L]
K = hydraulic conductivity in [L/T]
i = hydraulic gradient in [1]
n_e = effective porosity in [1]

The studies of Darcy (1856) can be regarded as the first groundwater model study. Porous media models such as sand models are still used extensively in research. They allow the study of special aspects of groundwater flow and transport under almost natural conditions. Large-scale physical models are used in practical application to evaluate particular flow or transport problems, such as three-dimensional flow toward monitoring wells in complex geological settings or multiphase transport. Otherwise, porous media models are used today only as a demonstration tool for students in teaching. In the study of field problems, porous media models have largely been replaced by numerical models.

1.1.3 Using Viscous Fluid Models

Hele-Shaw or parallel plate models are synonyms for the viscous fluid model. Figure 1.1 shows a Hele-Shaw model in its simplest form. The Hele-Shaw model makes use of the analog of the movement of a viscous fluid, such as glycerine between two closely spaced parallel plates, and two-dimensional groundwater flow. A similar analogy is Ohm's law in an electrical system and Fourier's law in a thermal system. The most prominent equation of this type is Newton's law of motion. Figure 1.2 presents these and other points of similarity in analog models.

The specific flux q in the viscous fluid model is given by Poiseuille's equation:

$$q_m = -\frac{ga^2}{12\nu_m} i_m \tag{1.6}$$

Equation 1.6 is analogous to Darcy's law (see Section 2.2), describing groundwater flow:

$$q_n = -\frac{\rho_n g}{\mu_n} k i_n = -\frac{g}{\nu_n} k i_n \tag{1.7}$$

where

n = subscript for groundwater system
m = subscript for viscous fluid system

Model Task \ Model Type	Porous Media Model	Numerical Model	Viscous Fluid Model	Electrical Analog Model: Electrolytic	Electrical Analog Model: RC-Network	Membrane Model
DIMENSIONALITY						
two-dimensional	●	●	●	●	●	●
three-dimensional	●	●	○	●	●	○
FLOW PROBLEM						
steady	●	●	●	●	●	●
unsteady	●	●	●	○	●	○
phreatic	●	●	◑	◑	○	○
anisotropy	●	●	●	●	●	●
heterogeneity	●	●	●	●	●	○
variably saturated	●	●	◑	○	○	○
TRANSPORT PROBLEM						
stream-/pathlines	●	●	●	○	○	○
advection	●	●	●	○	○	○
dispersion	●	●	○	○	○	○
sorption	●	●	○	○	○	○
decay/reactions	●	●	○	○	○	○

● Yes ◑ With certain constraints ○ No

Figure 1.2 Applicability of models and analogs (after Bear 1972).

g = acceleration due to gravity in [L/T^2]
ρ = density in [M/L^3]
μ = dynamic viscosity in [M/LT]
ν = kinematic viscosity in [L^2/T]
i = hydraulic gradient in [1]
k = permeability of porous medium in [L^2]
a = distance between plates in [L]

The corresponding scaling factors are

Geometry (linear scale ratio L_r):

$$L_r = \frac{L_n}{L_m} \tag{1.8}$$

Flux (q):

$$q_r = \frac{12k}{\nu_r a^2} i_r \tag{1.9}$$

Time (t):

$$t_r = \frac{L_r}{q_r} \tag{1.10}$$

where r represents the subscript for nature-model ratio. Thus the plate spacing and fluid can be selected to correspond to a required permeability.

Although the viscous fluid model can be of a horizontal or vertical type, the latter has found greater application. Hele-Shaw models have been introduced to study flow and seepage through dams, saltwater intrusion, and other phenomena. They have been applied by Schwille (1988) to model groundwater flow and transport in fractured rock. One major advantage of the Hele-Shaw method is that it solves problems concerning the phreatic surface for steady and transient flow situations. The water table can be observed and path lines produced if dye is added to the viscous liquid at discrete points. Permeability represented by viscous plate models is isotropic, but local variation of permeability can be achieved either by varying the width of the interspace or by placing obstructions between the plates. Storage can be simulated by connecting small storage reservoirs to the interspace. Numerical models have decreased the importance of Hele-Shaw models. However, since the method is very illustrative, Hele-Shaw models are still used for demonstration purposes.

1.1.4 Using Membrane Models

Membrane models were used in engineering laboratories before computers were available. A membrane model consists of a mechanism for depressing a membrane covering a frame, and a device to measure deformations precisely. For example, to represent a single well, a nail can be applied to depress the membrane at a discrete point. The strength of the well can be adjusted by pulling in on the nail. For a pumping well the top of the membrane is depressed, while for an injection well the nail is placed beneath the membrane and is pushed upward as shown in Figure 1.1. Sinks and line or area sources can be represented by depressing the membrane by line or area loads. The deformation of the membrane is then directly related to known equations from groundwater hydraulics. V. E. Hansen (1949) describes the analogy of the deformation of the water table due to water extraction from wells and the deflection of thin membranes due to vertical displacement at discrete point. The equation describing the deflection of a thin stretched membrane is

$$\frac{d^2z}{dr^2} + \frac{1}{r}\frac{dz}{dr} = -\frac{W_m}{T_m} \tag{1.11}$$

where

z = elevation of membrane surface in [L]
r = radial distance from the coordinate origin in [L]
W_m = weight of the membrane per unit area in [M/L^2]
T_m = uniform membrane tension in [M/L]

The equation describing steady, axially symmetrical flow in a homogeneous, isotropic groundwater system is

$$\frac{d^2h}{dr^2} + \frac{1}{r}\frac{dh}{dr} = -\frac{N}{Kh} \tag{1.12}$$

where

h = water-table elevation in [L]
N = groundwater recharge per unit area in [L/T]
K = hydraulic conductivity in [L/T]

The analogy is also true for groundwater flow without natural recharge if the weight of the membrane is small. Today the value of membrane models lies more in the illustration of the deflection of the water table in the vicinity of wells rather than in the solution of real field studies. It is, however, an inexpensive tool to use to visualize groundwater stress, helping the inexperienced earth scientist to understand groundwater hydraulics.

1.1.5 Using Electrical Analog Models

Among the analog models, those types based on the analogy between electrical and groundwater flow were very popular before the dominance of numerical models. The electrical conductive medium in which flow is studied can be conducting paper, an electrolytic tank, or a resistance capacitance (RC) network. Conducting paper is a simple tool for simulating two-dimensional groundwater flow. The paper is cut into the form of the groundwater system under investigation. A simple model to simulate groundwater flow underneath a hydraulic structure with a cutoff wall is shown in Figure 1.1. Free edges of the conducting paper represent the impermeable boundaries. Equivalent voltage E is imposed on boundaries with prescribed heads by clamping copper strips with controlled voltage to the relevant edges of the conducting paper. Lines of constant potential drop represent lines of constant heads and can easily be measured. Regions with different hydraulic conductivities are represented by different types of conducting paper or by perforating the paper. In RC networks the groundwater system is represented by electrical elements. Each element represents a specific volume of the groundwater system. The elements are connected at nodes and even complex three-dimensional systems can be reproduced. Voltage measuring devices assess the voltage distribution within the network, which again reflects the head distribution in the simulated groundwater system.

According to Walton (1970) four scale factors show the relation of electrical units associated with the analog model to hydraulic units associated with the groundwater system. These scaling factors are given by Walton (1970) as

$$k_1 = \frac{\text{quantity of water (gallons)}}{\text{resistance (coulombs)}} \tag{1.13}$$

$$k_2 = \frac{\text{potentiometric head (feet)}}{\text{voltage (volts)}} \tag{1.14}$$

$$k_3 = \frac{\text{flux (gallons per day)}}{\text{current (amperes)}} \tag{1.15}$$

$$k_4 = \frac{\text{time}_n \text{ (days)}}{\text{time}_m \text{ (seconds)}} \tag{1.16}$$

with

$$k_1 = k_3 k_4 \tag{1.17}$$

The relation between hydraulic conductivity K and electrical conductivity σ—which is equivalent to the reciprocal of electrical resistance R—can be calculated by substituting Ohm's law and Darcy's law in these four scaling factors. This relationship calculation is written as

$$\sigma = \frac{1}{R} = \frac{k_2}{k_3} K \tag{1.18}$$

For discontinuous electrical analog models (network), the hydraulic conductivity in Equation 1.18 is replaced by the transmissivity of the volume of the groundwater system, represented by electrical resistance in the model.

1.1.6 Using Empirical Models

Also referred to as *lumped parameter models*, empirical models are used to represent physical or chemical processes by generalities, simplifications, or at a scale larger than the process itself. These models fill a useful gap between the simpler analytical models and the more sophisticated numerical models. Empirical models are of two types: models representing individual processes or mechanisms and models representing an entire groundwater problem.

Examples of the first type of model include Darcy's law described in Section 2.2, Fick's law described in Section 3.2.3, and adsorption isotherms described in Section 3.2.4. These empirical models are embedded in analytical and numerical models and impact the accuracy of the model predictions. They are used when detailed site-specific data are lacking, or when it is impractical to simulate fine-scale processes. This type of empirical model indicates a lack of understanding of the processes involved and is a temporary solution to aid analysis.

Empirical models representing an entire groundwater problem invoke a series of physical laws, empirical laws, and conservative assumptions to represent a flowpath of interest. Examples of such models include the organic leachate model (51 Fed. Reg. 21,653, 1986), the landfill analysis model HELP (Schroeder 1994), and the exposure assessment model, MULTIMED (USEPA). These models can be misused or misunderstood because they are easy to use; they are applicable only in limited circumstances, and they mask their limitations by lumping processes together.

1.1.7 Using Mass-Balance Models

The mass-balance model, also known as the black box or single-cell model, is regarded as a numerical model in its simplest form. Mass fluxes, either of groundwater (water balance) or of some constituent (solute balance) are balanced over large volumes as schematically shown in Figure 1.1. For illustration, consider a groundwater basin of horizontal area A, bounded by impermeable boundaries. Groundwater recharge is from natural replenishment N, and groundwater discharge is from pumping wells with a total discharge rate Q. The difference of total N and Q for a given time period causes either an increase or decrease of the average water level in the cell. No variations of groundwater levels within the groundwater system are calculated except the average water level. Due to the simplicity of black box models, evaluation of field data is only concerned with fluxes in and out of the system. Except for storativity, no other feature of the groundwater system needs to be considered. Therefore averaging over the entire area is a crude approximation, especially for solute balances. It means, for example, that as various sources contribute solutes to the groundwater system, complete mixing of the solute within the entire system takes place.

Despite its simplicity, the black box model is useful, since it leads to an examination of the global mass balance. In numerical modeling, single-cell models are best fitted to complement modeling efforts, providing a comparison of mass balances in the early stages of the calculations. Single-cell models are discussed in Chapters 5 and 6.

1.1.8 Using Numerical Models

All groundwater models discussed in the previous subsections are strongly restricted in their scope of application, as shown in Table 1.2 which compares the model tasks and model types. With the exception of the analytical solution methods and possibly the conducting paper models, the discussed groundwater models are not easily applied. For example, constructing an RC network for a given scenario is time-consuming and the hardware model is voluminous. Then comes the tedious simulation of different scenarios that further requires extensive experience in interpreting the results.

The simulation of groundwater flow and transport by numerical models is a relatively recent development dating from the early 1970s. Today numerical models dominate the study of complex groundwater problems. Numerical models basically represent an assembly of many single-cell models. Tremendous advances in computer technology have made them the standard procedure for the solution of groundwater flow and mass transport problems. Computer programs for most common flow and transport problems are available and the model user can apply a relevant computer program to an investigated scenario without writing any computer code. The numerical model solves both simple and complex problems. Theoretically numerical models impose no restrictions on the boundary type, the initial conditions, the characteristics of the groundwater system, or the characteristics of the investigated solute. Once the numerical model is completed, various scenarios can be realized without undue effort. The dominance of the numerical models has led to the use of "groundwater model" as a synonym for numerical groundwater models.

TABLE 1.2 Analogy in Physics

Physical Process		Law	Conservation Law	Quantity	Potential	Proportionality Factor
Groundwater flow	Darcy	$q = -K \nabla h$	$\nabla^2 h = 0$	Darcy flux q	Potentiometric head h	Hydraulic conductivity K
Viscous fluid flow	Poiseuille	$v = -f_r \nabla h$	$\nabla^2 h = 0$	Velocity v	Potentiometric head h	Conductivity of fracture f_r
Electricity flow	Ohm	$I = -\sigma E$	$\nabla^2 E = 0$	Current I	Voltage E	Electrical conductivity σ
Heat flow	Fourier	$Q_\theta = -\lambda \nabla \theta$	$\nabla^2 \theta = 0$	Heat flow Q_θ	Temperature θ	Thermal conductivity λ
Force field	Newton	$f = m \nabla U$	$\nabla^2 U = 0$	Force f	Potential U	Mass m
Diffusion	Fick	$q_0 = -D_0 \nabla c$	$\nabla^2 c = 0$	Diffusive flux q_0	Concentration c	Diffusion coefficient D_0
Incompressible flow of a frictionless fluid		$v = -\nabla \phi$	$\nabla^2 \phi = 0$	Velocity v	Velocity potential ϕ	1

In the following text we will refer to numerical groundwater modeling. The term "groundwater model," is used equivalently to the term "numerical model," which is the combination of the mathematical description, the numerical computer code, and its application to the specific groundwater problem. The same meaning is accepted when we discuss "groundwater modeling."

The numerical computer code is a tool for solving the governing equation of flow or transport. The numerical computer code is transformed into a groundwater model by incorporating the site-specific geometry and boundary conditions, by introducing the actual flow and transport parameters, and by calibrating and verifying the model.

1.2 DESIGNING NUMERICAL MODELS

Alternate numerical theories for solving partial differential equations lead to different types of numerical models, but there are no fundamental differences in the overall model approach to a given problem. The modeling components listed in Table 1.1 result in the following main phases in model design:

- Compiling and interpreting field data.
- Understanding the natural system.
- Conceptualizing the groundwater system.
- Selecting the numerical model.
- Calibrating and validating the model.
- Applying the model.
- Presenting the results.

This section provides some introductory information on each phase in the design of numerical models. Details are discussed in the following chapters.

1.2.1 Compiling and Interpreting Field Data

Field data are essential to understand the natural system, to specify the investigated groundwater problem, to facilitate selection of computer code, and to derive model input data. The computer code of a numerical model is a tool for solving the governing flow or transport equation. It can be used for any model study as long as the flow or transport follows the mathematical model approximated by the computer code. The numerical model actually develops into a site-specific groundwater model when real field parameters are assigned. The accuracy of the modeling results is not primarily a question of the sophistication of numerical code, long calculation time, fine discretization, or large memory requirements. The quality of the simulations depends in large part on the validity of the model physics and on the quality of the input data. The gigo rule, "garbage in, garbage out," also applies to groundwater modeling.

Generally, field data do not directly provide the parameters required for the model such as transmissivity or groundwater recharge per model segment. Model parameters must be derived from field data. Consider a pumping test. The measured field data are well discharge, time, and drawdown. The calculated model parameters are transmissivity and storage coefficient. In addition the same regional parameters must be assigned to each model segment. The time needed to perform a model study will depend significantly on the time required to collect and prepare the model input data.

It is wrong in many groundwater studies to first evaluate a huge volume of field data and then to design the groundwater model. Due to its strength in combining field data, a groundwater model will help to guide the evaluation of new meaningful data.

1.2.2 Understanding the Natural System

To ensure accurate modeling, the model user should make an effort to gain an appreciation of the natural system. This is the next phase in model design. Distribution of geological parameters and boundary conditions are identified as shown in Figure 1.3. Having a clear definition of the flow or transport problem is important. In only a few investigations will the model user apply a general groundwater model to simulate flow or transport in detail. More often the simplest, most appropriate approach is acceptable, depending on the problem to be solved. Stating the problem properly is often half the battle. Quite often, this will allow the model user to accomplish the analysis of a complex groundwater system with individual solutions for well-defined tasks. These tasks may then be solved independently assuming simplified configurations (two-dimensional instead of three-dimensional, analytical instead of numerical, etc.). Solutions to groundwater problems do not necessarily require the most sophisticated model. In each case the most appropriate model is the one that addresses the investigated problem with as little effort as necessary to represent the real system. The model should be simple enough to facilitate model efforts but not too simple so as to exclude features dominant to the investigated groundwater problem. In summary, the natural system must be well understood to design the best-fitted model in view of needs, cost, and the availability of data to develop and calibrate the model.

1.2.3 Conceptualizing the Natural System

In each model study the natural system is represented by a conceptual model, as shown schematically in Figure 1.3, for which an approximate solution is applied. To design and construct equivalent but simplified conditions, extensive information is required on the natural system. Transferring the real world

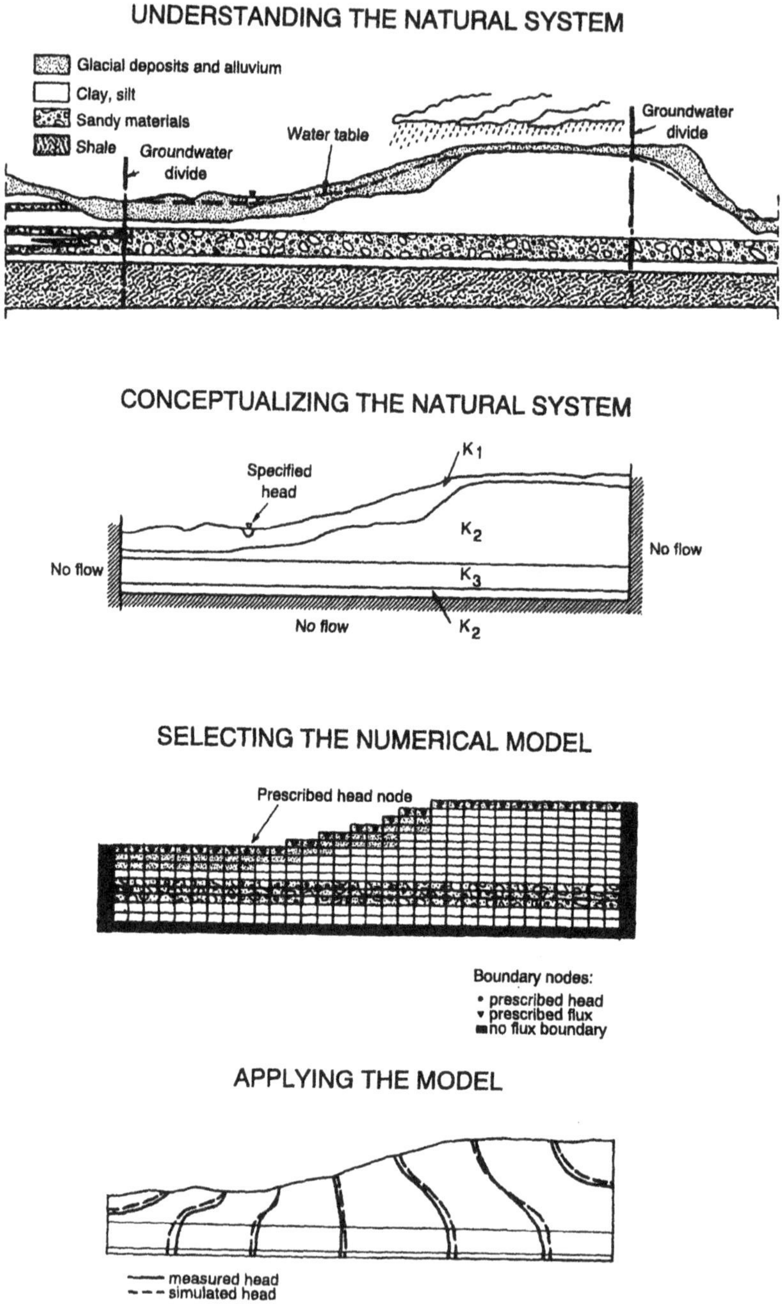

Figure 1.3 Examples of some of the main phases in model design.

into an equivalent model system, which can then be solved using existing program codes, is a crucial step in groundwater modeling. Errors in the conceptual model cannot be corrected during the model calibration or at any later stage of the model study without major revisions.

To illustrate the importance of an appropriate conceptual model, consider a flow problem with significant three-dimensional flow components such as the flow in the vicinity of a partially penetrating well. In approximating the flow field close to the well with a depth-integrated model such as a two-dimensional horizontal model, strong discrepancies between the real and the simulated groundwater flow must be expected.

Problems with approximating reality are magnified in transport modeling. Besides simplifying flow, solute transport processes are reduced to a few transport mechanisms considered dominant. It is the number and complexity of controlling processes in solute transport that generally make transport modeling so difficult in comparison to flow modeling and, in particular, require a sound and solid understanding of the relevant natural transport processes at this stage of modeling. The numerical model is only as good as the underlying assumptions or as the conceptual model allows.

1.2.4 Selecting the Numerical Model

Consider numerical models in the same way as analytical solutions. Both give meaningful results only for well-defined questions. No existing model is applicable to all types of flow and transport problems. Consequently, by this stage of the model study, the model user must decide which computer code to use for the calculation of a particular groundwater problem. Sometimes model users are not aware of what models are available. For sources of groundwater models, see Chapter 9.

Most numerical models are similar in that the investigated area is subdivided into either rectangular or irregular polygonal segments, as shown in Figure 1.3. The handling of the input and output data, which are referred to as pre- and postprocessing, determines if the computer code is "user friendly."

Today all popular and well-tested programs provide results of a similar accuracy for their range of application. Some models like the ones based on the finite element theory have proved to be more versatile. In each model the choice of the discretization of space and time basically controls the computational accuracy.

1.2.5 Calibrating and Validating the Model

Model calibration and validation are required to overcome the lack of input data, but they also accommodate the simplification of the natural system in the model. Calibration and validation will become meaningless, fail, or yield inadequate results if significant features of the natural system are excluded from the model.

In model calibration, simulated values like potentiometric surface or concentrations are compared with field measurements. The model input data are altered, within observed ranges, until the simulated and observed values are fitted within a chosen tolerance. Input data and comparison of simulated and measured values can be altered either manually (trial-and-error adjustment) or automatically (inverse or parameter estimation models). Be prepared to quantitatively assess model calibration. The more convincing the calibration, the more useful will be the model validation. Model calibration is time-consuming and can easily take up half of the time required for the whole study.

Model calibration represents a crucial stage in groundwater modeling. Model calibration tries to demonstrate that the site-specific groundwater model is capable of reproducing observed responses by the natural system. The observed responses of the natural system or the behavior of a particular solute in the subsurface environment may have different causes. The bend of a streamline, for example, may be caused by a change in hydraulic conductivity or by areal discharge or recharge. Relationships are much more complex in transport problems. Decline of concentration may be due to dispersion, decay, or variation of the source strength. To calibrate by trial-and-error adjustment or by inverse models, the model user has to judge, based on the understanding of the investigated problem, if calibrated data represent the natural system. A good fit to historical data based on unrealistic input data is not only wrong but misleading if the same model is applied for predictions.

Model validation has different meaning for different people. A model can sometimes be calibrated to match observed conditions based on arbitrary input parameters. Model validation is required to demonstrate that the model can be reliably used to make predictions. At present, no standard criteria exist on how to demonstrate model accuracy, neither will they be established easily in the near future due to limited data collection in field investigations. A one-time fit of calculated and measured values does not guarantee accuracy. In most model studies there are a large number of adjustable values compared with a small number of field observations. A common practice in validation is the comparison of the model with a data set not used in model calibration. This procedure is more useful when the simulated conditions differ significantly from the one used in calibration. If the calibrated groundwater model does not reproduce accurate results in model validation, model data are recalibrated using both data sets. Then other approaches to validation must be used. Calibration and validation are accomplished if all known and available groundwater scenarios are reproduced by the model without varying the material properties or aquifer characteristics supplied to the model. Model calibration and validation are two of the critical steps that precede model application and are discussed in more detail in Chapter 8.

1.2.6 Applying the Model

Model application is the part of the study in which the numerical model demonstrates its dominance over all other models. Alternative scenarios for a given area may be assessed efficiently. When applying numerical models in a predictive sense, however, limits exist in model application. Predictions of a relative nature are often more useful than those of an absolute nature. The outcome of a numerical model, must be reviewed critically.

It makes a difference whether model users forecast water levels or concentrations for one week or some years into the future. Predictions into the far future are generally more uncertain. As in statistics, extrapolations into the future are more accurate if they are based on a long-term series of observed events in the past. There is a saying in geology that applies here: "The past is the key to the future." The larger the historical "key," the better is our ability to predict the future. Groundwater models calibrated on field data observed within a couple of weeks are not suitable for predictions over 100 years and more.

Another kind of limitation on predicting responses of a groundwater system exists. Flow or transport conditions during predictive runs may stress system parameters that have been irrelevant for model calibration or validation. Assume that a model calibrated to a steady-state condition is applied for predicting system behavior under unsteady conditions. Specific yield or storativity would not be a concern. Predictions would have to be based on best-guess values, and model results must be reviewed carefully. Thus sensitivity analyses become important. Sensitivity analysis helps to rank the input data in terms of its influence on model predictions and gives answers to "what if" questions. Sensitivity analysis also allows one to assess unforeseeable groundwater stresses in the future. Unknown future landuse may, for example, change natural groundwater recharge.

1.2.7 Presenting the Results

When reviewing modeling results, appreciate the presentation but trust the numbers. The output of numerical models is numbers such as hydraulic heads at discrete points of the solution domain at a given time. Model output normally undergoes postprocessing to produce modeling output that is understandable to nonmodel users. Such effort is necessary but postprocessors are not a substitute for lack of modeling experience. Modern postprocessors such as plot software packages, commercial or in-house codes, are unique tools in presenting modeling results illustratively and informatively. However, there is the chance of misuse or misinterpretation. Professional presentation of output-model data may infer an accuracy that does not exist. Interpolation and extrapolation of data in order to facilitate data interpretation are necessary, but manipulating output data to cover uncertainty in model results should never occur. Graphical interpolation programs actually weaken the accuracy of data presentation. Plots of measured potentiometric surface presented as isolines should include the number and location of observation points used for the interpolation. Section 8.10 is devoted to the topic of presenting modeling results.

2 Reviewing the Mathematical Model for Flow

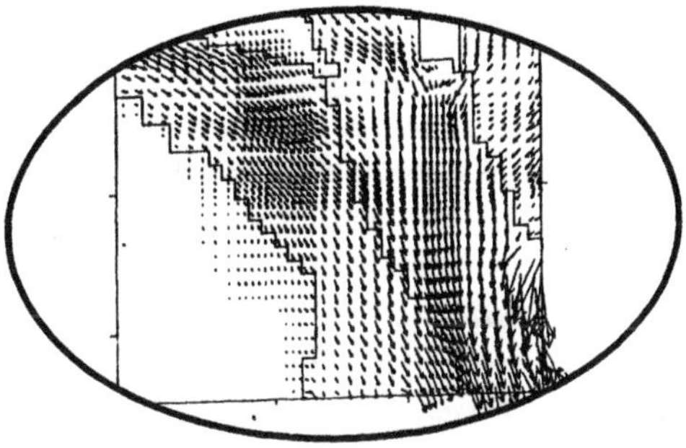

This chapter deals with the description of groundwater flow. Topics are presented from the model user's view, to create a good understanding for applying groundwater flow models and for grasping solute transport caused by groundwater movement. After introducing subsurface water and porous bodies, emphasis is on the mathematical model for the description of groundwater flow: Darcy's law, the water balance equations, and the definition of suitable hydrogeological boundary conditions. Acknowledging its growing importance in groundwater modeling, multiphase flow is briefly discussed in Section 2.5.

2.1 IDENTIFYING GROUNDWATER AND AQUIFERS

To use models, the several forms of occurrence of water in the subsurface and the various openings in soil and rocks that allow water movement must be understood. The model user must also be able to identify characteristic regions of the groundwater system. The following sections are devoted to these topics:

- Defining water in the subsurface
- Classifying voids in soil and rock
- Identifying hydrologic units and aquifers.

2.1.1 Defining Water in the Subsurface

Subsurface water occurs in the following several forms:

- Mobile water. Water in interconnected pore space. The water is transmitted freely through the pores of rock or soil or, alternatively, through fractures in rocks.
- Immobile water. Crystalline water contained in minerals. Such adsorbed water is referred to as hygroscopic water; it is bound by electrostatic and van der Waal's forces on the grain or water enclosed in disconnected voids.
- Water vapor. Gaseous water distributed in the unsaturated groundwater zone.

The model user is particularly concerned with mobile water. One exception is in transport simulations where immobile water may influence the transport behavior of the solute under investigation due to diffusive mass exchange between the mobile and immobile water zone.

To allow significant horizontal water flow, groundwater must occupy the entire pore space forming one continuous water body called the *saturated zone.* The saturated zone may be bounded from above by the water table which is the interface with the unsaturated zone where water and air fill the pores. The water table is defined as the surface of the zone of saturation. Even though there is no strict relationship, the water table often follows a pattern flatter than the profile of the land surface. If groundwater is restrained in its movement by an impervious layer on the top, the water table in the above sense does not exist. Groundwater movement is confined and is described by defining an imaginary surface called the *potentiometric* surface to which water will rise when tapping the confined or semiconfined groundwater zone.

In the unsaturated zone, gravity is the driving force causing predominantly vertical flow and, consequently, vertical transport. This book primarily focuses on flow and transport in the saturated zone where the model study is not restricted to a vertical profile. The importance of flow and transport in the unsaturated zone in specific groundwater problems is briefly discussed.

2.1.2 Classifying Voids in Soil and Rock

Voids in soil and rock are fundamental hydrologic parameters. The fraction of voids in soil and rock are expressed by porosity and defined as a ratio of void volume to total soil or rock volume. Soil and rock with little void volume have little capacity to hold water. Openings in solids are necessary to allow water movement. The model user classifies voids in soil and rock into the following four groups, as shown in Figure 2.1:

- Matrix porosity. Soils and rocks are granular and yield intergrain pores (gravel, sand, silt, clay, sandstone, etc.). The rock or soil grains themselves may be porous (leading to higher overall porosity), or the aggregate porosity may be reduced by mineral deposition in pores.
- Fractured porosity. Rocks contain joints and fractures, regular or irregular.
- Karstic fractures and solution cavities. Rocks are characterized by distinguishable solution cavities.
- Fracture and matrix porosity (dual porosity). Such formations are porous units of the granular type such as clay or sandstone units, but they also contain fractures.

Figure 2.2 shows the typical range of porosity of various geological formations. Additionally porosity values of different soils and rocks are compiled in Appendix B.1. Even though porosity is essential for groundwater flow, the capacity of soil or rock to transmit water is not only a function of porosity but also a function of void geometry and internal forces (see also Section 2.2.4). For practical purposes the percentage of void volume that contributes to percolation (called *effective porosity*) is equivalent to the specific yield. This yield is the volume of water released from a unit volume of saturated soil or rock material drained by a falling water table of a unit height. Specific yield, or the effective porosity, is always less than the total porosity as indicated in Figure 2.2. For a compilation of values of specific yield see Appendix B.2.

Traditionally groundwater mathematics and numerical models have focused on the description of flow and transport in matrix porosity or in sand and gravel. Frequently the model user investigating a groundwater problem in groundwater systems other than the granular type tries to adopt these solutions established for granular medium flow and transport. This is an assumption that approximates reality with varying degrees of success and should be justified on a case-by-case basis. If model calculation cells contain many well-connected fractures and no large-scale discontinuities, then the assumption of porous-media-like behavior may be justifiable.

2.1.3 Identifying Hydrologic Units

Groundwater modeling represents an approximation of natural features and observations. The groundwater system is not described in detail but as units with similar characteristics. The ability to identify main hydrologic units is essential in the conceptualization of the groundwater system. Hydrologic units fall into the following three groups:

- Aquifer. A permeable unit that can yield water in usable quantities when tapped by a well. Common aquifers are geological units of unconsolidated sand and gravel or fractured rock.

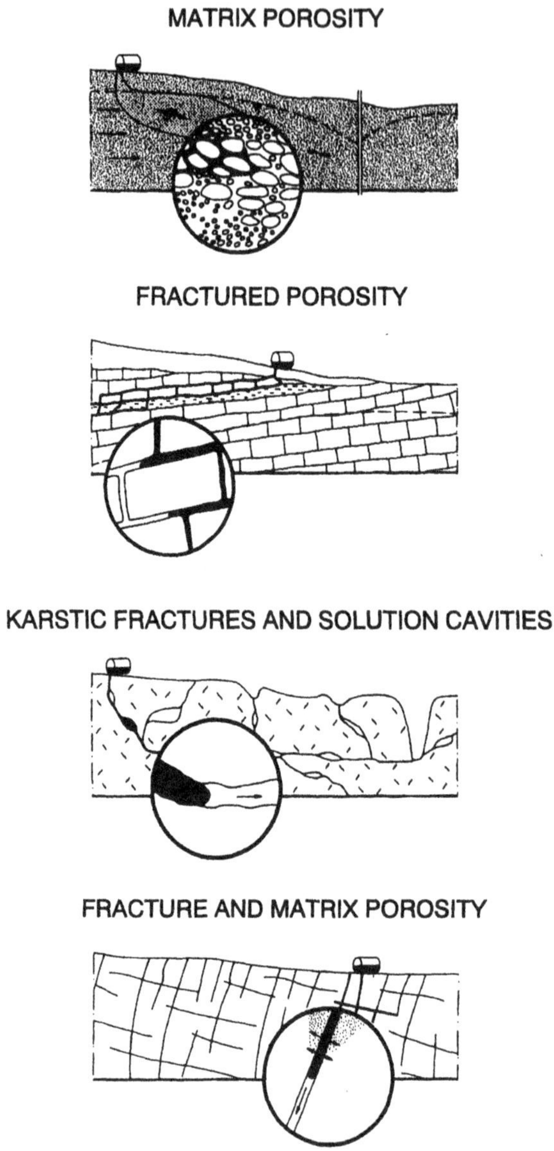

Figure 2.1 Classification of voids in soils and rocks according the type of porosity.

- Aquiclude. A very low permeability unit that will not transmit water freely. In an aquiclude, groundwater flow is often assumed to be zero. Solute transport, however, may not be zero. Common aquicludes are thick clay layers or solid rocks.
- Aquitard. A low permeability unit that falls between aquifers and aquicludes. Water cannot be produced economically through wells, but flow is

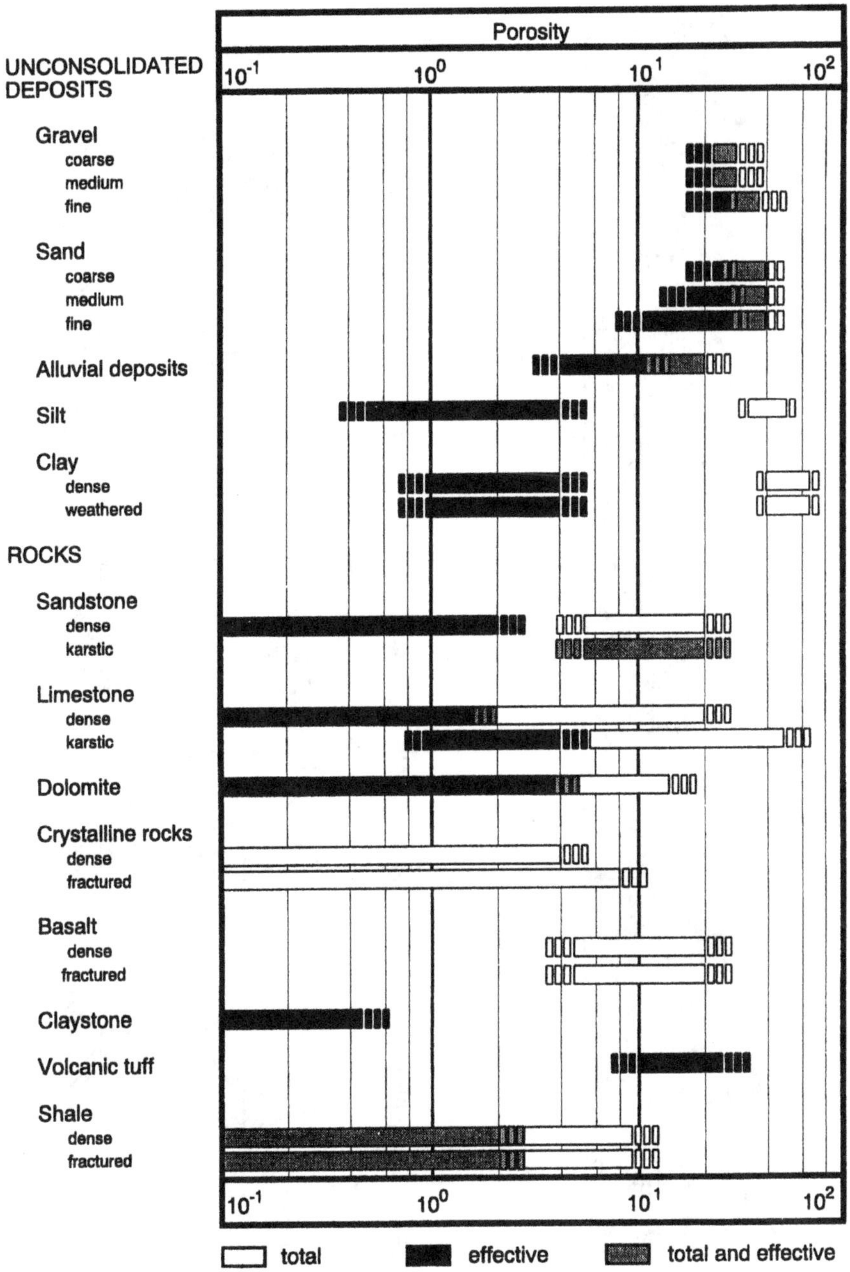

Figure 2.2 Range of porosity of geological formations.

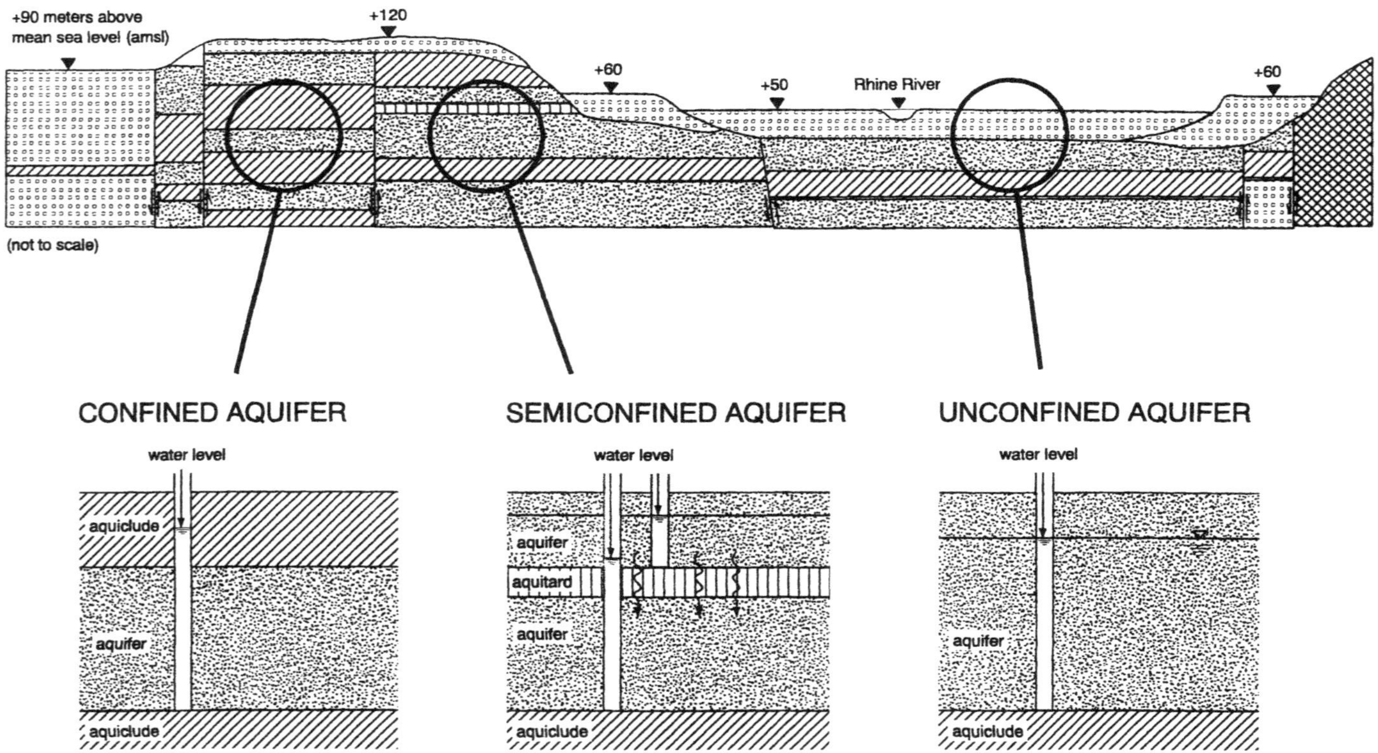

Figure 2.3 Different types of aquifers illustrated on a cross section of the Rhine River Valley at Cologne.

significant enough to feed adjacent aquifers. It is generally assumed that groundwater flow in an aquitard is predominantly vertical. Common aquitards are clay layers.

Figure 2.3 gives an example of how the model user can simplify the groundwater system in the conceptual model by identifying main hydrologic units. The given cross section of the Rhine valley at Cologne identifies schematically individual characteristic units relevant for a model study. The group in which a rock or sand formation falls (aquifer, aquiclude, or aquitard) depends on total porosity, specific yield, and hydraulic conductivity.

Modeling is often primarily concerned with flow in aquifers, even though flow, and transport, in aquitards may also on occasion become the subject of groundwater studies. In view of their hydraulic behavior, aquifers can be classified into three types shown in Figure 2.3:

- Confined aquifer. The aquifer is confined at the top and base and, in terms of its steady-state hydraulic behavior, is comparable to a pipeline. The saturated thickness is independent of flux or boundary conditions, unless these conditions cause the aquifer to become unconfined.
- Leaky or semiconfined aquifer. The aquifer receives or loses water to adjacent aquifers through an aquitard. In the case of an overlying aquitard, the aquitard may be only partly saturated. The leaky aquifer can, in many cases, be visualized as a pipeline with porous walls.
- Unconfined or phreatic aquifer. The unconfined aquifer has a free water table. A change in flux results in a change of the saturated thickness. Flow can often be compared to flow in an open channel. Leakage may, or may not occur across the base of a phreatic aquifer.

2.2 REVIEWING DARCY'S LAW

Darcy's law allows us to assess groundwater flow by means of graphical, analytical, and numerical models. The following sections illustrate Darcy's experiment and discuss the general form of Darcy's law.

2.2.1 Remembering Darcy's Experiment

The first groundwater model, a porous media model, was developed in a scientific application in 1856 by the French hydraulic engineer Henry Darcy (Darcy 1856). Figure 2.4 illustrates Darcy's original model design and its extension to study flow in each direction. Darcy had been enlarging and modernizing the water works in Dijon, France. While designing sand filters, he encountered a problem with the physics of flow through porous media for which no published information existed. The detailed description of flow at

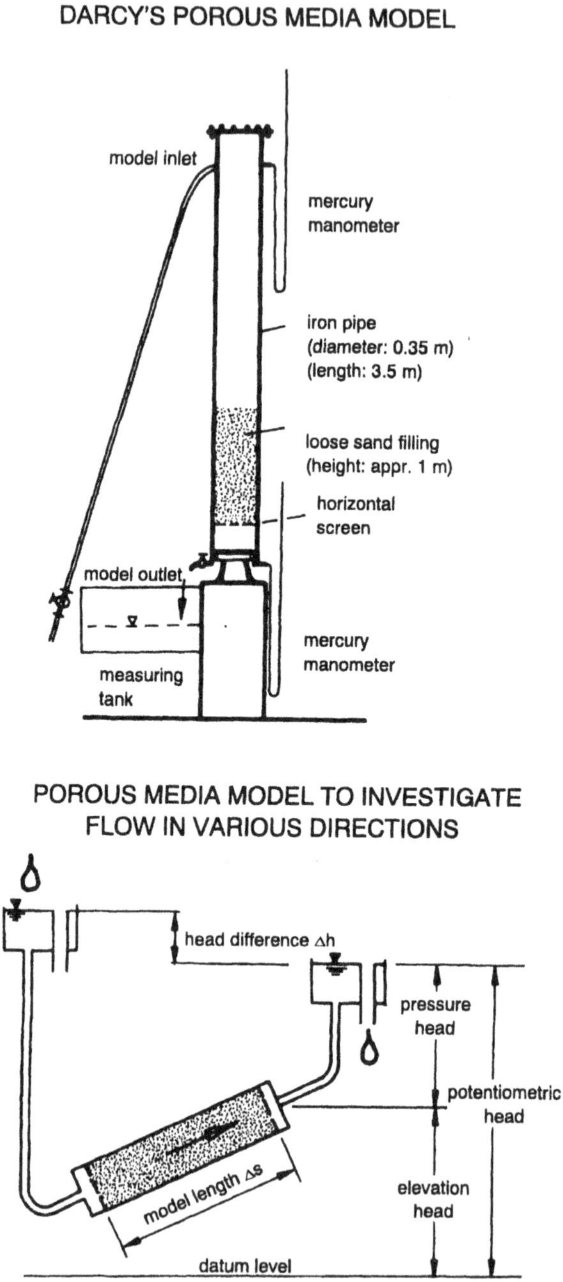

Figure 2.4 The first groundwater model: Porous media model to elaborate Darcy's law.

pore scale in a porous media was too complex to be practically applicable. To gain information about groundwater flow, Darcy constructed a porous media model.

In Darcy's experiment, fully saturated flow through a vertical experimental tank filled with porous material is generated by imposing different pressures on the model inlet and outlet. Darcy's law in its simplest form shows the relationship between flux, pressure gradient, and an empirical coefficient called the *hydraulic conductivity*, that depends on the characteristics of the porous material and of the water. This relationship is shown in Equation 2.1:

$$q = -K \frac{\Delta h}{\Delta s} = -Ki \tag{2.1}$$

where

q	=	Darcy flux or specific discharge in [L/T]
K	=	hydraulic conductivity in [L/T]
Δh	=	head difference in [L]
Δs	=	length in [L]
i	=	hydraulic gradient in [1]

Hydraulic conductivity is a measure of the capacity of the porous media for transmitting water. Since the hydraulic gradient is dimensionless, K has the dimension of velocity. Flux is proportional to hydraulic gradient. The simple relationship illustrates, for example, that flux is doubled either by doubling the hydraulic gradient or doubling the hydraulic conductivity.

In the literature the Darcy flux q is often referred to as the *Darcy velocity*. Velocity is the rate of unit flow over unit cross-sectional area perpendicular to the flow direction. In groundwater flow, however, only a portion of the unit cross section of the aquifer has the capacity to transmit water while the remainder is solid material. Therefore the term velocity is misleading. Darcy velocity does not represent the velocity of water molecules. In groundwater flow studies which by nature are concerned with water quantities, the actual velocity of water in the irregular pore space is of no interest. In contrast, the true velocity of water-carrying solutes is necessary when studying solute transport. Thus in transport modeling the problem of approximating the true velocity based on known Darcy fluxes arises. This is discussed in more detail in Section 3.2.1.

2.2.2 Generalizing Darcy's Law

The hydraulic conductivity K depends on the permeability of the porous media k, on the physical properties of the fluid, on the density ρ, and on the dynamic viscosity μ or the kinematic viscosity ν:

$$K = \frac{\rho g}{\mu} k = \frac{g}{\nu} k \tag{2.2}$$

where

K = hydraulic conductivity in [L/T]
k = permeability in [L^2]
ρ = density in [M/L^3]
g = acceleration due to gravity in [L/T^2]
μ = dynamic viscosity in [M/LT]
ν = kinematic viscosity in [L^2/T]

In most cases the permeability k varies from point to point in an aquifer; this is called *heterogeneity*. In addition, if permeability also depends on the direction of measurement, the aquifer is called *anisotropic*. The aquifer is called *homogeneous* if k is independent of the position of measurement. If k is independent of the direction of measurement at any point of the investigated aquifer, the aquifer is called *isotropic*. Mathematically speaking, the permeability, or respectively, hydraulic conductivity, is a second rank tensor with nine components.

Incorporating Equation 2.2 into Darcy's law, and expressing the hydraulic gradient in terms of elevation and pressure components, the general Darcy law is represented as

$$q_i = -\frac{k_{ij}}{\mu}\left(\frac{\partial p}{\partial x_j} + \rho g \delta_j\right) \tag{2.3}$$

where

i,j = 1, 2, 3 (principal coordinate directions)
q = Darcy flux in [L/T]
p = pressure in [M/LT^2]
x = space coordinate in [L]
δ_j = 0 in horizontal flow direction
δ_j = 1 in vertical flow direction

In Equation 2.3, and hereafter, the summation using double indexes is used (Einstein's convention of summation).

In modeling studies where the properties of water change with time or space (saltwater intrusion, migration of highly contaminated groundwater, warm water injection, etc.), the numerical formulation of groundwater flow must be based on the general Darcy law. The only remaining parameter in the general flow equation that is independent of water properties is permeability, characterizing the shape of the passages through which flow occurs.

2.2.3 Underlying Assumption of Darcy's Law

Darcy's law is established by experiments with various sands and gravels. The Darcian approach is to replace the actual aquifer with a representative continuum in order to use macroscopic law to describe flow at microscopic scale (Bear 1972). Groundwater flow is expressed by a simple relationship ignoring the complex flow configuration at pore scale. The macroscopic equation is also applicable for fractured or karstic material, assuming that the scale applied is large enough to eliminate the dominance of single fractures within the system. This restriction is important when describing flow in fractured or karstic aquifers. While numerous experiments demonstrate that Darcy's relationship holds true for unconsolidated media, as soon as one averages over a certain distance, similar rules do not exist for fractured or karstic formations. Averaging distances as large as several kilometers, beyond the size of the aquifer under investigation, may be required.

2.2.4 Summarizing Hydraulic Conductivities of Soils and Rocks

Since the first experiment by Darcy (1856), numerous laboratory and field studies have contributed to sketch a picture of hydraulic conductivities for natural rocks and unconsolidated deposits, facilitating the estimation of the hydraulic conductivity for a given aquifer. Figure 2.5 gives the range of hydraulic conductivity of various geological formations. A detailed compilation of relevant data concerned with the hydraulic conductivity is presented in Appendixes B.3 and B.4.

The hydraulic conductivity is a key factor when solving for groundwater flow. Although numerous approximate formulas exist to estimate the value theoretically, it is always preferable to rely on accurate field measurements rather than on calculation. When modeling at a regional scale, aquifer pumping tests giving conductivity values integrated over larger distances provide more helpful estimates of the hydraulic conductivity than local point measurements such as slug tests. Hydraulic conductivities of unconsolidated materials are considerably higher than those of most rocks. For rocks, K depends either entirely on the secondary porosity of the rock (fractures, weathering, etc.) or, in the case of sandstone, on the degree of cementation of pore space. High porosity environments such as karstic systems yield high conductivity values (as long as Darcy's law holds). Generally, horizontal hydraulic conductivities in sandy aquifers are in the region of a factor 10 to 100 higher than the vertical ones. The reason for larger horizontal conductivities lies in the bedding of grains during the geological formation of the aquifer.

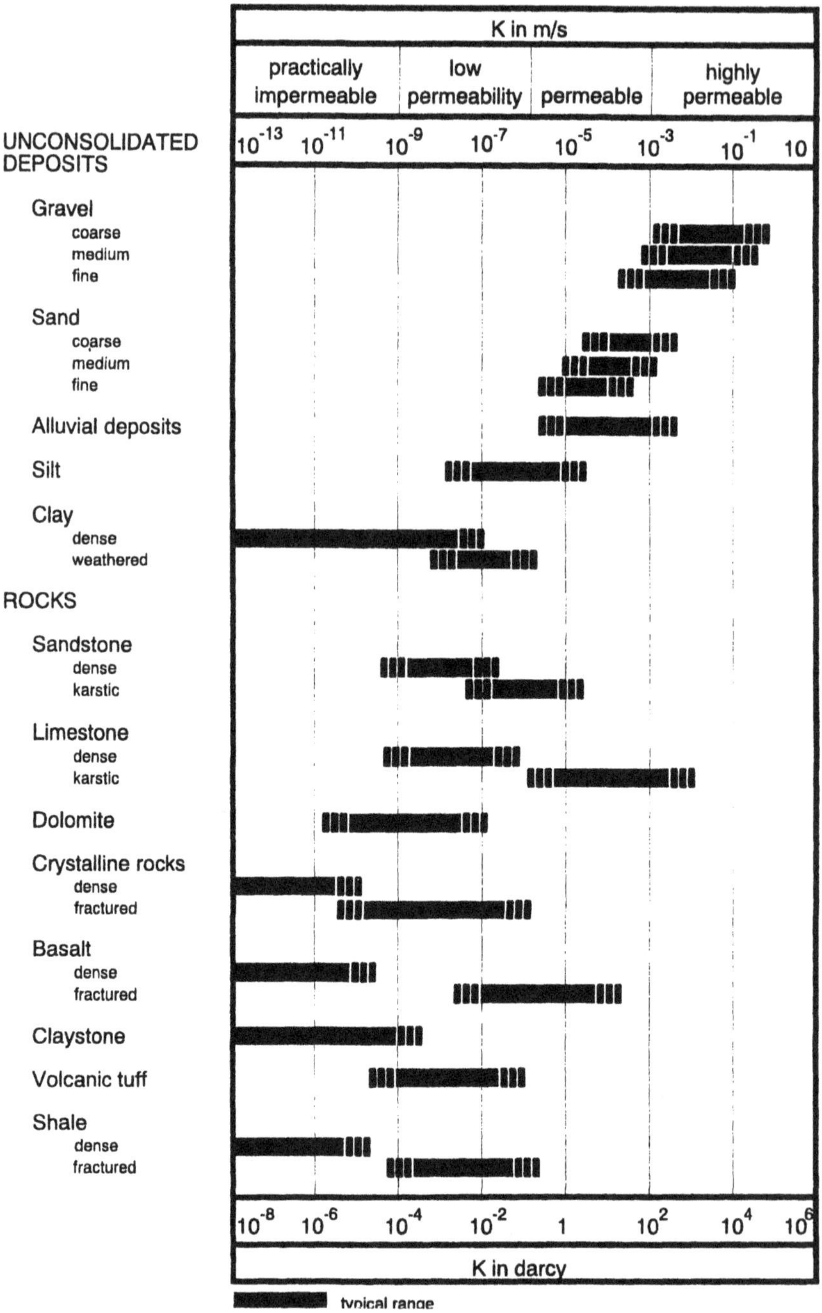

Figure 2.5 Range of hydraulic conductivity of geological formations.

There is no strict relationship between porosity and hydraulic conductivity despite the fact that rock of very low porosity is likely to have low hydraulic conductivity. Such correlation, however, does not necessarily hold true in the opposite case of materials with high porosity. Apart from the total amount of pore space, hydraulic conductivity depends on the openings or the geometry of pore space. Clay is commonly considered impermeable, even though its porosity is high. In clay, interconnecting tubes are small and hydraulic conductivity is low due to molecular attraction of water on the solid materials. In gravel, or reasonably coarse sediments, the force of molecular attraction between water and soil grains does not encompass or bridge the wide pore opening, and water is free to move in response to differences in potential. As long as the influence of molecular attraction is of the second order, hydraulic conductivity in unconsolidated deposits and sandstones follows trends in relation to porosity. Hydraulic conductivity increases with an increase of porosity, as can be seen by comparing Figure 2.2 and Figure 2.5.

Limestone reveals characteristics similar to sandstone. Dense limestone has low porosity and low hydraulic conductivity. However, limestone formations altered due to dissolution along fractures (karstic limestone) may transmit large quantities of water. Flow velocities in karstic systems may be on the order of flow velocities found in surface water.

2.3 FORMULATING THE GENERAL FLOW EQUATION

Darcy's law alone is not enough to describe groundwater flow, unless the head distribution within the entire flow system under review is measured. When modeling groundwater flow, however, the aim is to predict the head distribution under various groundwater stress situations. The head distribution known beforehand serves as initial conditions in a transient groundwater study.

The general flow equation for saturated groundwater flow is derived in numerous excellent textbooks such as Bear (1972). In most analyses the general flow equation is formulated by applying the law of conservation of mass over a control volume of an aquifer situated in the flow field. The net inflow into the volume must equal the rate at which water is accumulating within the volume under investigation, which leads to

$$\frac{\partial}{\partial x_i}\left(K_{ij}\,\frac{\partial h}{\partial x_j}\right) = S_s\,\frac{\partial h}{\partial t} + Q \tag{2.4}$$

where

i,j = 1, 2, 3 (principal coordinate directions)
K = hydraulic conductivity in [L/T]
h = head in [L]

S_s = specific storage in $[L^{-1}]$
Q = local sources and sinks per unit volume in [1/T]
x = space coordinate in [L]
t = time in [T]

Darcy's law gives three equations for flow in three directions. The flow equation yields the fourth equation for the four unknowns: the three components of groundwater flux and the head. The specific storage, or specific storage coefficient S_s in Equation 2.4, represents the amount of water released from storage or stored per unit of volume in an aquifer, per unit of change in head. Under unconfined conditions the coefficient is set equal to the specific yield for practical conditions. Under confined conditions water is stored or released due to the compressibility of water and the grain matrix. In comparison to the values of the specific yield which are in the order of 10^{-1}, this storage coefficient is small and ranges in the order of 10^{-4} (see also Appendix B.5).

Specific storage must only be considered for transient flow. Release of water from the aquifer in response to a decline in head is then generally assumed to be instantaneous. In some cases, however, the aquifer response differs from this modeling assumption (Boulton 1963). In fine-grained aquifers, drainage may be slow and similar to that of a leaky aquifer for a prolonged period of time. In an aquifer with two distinct regions of porosity and hydraulic conductivity, such as fractured sandstone, initial release from storage may be relatively rapid, but it decreases with time. Whenever the characteristics of release or storage of water are a concern, the model user must be aware of these effects.

The term Q represents sources and sinks of water in Equation 2.4. Local sources and sinks are often discharge or recharge wells. Natural groundwater recharge due to precipitation or natural groundwater loss due to evaporation are examples of areal sources or sinks. Water leaking from a river, regardless of groundwater level, can be approximated as a line source.

Only in a few flow configurations is there a need to solve the general flow equation. Depending on the flow problem, Equation 2.4 can be simplified to solve for one- or two-dimensional, steady-state, or transient flow. Consider a groundwater flow regime where the horizontal flow components dominate the vertical ones. Neglecting vertical flow components and expressing fluxes as averages over depth allows one to approximate groundwater flow as two-dimensional and horizontal (see Section 7.3). The relevant flow equation for confined flow is

$$\frac{\partial}{\partial x_i}\left(T_{ij}\frac{\partial h}{\partial x_j}\right) = S\frac{\partial h}{\partial t} + Q \tag{2.5}$$

with ($i = 1, 2; j = 1, 2$). T is referred to as aquifer transmissivity and is calculated as the product of hydraulic conductivity times saturated thickness. Hence T has dimensions L^2/T and depends on the elevation of the water table for unconfined conditions. S is storativity, representing specific storage multiplied by depth. Storativity is dimensionless. Q represents local sources and sinks integrated over depth. Chapter 7 gives numerous examples of how and when to simplify the dimensionality of a groundwater problem.

To obtain the solution of the flow equation to a given flow problem, it is necessary to define additional information that does not appear in the equation. Mathematically speaking, the flow equation falls into the group of equations known as *boundary-value problems.*

The model user is concerned with defining this additional information by defining the area of investigation, finding the relevant boundary and initial conditions (the latter only in transient flow), and finding realistic values for the spatial distribution of the hydrogeological parameters such as hydraulic conductivities. The model user's main task is to translate the numerical code into a site-specific model.

2.4 IDENTIFYING BOUNDARY AND INITIAL CONDITIONS

One of the first and most demanding tasks in groundwater modeling is that of identifying the model area and its boundaries. In defining the model area, the model user distinguishes the investigated area from the adjacent groundwater system. Consequently a model boundary is the interface between the model area and the surrounding environment. Conditions on the boundaries, however, have to be specified. Boundaries occur at the edges of the model area and at locations in the model area where external influences are represented, such as rivers, wells, and leaky impoundments.

The decision criteria for selecting hydraulic boundary conditions are primarily topography, hydrology, and geology. The topography, geology, or both, may yield boundaries such as impermeable strata or potentiometric surface controlled by surface water, or recharge/discharge areas such as inflow boundaries along mountain ranges. The flow system allows the specification of boundaries in situations where natural boundaries are a great distance away (referred to as *infinite* aquifers).

Boundary conditions must be specified for the entire boundary, and all may vary, with time. At a given boundary section just one type of boundary condition can be assigned. As a simple example, it is not possible to specify groundwater flux and groundwater head at an identical boundary section.

Boundaries are classified into three mathematical boundary types:

- Boundary of prescribed pressure or hydraulic head. This is the boundary condition that most constrains the model solution and is therefore the easiest to solve.

- Boundary of prescribed flux of groundwater. This is the boundary condition that least constrains the model solution and is therefore the most difficult to solve.
- Semipermeable or head-dependent flux boundary.

Figure 2.6 illustrates how natural features may indicate which boundary type to use. The different boundary conditions are discussed in the following subsections and in Subsection 8.4.3. Also reference is made to initial conditions required when solving transient flow.

2.4.1 Using Prescribed Head Boundaries

The most common boundary condition is the type where the head or for unsaturated flow, the pressure head, is fixed. They are often referred to as the *first kind boundary*, since the groundwater heads for the parameter being solved, are given. *Dirichlet's condition* is the mathematical terminology for this type of boundary. To add to the confusing terminology, some authors also use the term *recharge boundary*. This latter expression, however, is more common for prescribed flux boundaries. There is no change of head along prescribed head boundaries due to predicted changes in groundwater conditions in the modeled groundwater system. This does not necessarily mean that the potentiometric surface is constant over time. Prescribed heads at the boundary may change with time, for example, due to water level fluctuation of surface water levels or due to factors affecting heads in a predefined way.

Prescribed head boundaries can occur when surface water bodies, such as rivers, lakes, canals, sea coast, impoundments, and drains, interact freely with the aquifer, as shown in Figure 2.6. It is assumed that the elevation of the surface water equals the elevation of the water table. In nature, few surface water bodies cut through the entire thickness of an aquifer and are completely freely connected hydraulically with the aquifer. Typically a stream or lake partially penetrates the aquifer. In addition layers of silt and clay may reduce free exchange of surface water and groundwater, often making it difficult to represent boundary conditions accurately in the model.

The model user also often chooses extrapolations of measured groundwater heads as prescribed head boundaries (also called *unnatural boundaries*) as long as these boundaries stay unaffected by stresses within or outside the model domain during the simulation. This is best managed by recording and checking the flux over the boundary. The model user must judge if the calculated flux is reasonable and unaffected by model simulations. The need for using measured groundwater heads as prescribed head boundaries arises if the investigated aquifer is of such a wide extent that natural boundaries are too remote for modeling purposes.

When interpreting modeling results, consider that streamlines and potential lines are orthogonal to each other in isotropic and homogeneous aquifers.

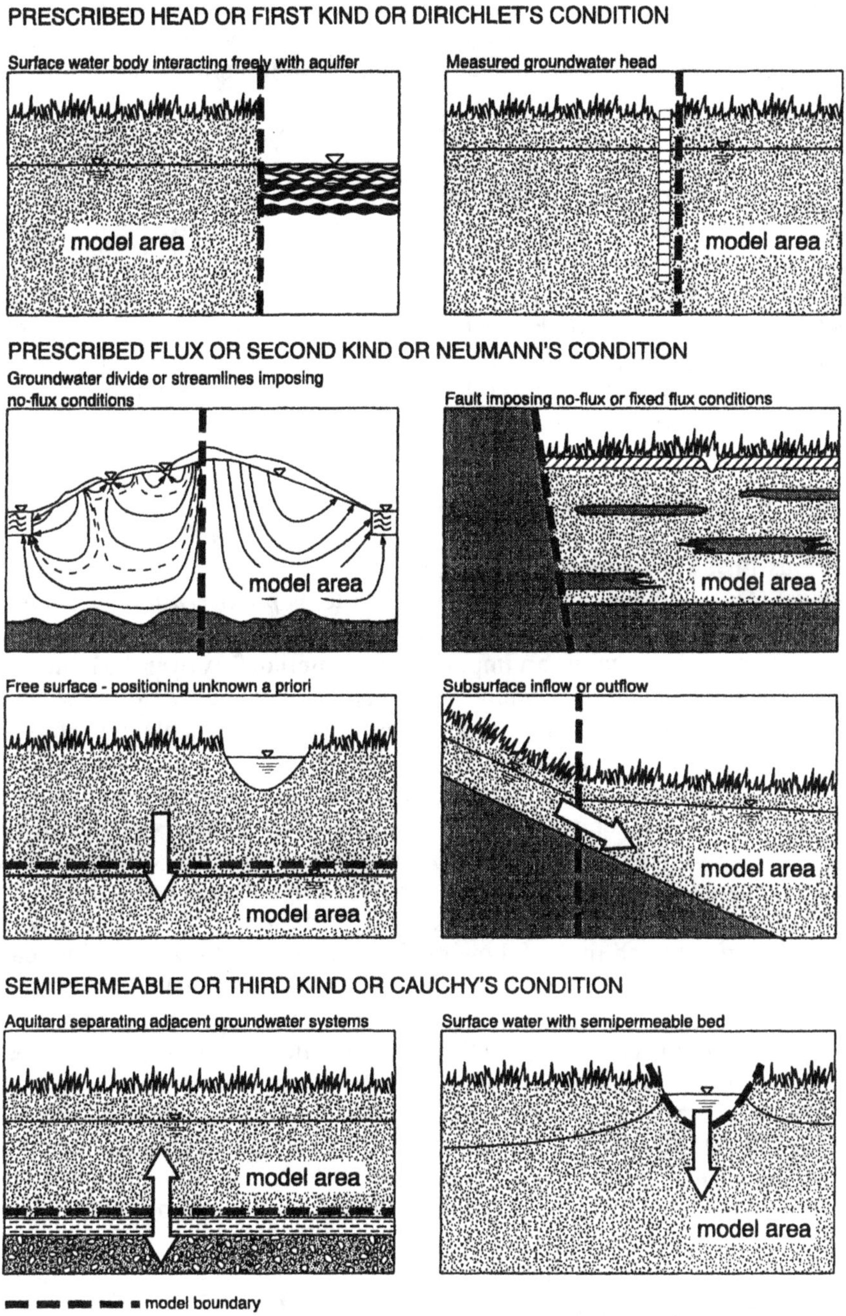

Figure 2.6 Boundary conditions in flow modeling.

Consequently streamlines generated by the model are perpendicular to prescribed constant head boundaries. Use of some prescribed head boundaries is preferred, since they facilitate the convergence of the numerical solution of flow.

In the modeling of steady-state groundwater flow, the potentiometric surface must be specified at least at one point in the model area. Otherwise, an infinite number of head distributions will be found that all differ by a constant value from each other at each point in the solution domain. That is, the steady-state solution needs a datum or, respectively, a "reference point" to work with. The transient solution requires the initial condition for a datum. To specify the head distribution, the model user normally applies at least one prescribed head boundary along the model boundary. Alternatively, the semipermeable boundary can serve as a datum. In this case numerical efforts increase significantly and stable solutions are not always ensured.

2.4.2 Using Prescribed Flux Boundaries

Various physical systems exist where groundwater flux at the boundary is known. These boundaries are also referred to as *prescribed flux boundaries, second kind boundaries, Neumann's condition*, or *recharge boundaries*. If the aquifer terminates against an impermeable material, flux over the boundary becomes zero. The no-flux boundary is a special type of the prescribed flux boundary and is also referred to as a zero-flux, impermeable, reflective, or barrier boundary. Groundwater flow is parallel to the no-flux boundary. In the analysis of model results, no-flux boundaries are identical to streamlines. Hence the models often represent natural groundwater divides or streamlines as no-flux boundaries, as shown in Figure 2.6. Groundwater divides or streamlines are unnatural boundaries. The model user must ensure that during the model simulation the chosen boundary stays unaffected by the calculations. This is demonstrated in the following field case (Spitz 1989). This field case also demonstrates how to define boundary conditions in a field application.

The investigated area is centered on the city of Semarang, Indonesia (Figure 2.7). In the north the Java Sea forms a natural boundary for the investigated groundwater basin and prescribes the potentiometric surface (section A–B). The southern boundary of the model area coincides with the border line of the hilly area (section C–D). A flux boundary simulates recharge from the mountain area. To the west the boundary is elongated along the estimated streamline under natural groundwater flow conditions. A zero-flux boundary separates the model area from the surrounding environment (section D–A). Only a small amount of data are available to support a choice for an appropriate boundary in the eastern part of the study (section B–C). It is assumed that subsurface drainage coincides with the surface water drainage system. Hence a flux boundary following the main river in this area is selected, and it is far enough away from the area of prediction. Figure 2.8 compares model results for natural flow conditions, with simulations for flow conditions im-

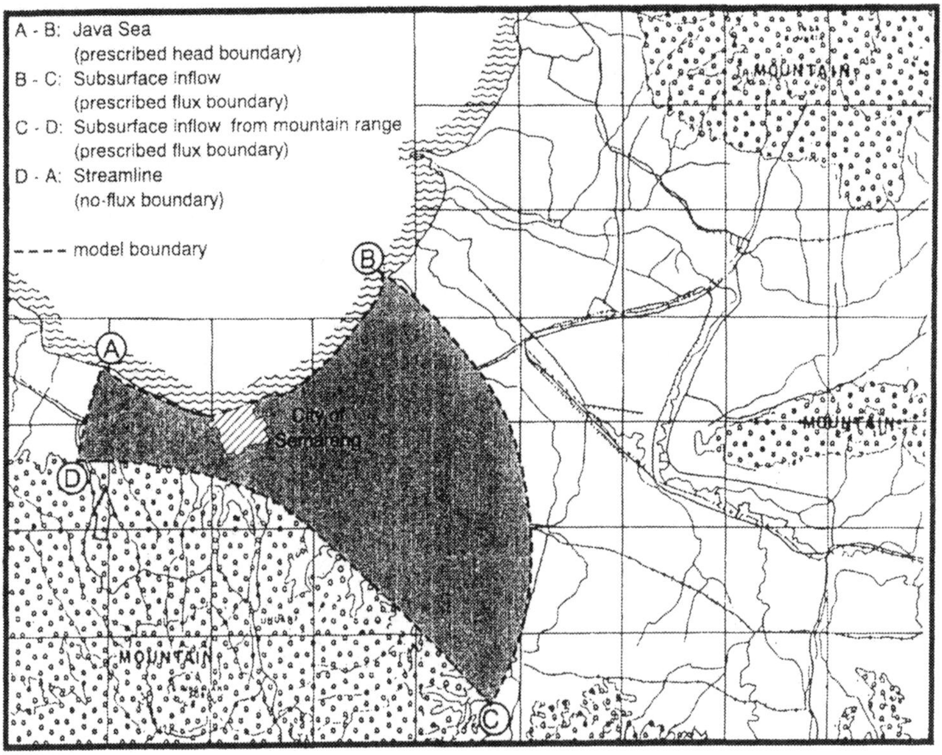

Figure 2.7 Defining boundary conditions in the model study of the Semarang basin (Spitz 1989).

posing a total groundwater discharge due to pumping of $Q = 9{,}500\ m^3/d$ and $32{,}000\ m^3/d$. The drawdown cone caused by the discharge of $9{,}500\ m^3/d$ does not significantly affect the eastern boundary with prescribed flux estimated in model calibration. Imposing a total pumping rate of $Q = 32{,}000\ m^3/d$ on the system, however, enforces significant drawdown along the eastern boundary. Groundwater heads in the model domain interact with the groundwater environment not accounted for in the model study. Hence the model user is forced to enlarge the model area or to estimate how this interaction affects model results.

Generally, natural flow boundaries, such as surface water and impermeable boundaries, that are independent of modeling objectives are preferred to unnatural boundaries.

2.4.3 Using Semipermeable Boundaries

Semipermeable boundaries are generally used to represent leakage from surface water bodies where the exchange of water between aquifer and surface water

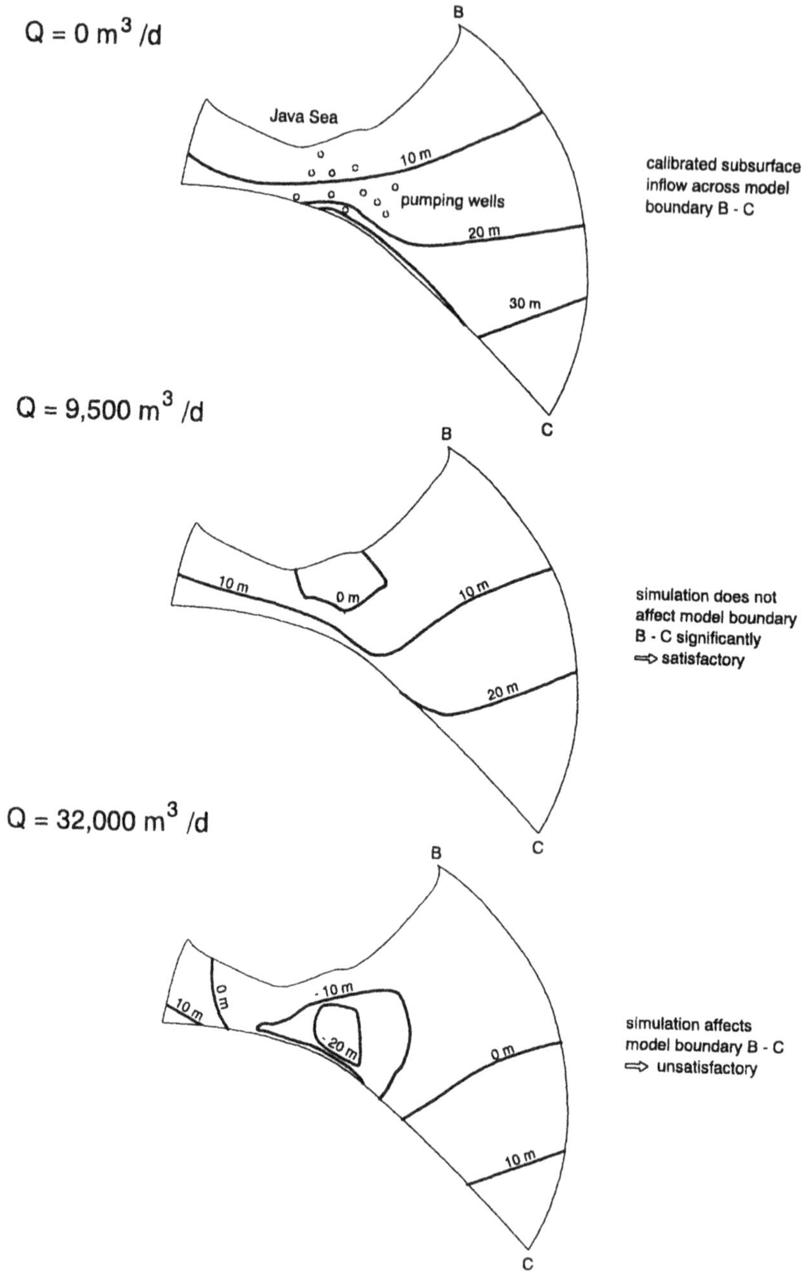

Figure 2.8 Unnatural model boundaries and their dependency on model simulations (Semarang field study).

depends on the head differences between surface water and groundwater (flux or groundwater head cannot be described in advance). Such flow conditions can occur if the river bed acts as a semipermeable membrane between the river and aquifer. Several synonyms are common for the semipermeable boundary. It is often also referred to as the *third kind, mixed, Cauchy's, Fourier's condition*, or *head-dependent boundary*.

It is convenient to introduce the semipermeable boundary in the model using the leakage principle. Flux is approximated by applying Darcy's law. The driving gradient is the head difference over the thickness of the semipermeable river bed. Since both the hydraulic conductivity K of the river bed and its thickness d remain unknown in most field studies, both parameters are commonly united within one parameter called the leakance c of the leaky layer:

$$c = \frac{K}{d} \tag{2.6}$$

where c is expressed in $[T^{-1}]$. Thus the semipermeable boundary is of the form

$$K \frac{\partial h}{\partial x} - c(h - H) = 0 \tag{2.7}$$

where

h = groundwater head in [L]
H = surface water elevation in [L]
K = hydraulic conductivity of aquifer in [L/T]

The semipermeable boundary expresses a linear relationship between the groundwater head and groundwater flux at the boundary. Obviously, if the resistance of the leaky aquifer tends to infinity, the head difference between surface water and groundwater must tend to zero to allow finite flux. Hence semipermeable boundaries act as prescribed head boundaries when choosing very large numbers for c. On the contrary, if c is equal to zero, the surface water is entirely separated from the groundwater system. No interaction occurs.

2.4.4 Using Initial Conditions

Initial conditions are vital for modeling transient flow problems. Initial conditions must be specified for the entire area. The initial condition is the head distribution within the model area at some initial time, usually denoted as $t = 0$. Generally, measured head distribution serves as the initial conditions. In areas where information on groundwater heads is missing, the model user relies on estimates.

The model user also needs an initial steady-state head distribution to solve steady-state flow problems. Here the initial head distribution acts as the starting distribution for the numerical calculation. A good estimate of the steady-state solution accelerates convergence particularly for unsaturated-flow problems. However, unlike the initial conditions for transient flow, initial steady-state conditions are a numerical and not a physical requirement.

2.5 MODELING MULTIPHASE FLOW

Most model users will apply models to study flow or transport in the saturated zone where water is the only phase in motion. There exist a number of problems relevant to the unsaturated zone in which water and air fill the pore space forming two distinct phases. Flow in the unsaturated zone is a special case of multiphase flow through porous media and is extensively studied in the petroleum industry (simultaneous flow of oil, gas, and water). An increasing demand for modeling studies focusing on multiphase flow has recently occurred in contaminant transport problems in order to predict the spreading of immiscible hydrocarbons and chlorinated solvents. Figure 2.9 illustrates typical multiphase flow configurations.

Modeling multiphase flow is more demanding than modeling saturated groundwater flow. In fact the strong nonlinearity of the governing flow equations together with the hysteresis of the hydraulic conductivity practically preclude exact analytical solutions. This fact has stimulated the development of numerous numerical models with emphasis on the liquid-phase transport of water rather than on air flow (Pinder and Gray 1977). Apart from the application in the petroleum industry, modeling of multiphase flow is in its infancy in groundwater hydrology and is mostly the object of research studies (Kueper and Frind 1991 a, b). Of all the models listed in Figure 1.2, with few exceptions only the porous media and the numerical model are applicable to multiphase flow. Because of scaling problems, however, porous media models fail to apply to practical studies, but they are an essential tool in the research of multiphase flow. This section presents unsaturated flow to highlight the specific characteristics and difficulties encountered in multiphase flow modeling. For a rigorous approach to multiphase flow, refer to classical textbooks (Muskat 1949; Bear 1972).

As in saturated groundwater flow Darcy's law is applied to describe water motion (or flow of any other liquid). Now, however, the hydraulic conductivity depends on the degree of saturation (or volumetric moisture content) θ, defined as water volume to total void volume in a unit aquifer volume:

$$q_i = -\frac{k_{ij}(\theta)}{\mu}\left(\frac{\partial p}{\partial x_j} + \rho g \delta_j\right) \tag{2.8}$$

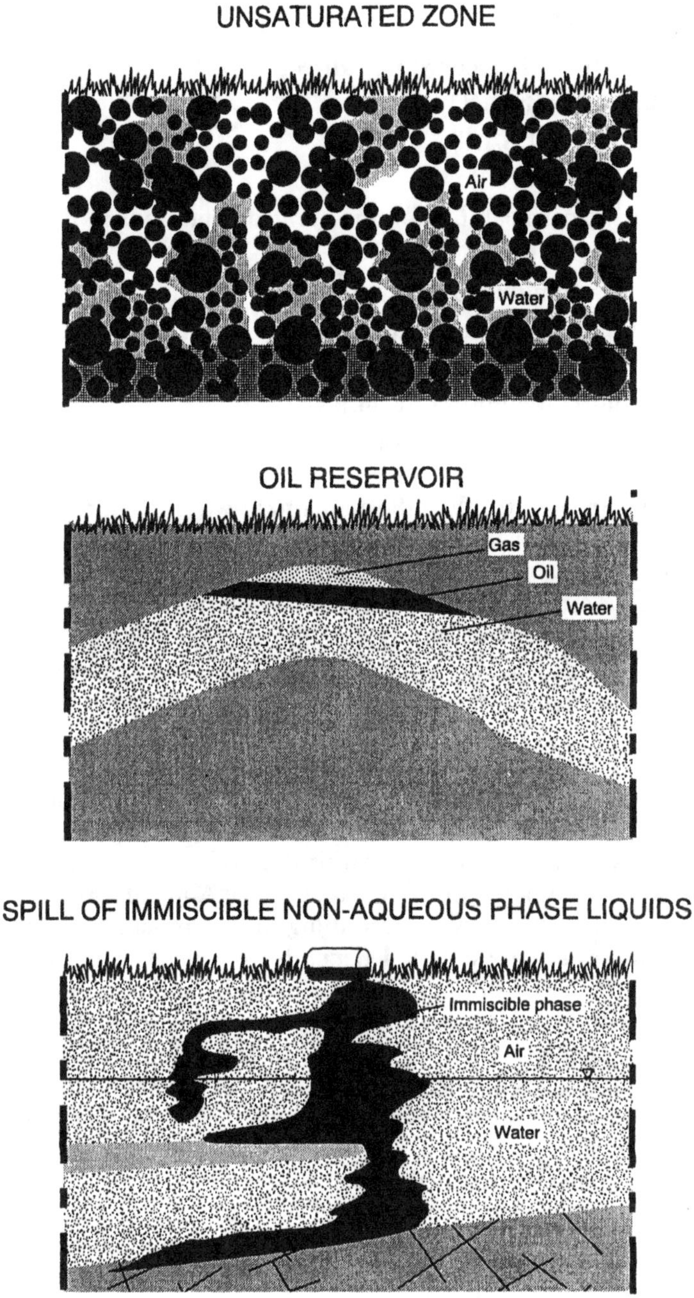

Figure 2.9 Examples of multiphase flow.

where

i,j = 1, 2, 3 (principal coordinate directions)
q = Darcy flux in [L/T]
μ = viscosity in [M/LT]
ρ = density in [M/L^3]
k = specific permeability for unsaturated conditions in [L^2]
p = pressure in [M/LT^2]
g = acceleration due to gravity in [L/T^2]
θ = volumetric moisture content in [1]
x = space coordinate in [L]
δ_j = 0 in horizontal flow direction
δ_j = 1 in vertical flow direction

In modeling unsaturated flow, the main unknown is unsaturated hydraulic conductivity. Typical unsaturated hydraulic conductivity relationships are given in Appendix B.6. For saturated flow, θ is equal to porosity n. As illustrated in Figure 2.4, the head h has two components; the elevation head (elevation z of the measurement point) and the pressure head ($p/\rho g$). In the unsaturated zone the head is still the algebraic sum of elevation and pressure head. Now, however, the pressure head is less than zero. Pressure head and degree of saturation are also not independent of each other but follow a typical functional relationship. Since hydraulic conductivity is a function of saturation, which in turn relates to the pressure head being calculated, nonlinearity is introduced to the flow equations. This significantly increases the modeling effort. For an assortment of typical relationships of pressure head and degree of saturation, or volumetric moisture content, see Appendix B.7.

There is another difficulty in describing of multiphase flow. When two immiscible phases are present, flow depends on whichever phase displaces the other. Interfacial tension and wettability differ when the water–air (or, generally, fluid–fluid) interface is advancing or receding. This phenomenon is referred to as hysteresis, and it is reflected in two different functional relationships for hydraulic conductivity and degree of saturation for the drying and wetting process. To add to the complications, the nonwetting fluid (e.g., air in the unsaturated zone) requires a certain pressure head, termed *excess pressure*, in order to enter water-filled pores.

As with the solution of saturated groundwater flow, the model user requires initial and boundary conditions for solving for a given multiphase flow problem. However, since the hydraulic conductivity is subject to hysteresis, in addition to the boundary conditions specified in single-phase flow, it is necessary to state whether a wetting or drying process is taking place along the boundary.

For many years, expansion of our society has enforced industrial growth without adequate consideration of environmental protection. One of the not immediately noticeable consequences of industrial growth is the decline of groundwater quality. The reasons for the decline of groundwater quality are:

- Continual use of many products considered harmful to the environment.
- High cost for the proper handling of these products from production to final disposal.
- Lack of legislation to protect groundwater.
- Lack of enforcement of regulations that do exist.

3 Reviewing the Mathematical Model for Transport

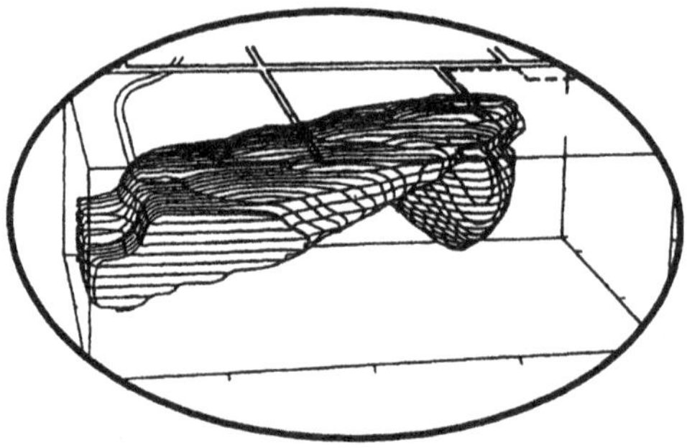

Chapter 3 provides an introduction to groundwater contamination issues from the model user's point of view. Contaminant transport through granular aquifers is briefly reviewed as a basis for the mathematical model for transport modeling. Although numerical modeling of contaminant transport can be achieved without in-depth knowledge of the underlying mathematical model, an understanding of the physics of contaminant transport is nevertheless required.

3.1 DISCOVERING SIMILARITIES IN GROUNDWATER CONTAMINANTS

This section provides an introduction to groundwater contaminants. Natural groundwater quality is analyzed to identify human impact upon groundwater quality. The section also distinguishes the sources from which groundwater contaminants may originate. The last subsection classifies contaminants by transport behavior.

3.1.1 Analyzing Natural Groundwater Quality

Groundwater is not pure water. It contains dissolved elements and suspended particles. Water quality can also be affected by bacteria, viruses, gases, and so on. The type and concentration of dissolved elements depends on the chemical composition of the rock formations in which groundwater flow occurs and on the travel time of water in these rocks. If groundwater movement is slow and minerals are relatively soluble, the dissolving process will finally result in chemical equilibrium between the water and the mineral. Hence the range of chemical compositions of groundwater, as shown in Figure 3.1, is great. Figure 3.1 compares concentrations with the sugar concentration if a sugar cube of 2.7 grams is dissolved in a bottle of water (2.7 liters) yielding a concentration of 10^{-3} mg/l, in a gasoline truck approximating a concentration of 1 mg/l, in an oil tanker representing a concentration of 10^{-3} mg/l, or in a lake as the approximate of a concentration of 10^{-6} mg/l. A more comprehensive discussion of the chemical constituents of groundwater can be found in Hem (1970).

In most cases the model user is concerned with the assessment of groundwater contamination or pollution. Both terms are generally reserved for adverse groundwater quality changes related to human activities, which introduce contaminants directly or indirectly to the groundwater. Information about the natural groundwater quality, however, is essential to quantify human impact on natural groundwater quality. High naturally occurring concentrations of nitrate, for example, which commonly exceed 80 mg/l in the inland areas of western Australia, would not be classed as contamination or pollution. Similarly high radioactive counts, salt levels, and heavy metal concentrations, as well as others, occur naturally.

Where saltwater intrusion occurs through the abstraction of groundwater, good quality groundwaters can become brackish to saline. Classifying this as contamination is debatable, but various model studies, such as the one discussed in Appendix A.4, address this groundwater quality problem.

It is proposed to define groundwater contamination as undesirable changes in groundwater quality that do not render the groundwater unfit for domestic consumption. Thus, if septic tank effluent increases the nitrate content in groundwater to 20 mg/l compared to the World Health Organization (WHO) maximum permissible drinking water standard of 45 mg/l, this would constitute groundwater contamination. If, however, the nitrate content were to rise above 45 mg/l so that the groundwater is unsuitable for drinking, the groundwater would then be classified as polluted.

The use of the term contaminants or pollutants is consequently arbitrary and what is contamination today can become pollution tomorrow if an increase in effluent disposal, leaching, or rate of leakage occurs, or if new more rigorous water quality standards are introduced. For example, in Germany the maximum allowable concentration of nitrate in drinking water prior to 1980 was 80 mg/l. The limit was then reduced to 50 mg/l in accordance with the European Economic Community (EEC) standards, so that overnight between 5 and 8 percent of groundwater production was reclassified as polluted.

In general, most model users will deal with groundwater pollution caused by chemicals. Among the most common groundwater contaminants are carbon tetrachloride (CTET), 1,1,1- trichloroethane (TCA), and trichloroethylene (TCE). Figure 3.2 lists common organic groundwater contaminants. Solubility and drinking water standards are given. For a complete list of U.S.

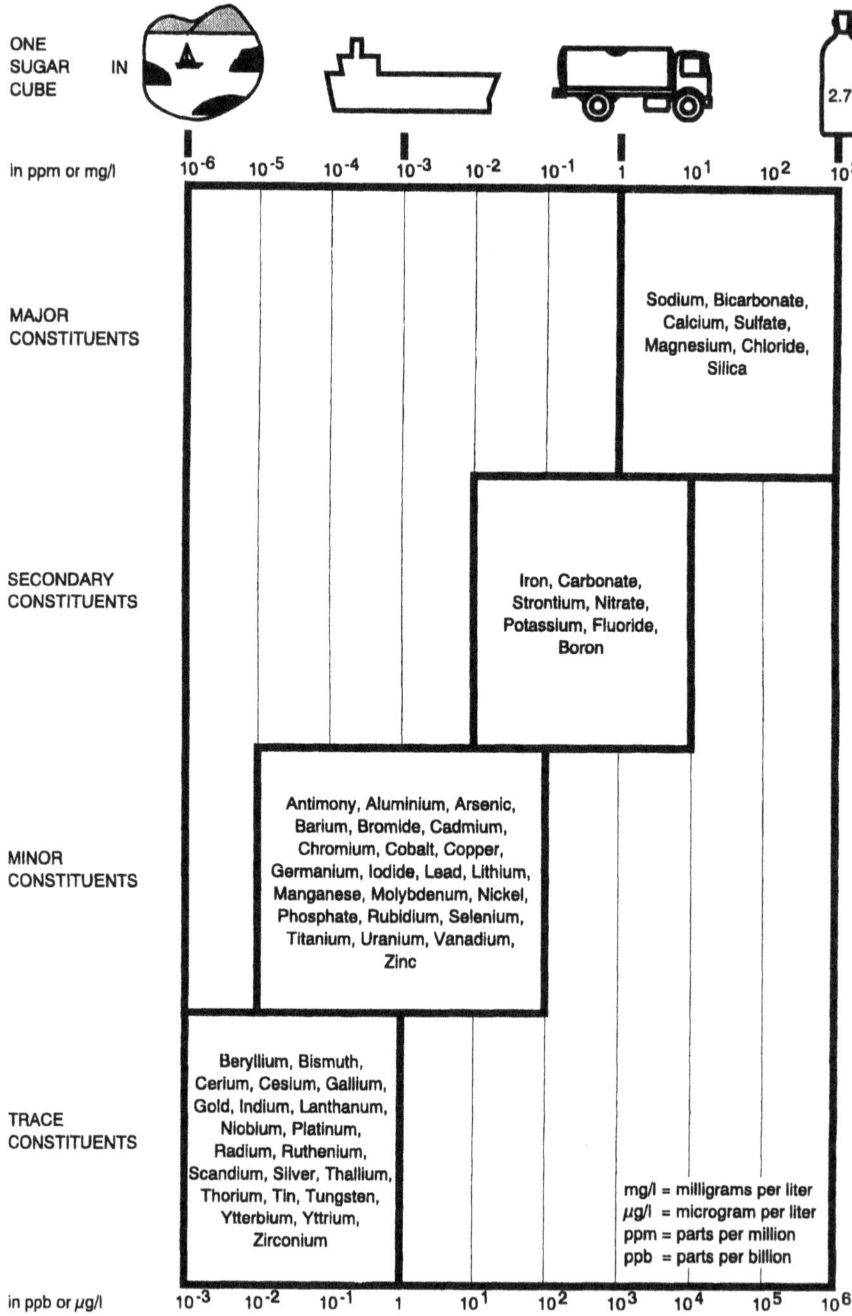

Figure 3.1 Dissolved constituents in groundwater (after Davis and de Wiest 1966).

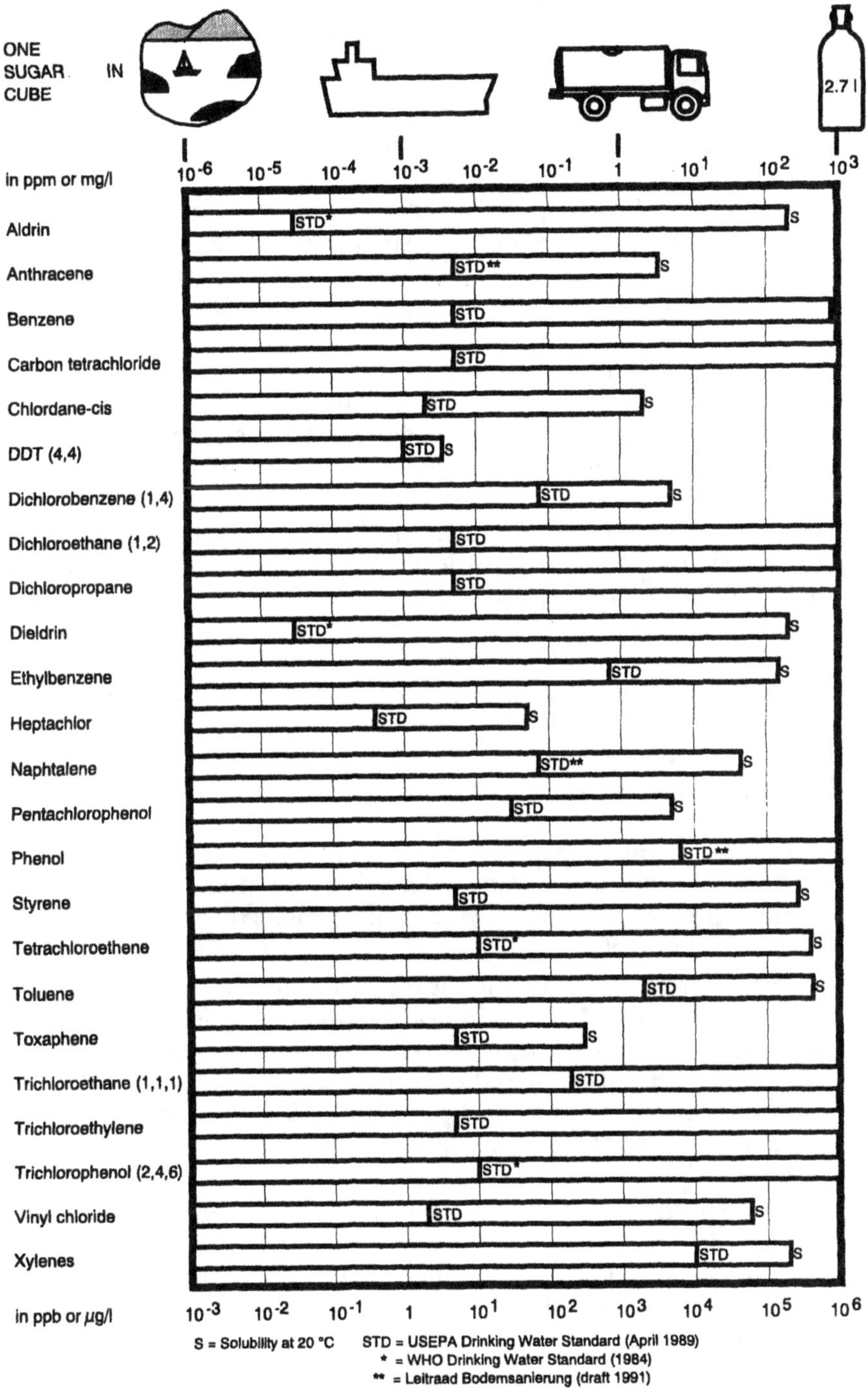

Figure 3.2 Common organic groundwater contaminants.

Environmental Protection Agency (USEPA) priority pollutants and their main characteristics, see Appendix C.1. Drinking water standards are listed in Appendix E.

3.1.2 Distinguishing Sources of Groundwater Contamination

The evaluation and cleanup of contaminated land is one of the highest priority environmental issues in the western world. Because concerns with groundwater contamination figure prominently in these cleanup operations, a dramatic increase in transport modeling activities has occurred in the last decade.

Groundwater contamination from domestic wastes has been with us since humans first walked the earth. But because the subsurface soils act as very effective filters, and the velocity of groundwater flow is relatively slow, viruses and bacteria do not usually travel too far from the source, so that groundwater contamination from domestic sewage was not recognized as contamination. Although the close proximity of waterwells to sewage pits has been responsible over the years for multitudes of deaths and illnesses, there was little or no awareness of the connection between domestic effluent and contaminated groundwater.

The nature and extent of groundwater contamination changed dramatically with the production of chemicals and toxic wastes, which began on a major scale with the Industrial Revolution in England in the eighteenth century and has continued unabated for 200 years. Several examples of early industrial contamination are:

- A gasworks plant was built at Norwich, England, in 1815 and abandoned in 1830. Phenolic compounds, originating from whale oil used by the plant, infiltrated to the underlying chalk and were intersected in a water well drilled in 1950.
- The manufacture of soda ash, caustic soda, and chlorine at Barberton, Ohio, began in the 1890s. Calcium and sodium chloride wastes were discharged to the adjacent river and also to holding ponds. These wastes have contaminated the groundwater adjacent to the river for a distance of over 300 kilometers downstream of Barberton. Several towns utilizing the groundwater had to abandon their well fields as early as 1926 due to excessive chloride levels.
- During World War I wastes from the manufacture of explosives near the Thames River, northeast of London, were placed in abandoned chalk pits. In the early 1920s water from a nearby well was reported to be a yellow color derived from the picric acid used in the explosive manufacturing process. By 1942 the contamination extended more than 1.5 km from the source.

There are innumerable similar instances of groundwater contamination, typified by the long residence time of contaminants in the groundwater. By 1988 the list of potentially hazardous waste disposal sites in the United States had grown to over 200,000. Both Canada and the United States now use a list of some 3,500 chemical substances that are considered hazardous waste.

Today similar regulation of contaminated land exists in many countries. In fact all industrialized areas are potentially emitting contaminants to the groundwater. Figure 3.3 illustrates common contamination sources, classified according to the source geometry.

Classifying contamination sources according to their geometry is helpful because the source geometry significantly influences the plume evolution. Usually a contamination source falls into the category of point, line, or areal (nonpoint) source depending on the source geometry and the ratio of source size to the size of the investigated area. The source geometry influences the required plume assessment as a one-, two-, or three-dimensional problem, the size of contaminated aquifer volume, the strength of contaminant dilution, the impact on the biodegradation capability of the aquifer, and the remedial measures. The classification scheme according to the source geometry is an attempt to categorize sources and cannot be fixed. For example, depending on the specific situation, agricultural areas may fall into the category of line sources or point sources.

3.1.3 Classifying Contaminants by Transport Behavior

The frequent discoveries of groundwater contamination highlight the need to understand the transport behavior of water contaminants. Density, viscosity, solubility, chemical stability and other characteristics of the contaminant affect its spreading in the subsurface, as schematically shown in Figure 3.4. This profile helps in classifying contaminants and in selecting a suitable modeling approach.

Contaminants may reach the groundwater either dissolved in water (miscible) or as a liquid phase that may be immiscible in water. Some contaminants such as heavy metals may also be adsorbed on fine particles and be transported in suspension. Depending on the particle size, transport of suspended particles may be equivalent to that of dissolved contaminants. Miscible liquids (or dissolved contaminants) migrate through the aquifer differently than immiscible fluids. Miscible fluids and the natural groundwater form one phase whose movement can be expressed by Darcy's law (single-phase flow). Immiscible liquids flow in separate phases through the pore space (multiphase flow). Modeling multiphase flow is by far more complex than simulating single-phase flow. Multiphase flow occurs for example in the unsaturated zone (water–air) or in hydrocarbon or chlorinated solvent spills (hydrocarbon–water or chlorinated solvent–water phases).

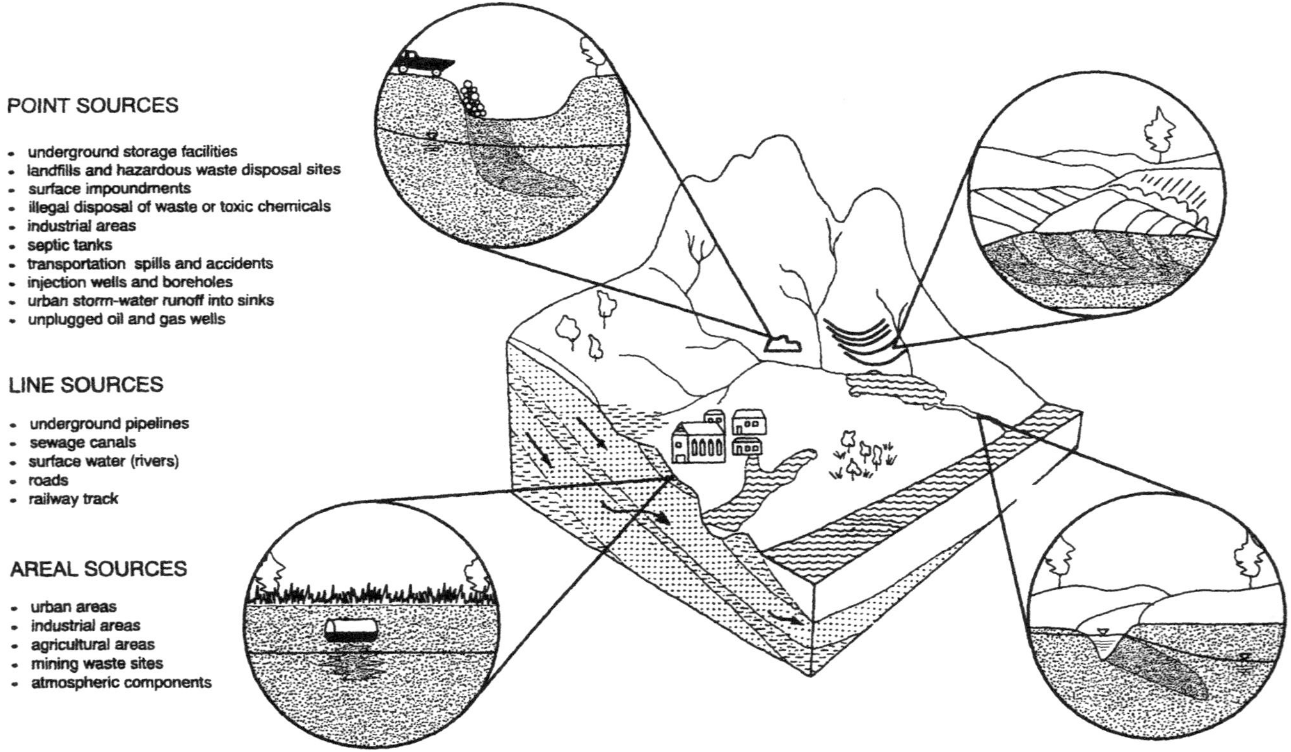

Figure 3.3 Classification of contamination sources according to source geometry.

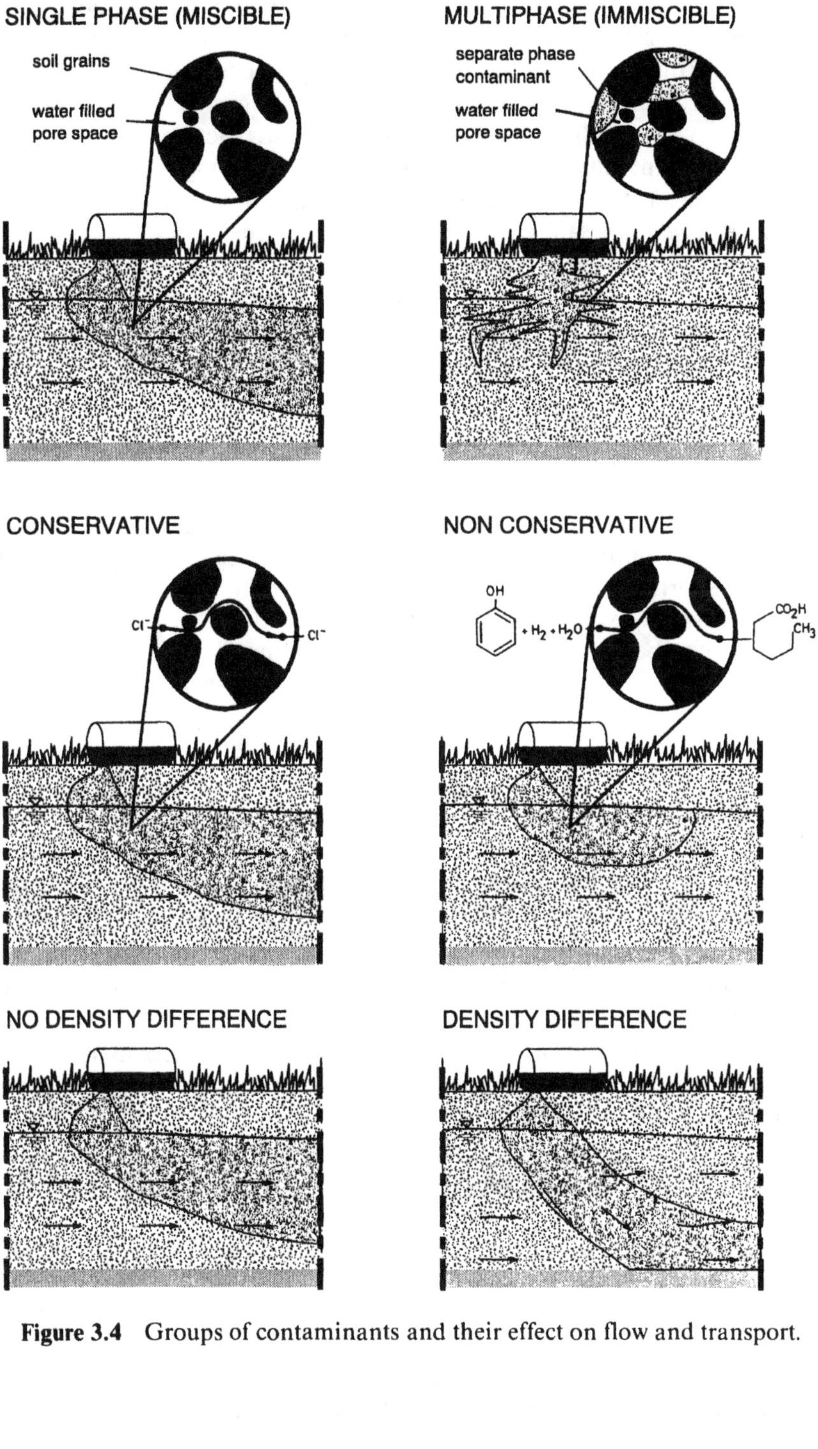

Figure 3.4 Groups of contaminants and their effect on flow and transport.

Solutes that remain stable in groundwater and do not change due to physical, biological, or chemical processes are described as conservative. Typical examples of conservative solutes are fluorescin, which is widely used in migration tests, and chloride. Describing the behavior of a specific water component as conservative depends on the ratio of migration to reaction time. For example, a contaminant that reacts and changes slowly, but migrates in an aquifer with high transport velocities, may be considered conservative.

Conservative solutes that do not influence the viscosity and density of water significantly can be used as tracers. Tracers are used in migration tests to determine the average transport direction and velocity, as well as the aquifer dispersivity. Normally the concentration determines whether a solute changes the physical characteristics of water. The best example of a solute that affects groundwater flow is salt in the transition zone between fresh- and seawater.

A groundwater contamination plume generally contains numerous contaminants. Each contaminant within the plume may migrate differently, and some will react with each other. In transport predictions one normally concentrates on the contaminant that is considered the most significant for the investigated transport problem. Consider the prediction of the maximum spreading of a plume. Here a conservative contaminant like chloride will give the largest plume and can be used as the contamination index. The maximum decontamination time will depend, however, on contaminants that adsorb on the soil grains.

Transport problems can be classified into six types:

- Heat transport.
- Tracer transport. Conservative contaminants, such as fluorescin or chloride in low concentration, that migrate with the natural unaffected groundwater flow.
- Coupled flow and transport without reactions. Conservative contaminants, such as chloride in high concentration (e.g., seawater intrusion), that influence the natural groundwater flow.
- Transport with reactions. Contaminants, such as trichloroethylene in low concentration, that react chemically or biologically but do not influence natural groundwater flow.
- Coupled flow and transport with reactions. Contaminants, such as ammonium sulfate, that react chemically or biologically and change viscosity and/or density of groundwater significantly.
- Multiphase flow and transport. Immiscible fluids, such as oil-water or transport in the unsaturated zone.

There is no practical numerical transport model that can be applied to all transport problems. Numerical models are normally programmed to solve only a given group of the transport problems. Hence the ability of the model user to apply a numerical model is directly related to the understanding of the processes controlling contaminant transport. Most transport model studies focus on transport of contaminants whose reactions can be approximated by linear equilibrium sorption or first-order decay. Complex chemical transport models, such as kinetic reaction or hydrochemical models, are restricted in their accuracy, due to both the lack of required model input data and the numerical difficulties experienced in solving the governing set of equations.

The physical, chemical, and biological processes that are thought to control contaminant transport, in most model applications, are discussed in the following section.

3.2 REVIEWING TRANSPORT MECHANISMS

The purpose of discussing transport mechanisms is to understand the processes that most strongly influence the migration of dissolved contaminants in saturated flow in granular aquifers. Since it is usually the dissolved contaminants that can affect billions of liters of groundwater and are the most mobile, we will concentrate on this class of contaminants. The transport mechanisms are:

- Advection
- Diffusion
- Dispersion
- Sorption
- Decay
- Hydrolysis, volatilization, and biotransformation
- Transport in aquifers with a pronounced bimodal permeability distribution.

To visualize the combined effect of advection, dispersion, sorption, and biodegradation on contaminant transport, consider two field tests in a sand aquifer at Borden, Canada. In both tests the extensive monitoring generates accurate pictures of the solute distribution at consecutive time increments. The distribution and extent of solutes migrating in groundwater is termed the plume. In the dispersion test (Mackay et al. 1985; Freyberg 1986; Roberts et al. 1986), about 12 cubic meters of water containing a number of solutes was injected. The concentrations of the solutes were low enough not to distort the natural flow regime. Spreading of the plume was observed over a period of about two years. Figure 3.5 shows measured plumes for chloride ions (Cl^-), carbon tetrachloride (CTET), and tetrachloroethylene (PCE). In a different test (Barker et al. 1987), about 1.8 m^3 of groundwater marked with benzene, toluene, and xylenes was injected in the same aquifer to study biodegradation.

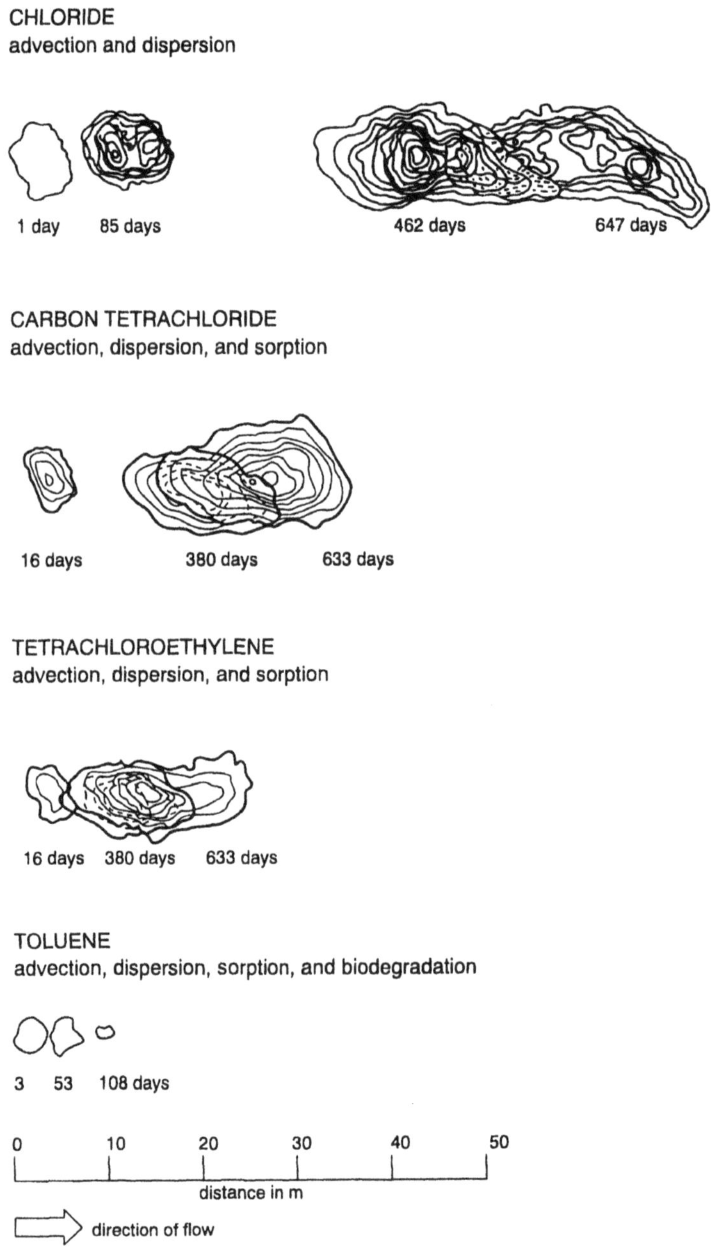

Figure 3.5 Combined effect of advection, dispersion, sorption, and biodegradation in a sand aquifer at Border, Ontario (adopted from Roberts et al. 1986 and Barker et al. 1987).

The spreading of toluene is also shown in Figure 3.5. Consider the growth of each plume:

- Cl^- is a contaminant that remains stable and does not change due to physical, biological, or chemical processes. Advection translates Cl^- with the advective flow velocity. If advection were the sole transport process, a case that does not occur in reality, the shape and concentration of the Cl^- plume would have been unchanged. The presence of dispersion, however, causes a spreading of the initial plume. Spreading due to dispersion also occurs transverse to the main flow direction, carrying Cl^- in aquifer regions that Cl^- would not reach in the absence of dispersion. Dispersive spreading in, and transverse to, the flow direction is at a cost of Cl^- concentrations along the centreline of the plume. Hence a decline of concentration with travel time is observed.
- CTET represents a contaminant that is subject to sorption. Adding linear adsorption to the transport process retards advection and dispersion, as shown in Figure 3.5. At any time, Cl^- moves significantly beyond CTET. This results in a separation of the initial plume containing several solutes, a phenomenon also referred to as a *chromatographic effect.*
- PCE represents a contaminant that is subject to strong sorption. Hence the growth of the PCE plume is less pronounced than the one of CTET. Compared to Cl^-, the average transport velocity of PCE was measured to be as little as $\frac{1}{5}$.
- Toluene represents a water-soluble and mobile gasoline component that is subject to biodegradation. Degradation reduces the contaminant mass and concentrations decrease. The mass of toluene being sampled in the field test at Borden was much less than the mass that was introduced. Little toluene was found on day 108.

A simple linear combination of transport processes does not always conform with the actual complex transport regime. For example, MacQuarrie and Sudicky (1990) conclude that large dispersion enhances biodegradation due to increased mixing with natural water.

The following Subsections 3.2.1 to 3.2.7 discuss the transport process observed in the field test at Borden, together with other processes commonly encountered in contaminant transport. Appendix C.2 provides information on the fate of priority pollutants in the subsurface environment allowing the identification of dominant transport mechanisms for the pollutant under review.

3.2.1 Understanding Advection

Advection is the mass transport caused by the bulk movement of flowing groundwater. The driving force is the hydraulic gradient. The average transport velocity v is calculated as the Darcy flux q divided by the effective porosity n_e:

$$v = \frac{q}{n_e} \tag{3.1}$$

The actual transport velocity within the pore space remains unknown. During the evolution of most plumes, advection is the important transport mechanism. Consequently the model user must concentrate on a fairly accurate description of groundwater flow in the transport simulation.

If advection is the sole transport process, then concentration changes in a control volume over time are equal to the difference between advective inflows and outflows. The following equation represents the one-dimensional transport equation in its simplest form, assuming the aquifer to be homogeneous:

$$\underbrace{\frac{\partial c}{\partial t}}_{\text{change in storage}} = \underbrace{-v \frac{\partial c}{\partial x}}_{\text{advective in/outflow}} \tag{3.2}$$

where

x = direction of flow in [L]
t = time in [T]
c = concentration in [M/L^3]
v = advective transport velocity in [L/T]

In highly permeable materials such as sand and gravel, advection is the most important transport process, and each transport prediction will only be as accurate as the flow description. The course by which groundwater and any accompanying contaminants travel to a discharge point is controlled by the geological setting and by the hydraulic gradient imposed by given boundary conditions. Water then follows the path of least resistance. In gentle topography, and in the absence of strong sinks and sources, the hydraulic gradient normally conforms with the topography. Under these conditions the typical magnitude of the gradient is a few tenths of a percent. Considering the hydraulic conductivities for sand and gravel aquifers, groundwater velocities typically range between 1 meter/year and 1,000 meters/year, the most common velocities being between 10 meters/year and 100 meters/year.

Groundwater flow, and consequently contaminant transport, in karstic or fractured aquifers is governed by a set of factors different from those for porous aquifers. Unknown preferred pathways, often on a large scale, will make transport predictions difficult if not impossible. The ability of groundwater models to develop numerical solutions for a problem that is not directly observable is limited in such aquifers.

Advective flow becomes more complex when the density and/or the viscosity of water change with solute concentration. Solutes with large density, in contrast with water, introduce a larger vertical component of flow movement. Depending on the extent of contamination, migration behavior can be drastically changed, allowing contaminants to flow against the natural groundwater flow direction or to flow "uphill." The effect of aquifer heterogeneities on flow is magnified in density-dependent groundwater flow.

3.2.2 Understanding Diffusion

Diffusion is the net flux of solutes from a zone of higher concentration to a zone of lower concentration. Diffusion does not depend on any bulk movement of the solution. The driving force is the random movement of ionic and molecular constituents under the influence of their kinetic activity called *Brownian motion.* Diffusion is irreversible. Once two solutions are mixed under natural conditions, they will form a single uniform solution. The global movement of solutes occurs in the direction of the concentration gradient and can be expressed by Fick's law:

$$\underbrace{q_o}_{\text{diffusive flux}} = \underbrace{-D_o \frac{\partial c}{\partial x}}_{\text{concentration gradient}} \tag{3.3}$$

Fick's law simplifies diffusive mass flux q_o as the product of concentration gradient and a coefficient D_o in $[L^2/T]$ called the *diffusion coefficient.* Typical values of diffusion coefficients in free solution are in the range of 10^{-9} m^2/sec. See Appendix D.1 for a detailed list of diffusion coefficient values. The coefficients are somewhat smaller in porous material because of the presence of the solid grain matrix. The diffusion coefficient in porous materials is termed *effective* to distinguish it from the diffusion coefficient measured in free solution.

Including diffusive contaminant transport in the one-dimensional transport equation (Equation 3.2), we now find that:

$$\underbrace{\frac{\partial c}{\partial t}}_{\text{change in storage}} = \underbrace{-v \frac{\partial c}{\partial x}}_{\text{advective in/outflow}} + \underbrace{D_o \frac{\partial^2 c}{\partial x^2}}_{\text{diffusive flux}} \tag{3.4}$$

Diffusion processes are relatively slow transport processes. This is illustrated by the following example addressing diffusive mass transport in clay (after Freeze and Cherry 1979). Assuming a constant source concentration at the boundary of a clay layer, with no advective groundwater flow but an effective diffusion coefficient of 5×10^{-9} m^2/sec, the concentration at 10 m from the source would show 10 percent of the source concentration after 500 years migration time. Hence the effect of diffusion will normally be masked by the effect of advection in groundwater zones with high flow velocities. Diffusion over geological time, however, can have a significant impact. Studies that address contaminant transport in fractured porous aquifers stress that diffusion in these geological settings may also significantly influence contaminant migration (see also Section 3.5).

3.2.3 Recognizing the Importance of Dispersion

In dealing with contamination transport, one of the main tasks is predicting dispersive spreading. Dispersive spreading, within and transverse to, the main flow direction causes a gradual dilution of the contamination plume. In some situations, when concentrations fall below the concentration values considered to be harmful to humans, dispersion may be a benefit. More often dispersion is an undesirable effect due to the fast spreading of contaminants, which increases the volume of contaminated groundwater. Dispersion is superimposed on the average transport velocity, and it also transports contaminants into zones that do not participate directly in the advective plume evolution. Dispersive spreading will lead to an increase in plume uniformity with travel distance.

Dispersive solute transport in aquifers has been intensively studied and a great amount of literature exists. In this section dispersion is approached by:

- Identifying factors that are likely to be interpreted as dispersion.
- Presenting Fick's law as the conceptual theory to describe dispersion.
- Explaining the scale dependency of dispersion.
- Reviewing methods that quantify dispersion.

When interpreting field observations of concentration patterns, be aware of the numerous factors that influence contamination transport. Variations in advective transport, caused by layers, density effects, flow field changes over time, as well as emission of contaminants from unknown contamination sources at changing rates, are likely to be interpreted as dispersion. Each factor may cause a breakthrough curve in a monitoring well that may be interpreted as an effect of aquifer dispersivity, as illustrated in Figure 3.6. In studies with insufficient data, the lack of field data is hidden in the dispersion term of the governing transport equation. However, dispersion stands exclusively for dispersive spreading due to aquifer heterogeneities.

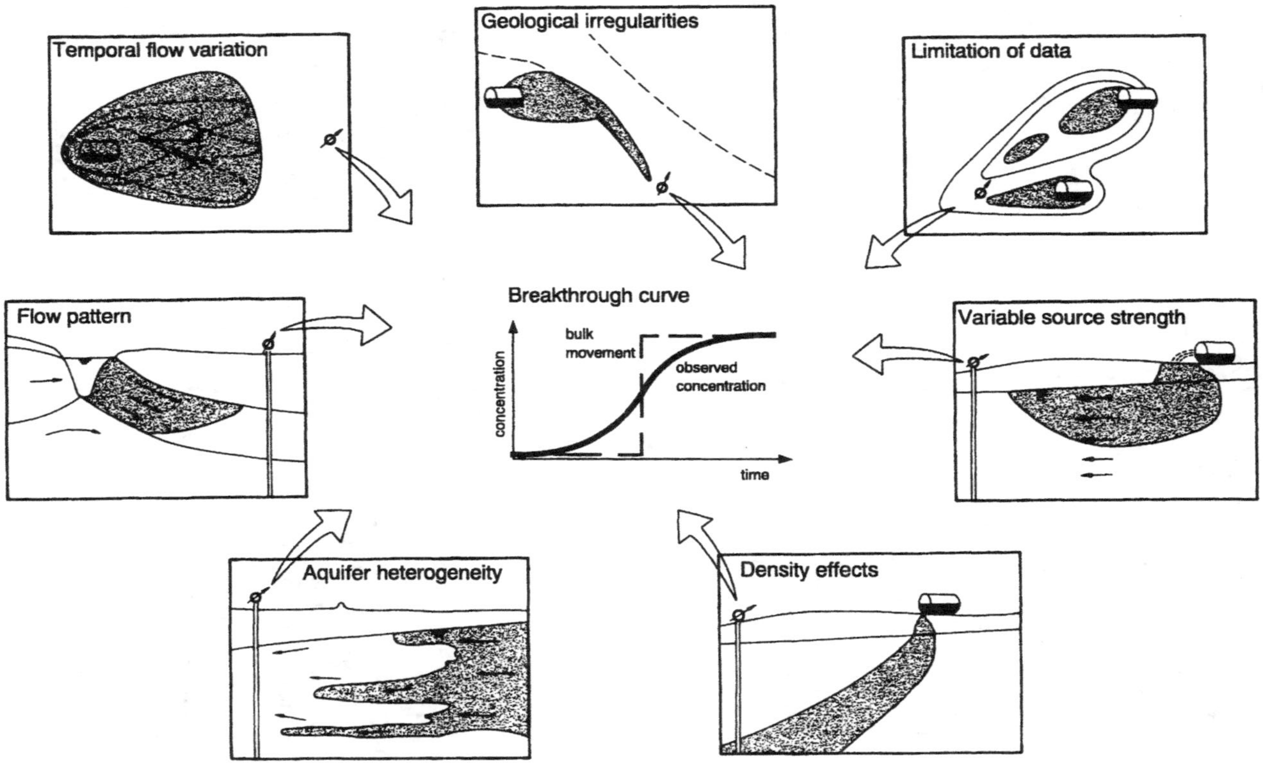

Figure 3.6 Factors which are likely to be interpreted as dispersion.

Groundwater transport models do not solve flow and transport in pores. The realistic alternative is to move on to a coarser scale of description by introducing measurable phenomenological coefficients such as hydraulic gradients. In the continuum approach (Bachmat and Bear 1964), the concept of the representative elementary volume (REV) is evoked. It is a theoretical approach in which representative values of flow and transport parameters are averaged over an appropriate volume. Using the diffusion analogy of Fick's law which summarizes the random motion of individual particles as an average flux, the unknown dispersive movement of contaminants through the pore-grain system is described by Fick's law. While Darcy's law combines average advective fluxes with average hydraulic gradients and hydraulic conductivities, Fick's law combines average diffusive fluxes with average concentration gradients and average dispersion coefficients as phenomenological coefficients. This leads to the well known advection-dispersion equation, which is similar to the advection-diffusion equation (Equation 3.4) with the exception that an unknown dispersion coefficient D_L now combines the effect of diffusion and dispersion in the flow direction due to aquifer heterogeneity. Consequently the advection-dispersion equation is written for the one-dimensional transport regime as

$$\underbrace{\frac{\partial c}{\partial t}}_{\text{change in storage}} = \underbrace{-v \frac{\partial c}{\partial x}}_{\text{advective in/outflow}} + \underbrace{D_L \frac{\partial^2 c}{\partial x^2}}_{\text{diffusive and dispersive flux}} \tag{3.5}$$

In accordance with the different orders of aquifer heterogeneities, dispersion can be visualized as taking place at several scales. These scales are shown in Figure 3.7.

At the pore-grain scale (heterogeneities of first order), contaminants are spread by mechanical dispersion. Diffusion within the pores, velocity variation in the pore channels, mixing at pore junctions, and similar mechanisms occurring during transport within the pore space contribute to dispersive spreading at the microscale.

At the pore-matrix scale, larger-scale heterogeneities within the pore matrix (heterogeneities of second order) contribute to the dispersive spreading. Permeability differences within the pore matrix cause a flow field with velocities varying in a complex irregular manner in three dimensions. Flow in highly permeable regions will carry the contaminant faster than the average transport velocity, generating dispersive spreading similar to that in the pore network.

At field scale commonly encountered geological structures influence contaminant transport drastically. Whether or not individual geological structures dominate the transport behavior depends mainly on the size of the contaminated aquifer volume, in comparison to that of the geological hetero-

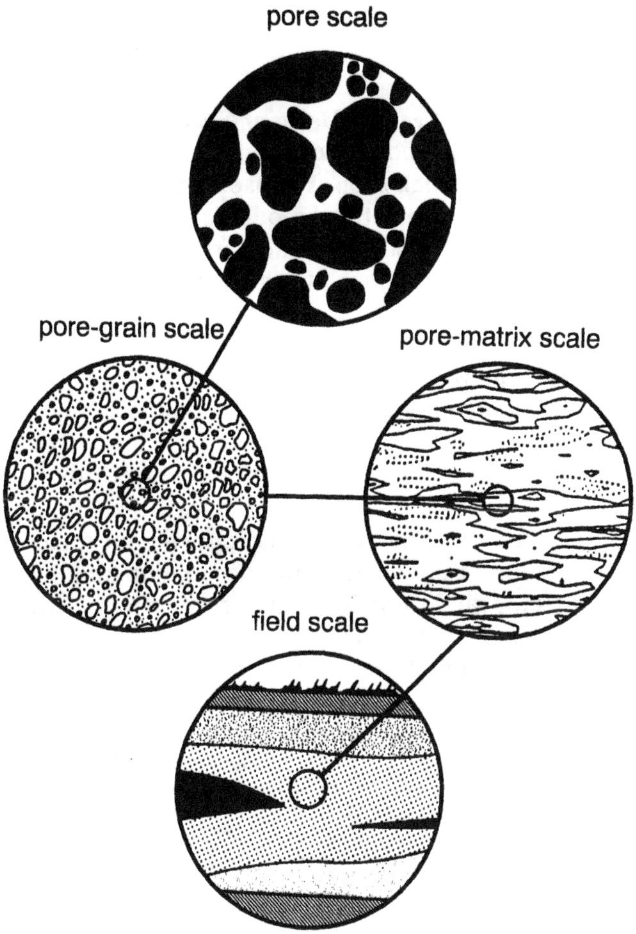

Figure 3.7 Scales of aquifer heterogeneities.

geneities. To predict dispersive transport at field scale is difficult. With the presence of geological large-scale heterogeneities, advective fluxes are still random at pore-grain and pore-matrix scale, but average Darcy fluxes on a large scale are related to the hydraulic conductivities of individual geological structures where there is a lack of accurate field data, dispersive spreading at field scale can still be approximated as a Fickian process, but the corresponding dispersion coefficients are several orders higher than those used to describe mechanical dispersion at pore-grain scale.

To predict dispersive spreading, the dispersion coefficient must be quantitatively related to known aquifer and fluid parameters. Attempts to quantify dispersion coefficients can be grouped in:

- Geometrical models
- Statistical models

- Hydraulic models
- Stochastic analysis

Quantifying Dispersion Coefficients by Means of Geometrical Models
Initial investigations of dispersion have aimed to quantify the dispersion coefficient using geometrical models such as transport through tubes (Taylor 1954; Aris 1956). However, since geometrical models cannot be easily extended to assess dispersion in natural materials, alternative methods to approximate dispersion under natural conditions also have been developed at the same time.

Quantifying Dispersion by Means of Statistical Models
Statistical theory does not provide a functional relationship between aquifer characteristics and the dispersivity values, but it forms the framework for classical dispersion theory. Statistical theory is developed directly from the physical processes commonly attributed to produce dispersive spreading. It is based on the random walk theory, which also provides the theoretical framework in the description of Brownian molecular motion and in groundwater modeling, the basis for random walk models (see Section 6.5). Scheiddegger (1954) developed a statistical model and introduced a constant of dispersion, which he called the medium dispersivity,

$$D = \alpha v \tag{3.6}$$

where

D = dispersion coefficient in $[L^2/T]$
α = dispersivity in $[L]$
v = transport velocity in $[L/T]$

The mechanical dispersion coefficient depends linearly on the transport velocity and a characteristic of the porous media referred to as *dispersivity*. As a consequence of the assumption of random steps independent of direction, the dispersion coefficient is considered to be scalar in Scheiddegger's model. De Josselin de Jong (1958) and Saffman (1960) showed that mechanical dispersion is more pronounced in the longitudinal direction than in the transverse direction. Dispersion coefficients for both directions were defined:

Longitudinal dispersion coefficient

$$D_L = \alpha_L v \tag{3.7}$$

Transverse dispersion coefficient

$$D_T = \alpha_T v \tag{3.8}$$

where

α_L = longitudinal dispersivity in [L]
α_T = transverse dispersivity in [L]

The general dispersion theory by Bear (1961b) and Scheiddegger (1961) expresses the dispersion coefficients as a second-rank tensor that is related to a fourth-rank dispersivity tensor. For an isotropic medium the dispersion coefficients can be reduced for two dimensions:

$$D_{xx} = \frac{\alpha_L v_x^2}{v} + \frac{\alpha_T v_y^2}{v} \tag{3.9}$$

$$D_{yy} = \frac{\alpha_T v_x^2}{v} + \frac{\alpha_L v_y^2}{v} \tag{3.10}$$

$$D_{xy} = D_{yx} = \frac{[\alpha_L - \alpha_T] v_x v_y}{v} \tag{3.11}$$

with

$$v = \sqrt{v_x^2 + v_y^2} \tag{3.12}$$

Diffusion is generally taken into account by adding the diffusion coefficients to the diagonal terms of the tensor. By rotating the coordinate system in the direction of flow, the coefficients are reduced to the one of longitudinal and transverse dispersion. Because there is no feasible method to determine the dispersivity tensor for anisotropic media (due to the number of components and the fact that the flow direction varies from point to point), the assumption of isotropy with respect to dispersion is generally used in application of the advection-dispersion equation in field studies.

Quantifying Dispersion by Means of Hydraulic Models

The magnitude of the dispersivities for homogeneous sands is traditionally obtained from hydraulic models by fitting the appropriate solution of the advection-dispersion equation with the observed concentration pattern. A comprehensive review of dispersion experiments is given by Spitz (1985; see also Appendix D.2). Typical values for the dispersivities at the pore-grain scale are

$$\alpha_L \approx \text{average grain diameter} \tag{3.13}$$

and

$$\alpha_T \approx \frac{1}{10} \alpha_L \tag{3.14}$$

Quantifying Dispersion by Means of Stochastic Analysis

Stochastic analysis enables the magnitude of the dispersion coefficients on the pore-matrix scale to be limited with the statistical properties of randomly distributed aquifer heterogeneities (Gelhar and Axness 1983; Dagan 1982; Sudicky 1986).

The stochastic analysis of macro-dispersion is widely accepted as the most general approach for assessing transport at pore-matrix scale. Local hydraulic conductivities are assumed to vary from point to point, but in a global sense the aquifer is assumed to be (statistically) homogeneous. The stochastic approach is basically equivalent to the averaging procedure used at the pore-grain scale. Since individual Darcy fluxes at every point in a heterogeneous aquifer are never known in detail, transport on the pore-matrix scale is translated into a coarser level of transport description by defining average aquifer properties (e.g. macroscopic hydraulic conductivities or macroscopic dispersivities) that have the effects of the small-scale variations embedded in them. In stochastic analysis the complex aquifer structure is described in terms of statistical properties of the aquifer parameters.

Macroscopic parameters are calculated for different configurations of aquifer heterogeneities by Dagan (1982) and Gelhar and Axness (1983). Mechanical dispersivities are added to the values obtained from stochastic theory. Theory indicates that transverse dispersion at a large scale remains controlled by mechanical mixing. This conforms to field studies on sand and gravel aquifers. The ratio of longitudinal and transverse dispersivities typically ranges from 10 to 100.

Despite the convincing stochastic theory, there are few methods to predict the magnitude of dispersion for a previously unstudied field situation. Generally, dispersive spreading in nature appears to be significantly stronger than dispersion observed in hydraulic laboratory tanks or calculated in theory. Dispersion coefficients are calculated to be several orders higher than the ones found for mechanical dispersion ($\alpha_L \leq 20$ m in controlled field experiments, $\alpha_L \leq 100$ m obtained from plume studies; Gelhar et al. 1992). This is not surprising since aquifer heterogeneities and therefore velocity variations occur at a larger scale in a field situation. In addition, in field study it is most likely that the plume will encounter large-scale heterogeneities that influence plume evolution beyond that predicted by stochastic theory for a statistically homogeneous aquifer.

Also the analogy between the diffusion process and dispersive spreading in a real aquifer cannot be strictly maintained. Velocities at field scale do not appear random but are linked to larger-scale heterogeneities. To use the Fickian theory to predict dispersive spreading requires that time-dependent or, respectively, travel-distance-dependent dispersion coefficients be introduced (Naff 1984). Published data summarized in Appendix D.3 suggests that, as a rule of thumb, longitudinal dispersivity is about 0.1 of the travel distance:

$$\alpha_L \approx 0.1 \text{ travel distance} \tag{3.15}$$

The actual value may however differ significantly from this simple approximation.

In conclusion, estimates of dispersion prior to the field study are characterized by a high degree of uncertainty. Dispersivities are fitted values in most model applications. One first tries to gain a knowledge of the variation of the flow field, the source strength, and the like, to obtain dispersivity values that approximately represent the dispersivity of the aquifer rather than the lack of input data. Equation 3.15 may be used in a first attempt to estimate dispersivities in field applications.

3.2.4 Interpreting the Effects of Sorption

Sorption refers to adsorption and desorption. Adsorption describes the adhesion of molecules or ions to the grain surface in the aquifer. The release from the solid phase is called *desorption*. Adsorption causes diminution of concentrations in the aqueous phase and a retardation of contaminant transport compared to water movement. The degree of sorption depends on a number of factors, including the concentration and the characteristics of the contaminant, the soil type and its composition, the pH value of water, and the presence of other water solutes. Each of these factors may vary in time and space, resulting in a variation of retardation in the natural environment. A source/sink term S, representing the rate at which solute mass is lost from or added to the solution phase (mass of solute per unit volume of solution per unit time), is introduced to the transport equation to represent sorption:

$$\underbrace{\frac{\partial c}{\partial t}}_{\text{change in storage}} = \underbrace{-v\frac{\partial c}{\partial x}}_{\text{advective in/outflow}} + \underbrace{D_L\frac{\partial^2 c}{\partial x^2}}_{\text{dispersion and diffusion}} - \underbrace{S}_{\text{source/sink due to sorption or decay}} \tag{3.16}$$

Equation 3.16 is the one-dimensional form of the advection-dispersion equation for the homogeneous aquifer, modified to account for the effect of sorption. Expressed in this manner, the equation describes any chemical or biochemical reaction involving the alteration of contaminants in groundwater. To be of practical use in model applications, the source/sink term must be expressed in a more specific form. In the following paragraphs the reaction term for sorption and decay are resolved.

The rates of adsorption onto the solid material as related to the concentration in the groundwater is expressed by adsorption kinetics. Which kinetic reactions apply and the way sorption is incorporated into the source/sink term S depends on the ratio of adsorption time to the typical time scale for flow. If adsorption processes are relatively fast compared to the advective transport time of contaminants, equilibrium between adsorbed concentration c_a and dissolved concentration c can be assumed (*equilibrium adsorption*). The relationship between the concentration of a solute in the adsorbed phase and in the adjacent water phase, at equilibrium, is called an *adsorption isotherm* and is expressed as:

$$c_a = f(c) \tag{3.17}$$

The simplest equilibrium relationship that can be found in transport problems with relatively small concentrations is the linear isotherm:

$$c_a = K_d c \tag{3.18}$$

where

c_a = adsorbed concentration measured as mass of contaminant per mass of dry matrix material in [1]
c = dissolved concentration measured as mass of contaminant per volume of water in [M/L^3]
K_d = distribution coefficient in [L^3/M]

The distribution coefficient K_d ranges typically between 0 and 30,000 ml/kg. Appendixes D.4 and D.5 present a compilation of K_d values for metals and organics respectively. However, the actual value will always be site specific and, in many cases, dependent on the local pH.

To establish the mass balance over the control volume to specify S, consider the total contaminant mass in a unit volume of aquifer given as

$$\Delta M = cn + c_a (1 - n) \rho_s \tag{3.19}$$

where

ΔM = total contaminant mass per unit aquifer volume in [M/L^3]
n = total porosity in [1]
ρ_s = density of dry matrix material in [M/L^3]
c = dissolved contaminants in [M/L^3]
c_a = adsorbed contaminants in [1]

Assuming constant mass in the aquifer volume over a given time period, the transfer of mass between the dissolved and adsorbed phase obeys

$$\frac{\partial c}{\partial t} = -\frac{(1 - n) \rho_s}{n} \frac{\partial c_a}{\partial t} = S \tag{3.20}$$

After substituting c_a in Equation 3.18, this becomes

$$S = \frac{(1 - n) \rho_s}{n} K_d \frac{\partial c}{\partial t} \tag{3.21}$$

Alternative expressions for c_a are used to describe more complex isotherms.

After a few rearrangements this source/sink term S for a linear isotherm leads to a one-dimensional transport equation identical in form to the advection-dispersion equation:

$$\underbrace{\frac{\partial c}{\partial t}}_{\text{change in storage}} = \underbrace{-\frac{v}{R}\frac{\partial c}{\partial x}}_{\text{retarded advective in/outflow}} + \underbrace{\frac{D_L}{R}\frac{\partial^2 c}{\partial x^2}}_{\text{retarded dispersion and diffusion}} \tag{3.22}$$

Thus equilibrium adsorption described by linear isotherm causes a retardation of contaminant migration compared to the advective transport with the average transport velocity. Such retardation is described by the dimensionless retardation factor R, defined as

$$R = 1 + \frac{(1 - n)\rho_s}{n} K_d \tag{3.23}$$

Sometimes R is also given as

$$R = 1 + \frac{\rho_b}{n} K_d \tag{3.24}$$

where ρ_b is the bulk density of the dry matrix material;

$$\rho_b = (1 - n)\rho_s \tag{3.25}$$

If linear, reversible, equilibrium sorption occurs, transport is identical to tracer transport (e.g., chloride in low concentration), except that the effective transport velocities are smaller than the bulk movement due to advection:

$$v_{\text{solute}} = \frac{1}{R} v_{\text{water}} \tag{3.26}$$

Thus contaminants would be expected to migrate at a lesser rate than the groundwater itself.

The K_d model may also be applied to describe transport with ion-exchange or hydrophobic sorption, although the mechanisms of hydrophobic sorption are still unclear. Experiments show that the greater the organic content of the aquifer, the greater is the sorption potential. Hydrophobic organics in an aqueous solution will leave the aqueous phase in preference to the solid organic phase present in the soil when added to the soil material. Solubility or, alternatively, the octanol-water partition coefficient K_{ow} are a measure of this preference. See Appendix C.1 for a compilation of solubility and K_{ow} values.

There are several methods of evaluating K_d. For hydrophobic organics, correlation equations determine the order of K_d. It has been found that the proportion of organics in the solid phase increases linearly with the organic carbon content of the soil:

$$K_d = K_{oc} f_{oc} \tag{3.27}$$

where

K_d = distribution coefficient in $[L^3/M]$
K_{oc} = organic carbon partition coefficient for organic solute under consideration in $[L^3/M]$
f_{oc} = weight fraction of organic carbon in soil material

The organic carbon weight fractions are usually low values and very difficult to measure accurately with f_{oc} typically varying between 0.01 and 0.001. See also Appendix D.6 for a compilation of f_{oc} values. The organic carbon partition coefficient K_{oc} has been tabulated for many environmentally important organic compounds and can be computed from the octanol-water partition coefficient K_{ow} by using correlation equations of the form

$$\log K_{oc} = a + b \log K_{ow} \tag{3.28}$$

where a and b are constants (Karickhoff et al. 1979; Schwarzenbach and Westfall 1981). A correlation exists between K_{ow} and solubility that can be approximated for many organics as

$$\log K_{ow} = 5 - 0.67 \log S \tag{3.29}$$

where S is expressed in [mmol/l] (Chiou et al. 1979). Since K_{ow} correlates with solubility, a relationship similar to Equation 3.28 can be defined between K_{oc} and solubility:

$$\log K_{oc} = a + b \log S \tag{3.30}$$

Empirical K_{oc} relationships based on Equations 3.28 and 3.30 are combined in Appendix D.7. In view of the high number of organics, it appears that there is no universal relationship between K_{oc} and K_{ow}, or solubility.

Whenever the K_d model applies, sorption processes can be introduced in transport simulation via the retardation factor defined in Equations 3.23 or 3.24. The higher the fraction of the contaminant sorbed, the more retarded is its transport. The retardation factor may range, for contaminants of concern, over four orders of magnitude, from slightly greater than 1 to more than 10,000. Figure 3.8 shows estimated retardation of organics. The formulas of Karickhoff et al. (1979) were used to produce these plots. Bulk density was assumed to be 2.65 g/cm^3. Porosity was 0.2. Figure 3.8 shows that all contaminants of water solubility less than the solubility of chloroform would be significantly re-

RETARDATION FACTORS BASED ON THE OCTANOL-WATER PARTITION COEFFICIENT K OW

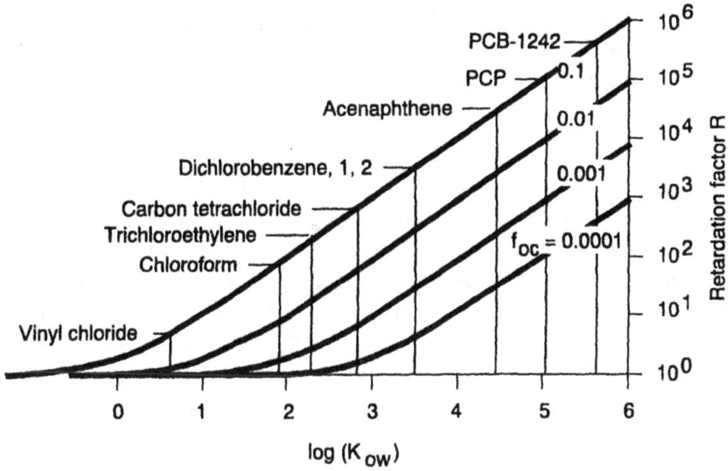

RETARDATION FACTORS BASED ON SOLUBILITY OF ORGANIC

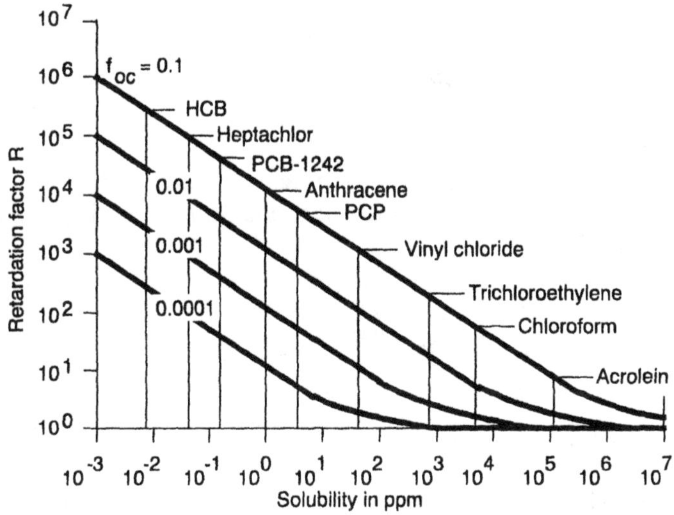

Figure 3.8 Generalized prediction of retardation factors assuming $\rho_s = 2.65$ g/cm^3 and $n = 0.2$.

tarded in a soil with an f_{oc} value of 0.001. In the same soil, vinyl chloride would be retarded by a factor of 10, whereas heptachlor would be retarded by a factor of 1000.

For a number of the adsorbed contaminants, the linear adsorption isotherm does not apply. Most reversible isotherms, however, can be fitted to the more general Freundlich equation:

$$c_a = Kc^{\alpha} \qquad (3.31)$$

where α is an empirical constant. (In the case of linear adsorption, α is equal to 1.) K factors are not defined, and the transport equation cannot be transformed into the conventional advection-dispersion equation. As well as causing a retardation of the plume movement, nonlinear isotherms also create an asymmetry in the concentration distribution. A tailing in the breakthrough curve may indicate a nonlinear adsorption isotherm as well as physical (immobile water) or chemical kinetics. Prediction of contamination migration then becomes more difficult, since the additional transport parameters (e.g., ratio of immobile to total pore space, etc.) must be determined.

3.2.5 Learning about Decay

Not all contaminants that are adsorbed or desorbed follow the principle of fast reactions. Reactions that are relatively slow in comparison to the average travel time of the contaminants are described by kinetics. Reactions of the first order follow the relationship

$$\frac{\partial c}{\partial t} = -\lambda c \qquad (3.32)$$

where λ is a decay constant in $[T^{-1}]$.

The above equation is identical to the source/sink term in Equation 3.16 and therefore the one-dimensional transport equation including the first-order reaction is as follows:

$$\underbrace{\frac{\partial c}{\partial t}}_{\text{change in storage}} = -\underbrace{\frac{v}{R}\frac{\partial c}{\partial x}}_{\text{retarded advective in/outflow}} + \underbrace{\frac{D_L}{R}\frac{\partial^2 c}{\partial x^2}}_{\text{retarded dispersion and diffusion}} - \underbrace{\lambda c}_{\text{decay and/or degradation}} \qquad (3.33)$$

Reactions of the first order are also applied to describe radioactive decay and/or simple degradation processes:

$$c = c_o e^{-\lambda t} \qquad (3.34)$$

where

$$\lambda = \frac{\ln 2}{T_{1/2}} \tag{3.35}$$

and

c = concentration at time t in $[M/L^3]$
c_o = concentration at time 0 (initial concentration or source concentration) in $[M/L^3]$
λ = rate constant that characterizes decay in $[T^{-1}]$
$T_{1/2}$ = half-lifes of radioactive isotopes or of the degraded contaminant in [T]

Appendixes D.8 and D.9 list some estimates of radioactive and organic half-lifes. Understand that the given half-lifes of organics are only crude estimates. The actual degradation will depend on numerous site-specific factors.

Since reactions of the first order are linear and do not change the characteristics of the transport equation, they can be easily built into transport simulations. Concentration values obtained without considering decay are simply multiplied by the factor $\exp(-\lambda/t)$, where t is the time for which concentrations have been calculated. Degradation processes also decrease the source of contamination with time.

3.2.6 Understanding Hydrolysis, Volatilization, and Biotransformation

We have briefly reviewed advection, diffusion, dispersion, sorption, and decay. There are a number of additional physical, chemical, and biological processes that may affect the fate of the contaminants. These processes include, for example, solution, dissolution, filtration, photolysis, hydrolysis, oxidation, chemical reactions with other species, volatilization, bioaccumulation and biotransformation, and biodegradation. Many of these transport processes are still poorly understood, considering the large number of potential groundwater contaminants. The following subsection briefly describes some of these processes. Deficiencies and limitations of information available in literature and model applications for complex transport simulation are identified.

Hydrolysis is the reaction of the contaminant in the water, usually with the introduction of a hydroxyl (OH) group into the contaminant's chemical structure, and commonly with the loss of an ion. The reaction rate may be "catalyzed" by metals, acids, and/or bases and microorganisms. It is also observed that hydrolysis is pH dependent (Mabey and Mill 1978). Hydrolysis may be understood as the pseudo first-order rate half-life of the contaminant. Thus it can be modeled in the same way as decay.

Volatilization of organic chemicals from water to the atmosphere can be an important pathway for chemicals with high vapor pressures. In fact soil and water decontamination measures by air extraction in vadose zone take advantage of the preference of many contaminants to be dissolved in gas phase rather than in the aqueous phase. The factors controlling the vapor concentration in the gas phase are the vapor pressure p of the contaminant under investigation (in atm) and Henry's law constant K_H (in atm m^3/mole):

$$p = K_H c \tag{3.36}$$

where c is the concentration of the contaminant in aqueous phase (in moles/m^3). Assuming equilibrium, Equation 3.36 yields the contaminant concentration in gas phase if gas phase is in contact with the contaminated aqueous phase. Thus Henry's law constant represents the equilibrium partitioning between water and vapor phases:

$$K_H = \frac{\text{concentration in vapor}}{\text{concentration in water}} \tag{3.37}$$

The vapor pressure helps to estimate the evaporation rate of the contaminant. The higher the vapor pressure, the more volatile is the contaminant. For a compilation of Henry coefficient K_H, values see Appendix C.1.

Biotransformation and biodegradation results from the enzyme-catalyzed transformation of the contaminant. Because biotransformations are carried out primarily by microbes, the rate of biotransformation is a function of the microbial mass in the studied aqueous environment. The aqueous and soil bacteria population is generally multispecies, and it is rare for a contaminant to be toxic to all microbes. Also a rapid adaptation of the native bacteria community to the contaminant is usual. In view of this, all organic contaminants are eventually biotransformed or biodegraded somewhere. The only remaining question is when the biotransformation occurs. In a natural environment, where biomass is relatively large and contaminant concentration is low, the degradation rate follows the pseudo first-order reaction. If the microbial populations change significantly, model predictions are considerably more complex.

3.2.7 Understanding Transport in Aquifers with a Pronounced Bimodal Permeability Distribution

Some groundwater systems have a large storage capacity for water or contaminants but a small pore volume that actually participates in flow and transport (unconsolidated aquifers with embedded fine-grained layers, karstic systems, or fractured clay). This results in field observations of concentration patterns that differ from predicted ones. In transport modeling, aquifer regions with little capacity for transmitting water but with a significant contribution to the total aquifer's pore volume may act as a reservoir storing or releasing contaminants. To simulate transient flow or transport accurately in groundwater systems with a pronounced bimodal distribution of porosity and permeability, the following two approaches are common:

- The entire transport regime is considered as an equivalent porous system. The governing equation for transport is extended, introducing the dual porosity concept by Coats and Smith (1964) to account for the storage and release of contaminants in less permeable aquifer regions.
- Contaminant transport is solved using the conventional transport equation (Equation 3.38 presented in Section 3.3) but accounting explicitly for transport in each aquifer region. This approach is recommended for assessing contaminant transport in fractured rock. Fractures allow advective transport in the groundwater system. Additionally contaminants spread into the rock by diffusion. Model studies on contaminant transport in fractured porous rock demonstrate that diffusion into the rock is significant and cannot be ignored in modeling efforts. For a detailed description of transport in fractured porous rock see Germain (1988).

3.3 FORMULATING THE GENERAL TRANSPORT EQUATION

In this section the equation governing one-dimensional migration of a contaminant in a groundwater system is extended to the general three-dimensional transport regime. The second-order partial differential equation describing contaminant transport is

$$\underbrace{\frac{\partial c}{\partial t}}_{\substack{\text{change of mass}\\\text{in aqueous phase}\\\text{over time}}} + \underbrace{\frac{(1-n)}{n}\rho_S\frac{\partial c_a}{\partial t}}_{\substack{\text{change of mass}\\\text{adsorbed over time}}} = \underbrace{-\frac{\partial}{\partial x_i}(v_i c)}_{\substack{\text{in/output}\\\text{due to advection}}} + \underbrace{\frac{\partial}{\partial x_i}\left(D_{ij}\frac{\partial c}{\partial x_j}\right)}_{\substack{\text{in/output due to}\\\text{diffusion and dispersion}}} \underbrace{-\lambda c}_{\substack{\text{loss due}\\\text{to decay}}} \underbrace{-\Sigma Q c_{in}}_{\substack{\text{other}\\\text{sources/sinks}}} \tag{3.38}$$

The set of differential equations that allow the study of contaminant transport in a given groundwater system is a combination of three types of equations:

- Groundwater flow equation (Equation 2.4).

- Contaminant transport equation (Equation 3.38).
- Auxiliary equations such as Equations 3.39 and 3.40 relating physical characteristics of groundwater to the contaminant concentration.

Sometimes groundwater flow is solved analytically or is known a priori due to extensive field investigation. In the latter case field data for groundwater flow serve directly as input data for contaminant transport predictions. More often, the numerical model for simulating solute transport contains a flow model that computes groundwater flow and then utilizes the groundwater flow velocities in a water-quality model.

In the process of describing saltwater intrusion, flow and mass transport simulations influence each other. The high salt concentration in seawater affects the density and the viscosity of groundwater, which in turn affects the flow pattern. For chloride, density ρ and viscosity η are approximately linearly related to the concentration

$$\rho(c) = \rho_0 + \frac{\rho_{\max} - \rho_0}{c_{\max}} c \tag{3.39}$$

$$\eta(c) = \eta_0 + \frac{\eta_{\max} - \eta_0}{c_{\max}} c \tag{3.40}$$

where indexes 0 stands for density and viscosity at concentration ($c = 0$), and "max" stands for density and viscosity at saturated concentration. Prediction of the movement and spread of saltwater requires interactive coupling between the flow and the quality submodels. Density-dependent transport modeling is often applied in treating saltwater intrusion problems; however, interaction between flow and transport is experienced in many contaminant studies. Groundwater flow is sensitive even to density differences as small as 1 percent (List 1965), which may occur due to a high concentration of solutes or due to temperature differences.

Numerically the interactive coupling of flow and transport is not a major problem and follows directly from the numerical approximation of time space. For each time step, pressure distribution is solved first using known concentrations at the beginning of the time step. After groundwater flow velocities are obtained from the pressure distribution, they are used as input parameters in the solute transport calculation. The new concentrations are then used to repeat the flow and transport simulation. These calculations are repeated until a final solution is obtained.

Coupled groundwater flow and transport is an example where physics rather than numerics controls the modeling success (assuming that the numerical code is error free). Let us apply one numerical code to two different density-dependent groundwater flow problems.

First, consider the case where highly concentrated leachate increases groundwater density on top of the groundwater system. The interface between contaminated water and natural groundwater becomes unstable, causing complex flow configurations. This is shown in Figure 3.9, which illustrates the numerical calculation of a simple experiment demonstrating density-dependent groundwater flow. The model aquifer is isotropic; the upper part is filled with saltwater overlying freshwater. The density difference initiates a vertical movement of saltwater in the form of a finger. A rotating flow field evolves indicated by arrows in Figure 3.9. The length of the arrow is a measure of the magnitude of velocity. The direction of the arrow coincides with the direction of flow. At later stages the flow field becomes more and more complex. Saltwater moves downward at several locations forcing freshwater to move upward. Modeling these unstable groundwater conditions exactly will require a high degree of modeling effort.

Second, consider the application of a numerical code to simulate the displacement of the saltwater interface in a coastal groundwater system due to a changing water table. Since flow is density stratified and stable, calculations may be less demanding. In summary, density-dependent flow is a predominantly vertical flow problem. Simplification of modeling efforts by neglecting the vertical space direction is not applicable.

3.4 IDENTIFYING BOUNDARY AND INITIAL CONDITIONS

As in the case of the flow equation, initial and boundary conditions are required for the solution of the transport equation. The initial conditions are the concentration conditions at the starting time of simulation. This implies that an existing contaminant plume must be investigated extensively, in terms of its extent and its concentration, prior to simulations.

As in the case of the flow problem, typical boundary conditions fall into the three categories of prescribed concentration (first-kind condition); prescribed concentration gradient (second-kind condition) and a linear combination of both (third-kind condition). Figure 3.10 illustrates commonly encountered chemical-species boundaries including leaky structures, immiscible contaminant sources, groundwater recharge/discharge, injection wells, and surface water bodies.

In general, boundary conditions are difficult to apply at the outflow end of a transport model, since concentrations at the outflow boundary are unknown. One usually circumvents this problem numerically by making the model area large enough so that contaminants do not reach the outflow boundary and by setting dispersive flux at the boundaries to zero.

The first-kind condition, the *Dirichlet condition*, specifies the concentration at the boundary. It is also referred to as a *concentration-type boundary*. Leaky structures, such as leaking landfills, ponds, drains, and infiltration beds can be represented by a prescribed flux boundary for flow modeling with as-

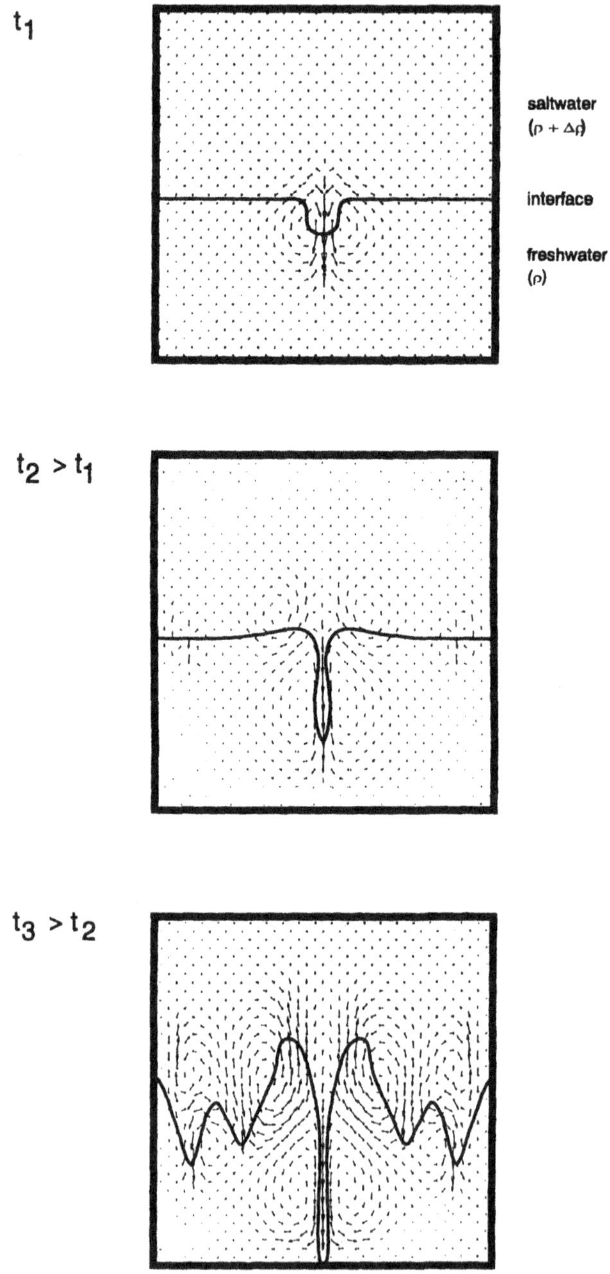

Figure 3.9 Numerical simulation of density-dependent groundwater flow for a simplified system.

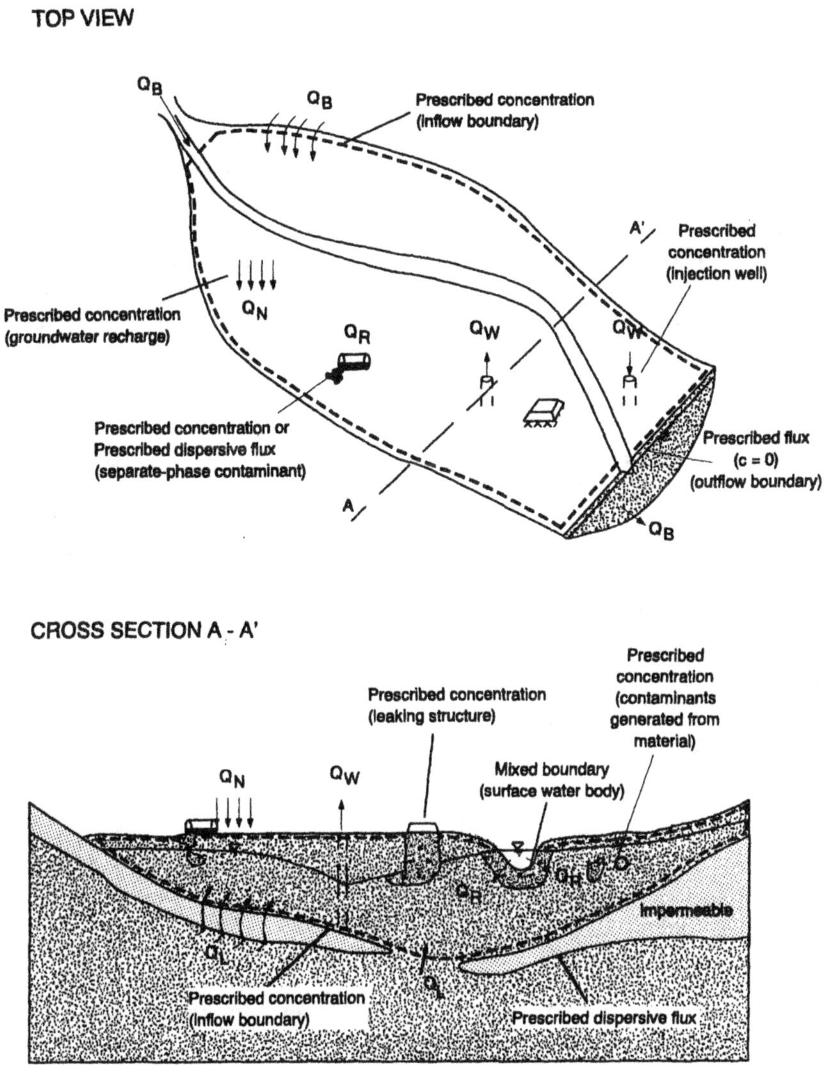

--- model boundary

Q_N = natural groundwater recharge or discharge
Q_R = exchange with surface water
Q_B = subsurface flow over boundary
Q_L = leakage component coming from adjacent aquifers
Q_W = local exfiltration or infiltration

Figure 3.10 Examples of boundary conditions in transport modeling.

sociated chemical-species concentration for transport modeling. Similarly injection wells can be simulated. The first-kind condition is usually applied at the inflow end of a model to represent known recharge with known concentration (advective flux). In this sense groundwater recharge, such as metal-laden recharge from mountainous areas, geothermal upwelling, recharge from agricultural irrigation, or recharge from domestic or industrial areas, can be represented by a prescribed flux boundary for flow modeling with known, associated chemical-species concentrations for transport modeling. Fixed concentrations inside the model area can also be used to approximate contaminated water around a body of pure contaminant. Usually discharge of contaminated groundwater into surface water where it is diluted can be incorporated by setting the concentration along the surface water boundary to zero.

The second-kind condition, the *Neumann condition*, specifying the concentration gradient along a boundary segment, prescribes dispersive flux. Contrary to the second kind of boundary in flow problems, it is less restrictive in the transport model, since it neither prescribes the total mass flux (advective and dispersive flux combined) nor prescribes advective flux alone. Setting dispersive flux equal to zero, the second kind of boundary is usually applied along impervious boundaries or at outflow boundaries where the concentration within the model area is zero. The third kind, the *Cauchy condition*, also referred to as *mixed* or as the *flux-type condition*, prescribes total contaminant flux in a linear combination of concentration (advection) and concentration gradient (dispersion) along the boundary. Mass conservation across the inlet boundary requires:

$$\underbrace{vc_o}_{\text{flux entering model area}} = \underbrace{vc - D\frac{\partial c}{\partial n}}_{\text{advective and dispersive flux inside model boundary}} \qquad (3.41)$$

The left side of the equation represents the flux entering the model area; the right side gives the mass flux just inside the boundary.

This third boundary condition is valid if concentration in the outside region adjacent to the model boundary (e.g., surface water) is perfectly mixed. In this case concentrations across the inlet boundary are discontinuous, in contrast to first-type boundaries. If there is a porous medium outside the boundary, then one should also include a dispersive flux on that side. But this creates problems unless the dispersive flux across the boundary is known in advance. Hence dispersive inflow is normally neglected. Note that at large times (as mixing across the boundary occurs), the concentration gradient or dispersive flux diminishes and the third kind of boundary condition approaches the first kind of condition. When advective flow is zero, the boundary condition reduces to the second type. Third-kind boundary conditions are easier to solve numerically, since no initial sharp concentration fronts exist.

Initial sharp concentration fronts may cause an overshoot in the numerical solution.

3.5 DESCRIBING TRANSPORT IN DIFFERENT GEOLOGICAL SETTINGS

The mathematical model for transport modeling is more demanding than for flow modeling. The physical processes in plume spreading are advection, dispersion, molecular diffusion, filtration, volatilization, and transport in the gas phase (unsaturated zone). The complexity of these processes, together with the many different chemical/biochemical and biological processes involved in contamination transport such as decay, dissolution, sorption, biodegradation, and biotransformation, restricts complete assessment of a real contamination problem to only a few field investigations.

Fortunately the type of transport problem and the geology often allow the model user to neglect transport components with little impact on the overall plume migration and to facilitate solving the mathematical model. Physical transport processes have different weights in the different geological settings shown in Figure 2.1. While diffusion may be the controlling transport mechanism in unfractured clay, it can generally be neglected in gravel aquifers with high transport velocities. There is, however, no general rule to determine in a particular situation which physical transport processes will control plume spreading. The following general aspects are important:

- Porous (granular) groundwater systems. In sand/gravel aquifers where significant natural groundwater now exists, the overall plume movement is controlled by advection. The role of diffusion increases with decreasing flow velocity. Dispersion causes a dilution of the plume. The unknown variability in the advective flow can be interpreted as dispersion due to the aquifer heterogeneity.
- Fractured/karstic groundwater systems. Contaminants migrate along the fractures and channels. If the fractures are regularly distributed, a theoretical assessment of the average advective transport is possible with reasonable effort using the continuum approach. Numerical models applicable for the simulation of mass transport in granular groundwater systems can be used as an approximation. The effects of diffusion and dispersion can be only estimated. In an irregular fractured system (karstic system), transport depends on preferred flow paths that are normally not known. Advective transport is dominant but difficult to evaluate. The influence of dispersion is secondary. In highly fractured systems, turbulent flow (which does not obey Darcy's law) may occur.
- Unfractured, low permeability groundwater systems. The process of contamination migration is small. Diffusion controls contaminant trans-

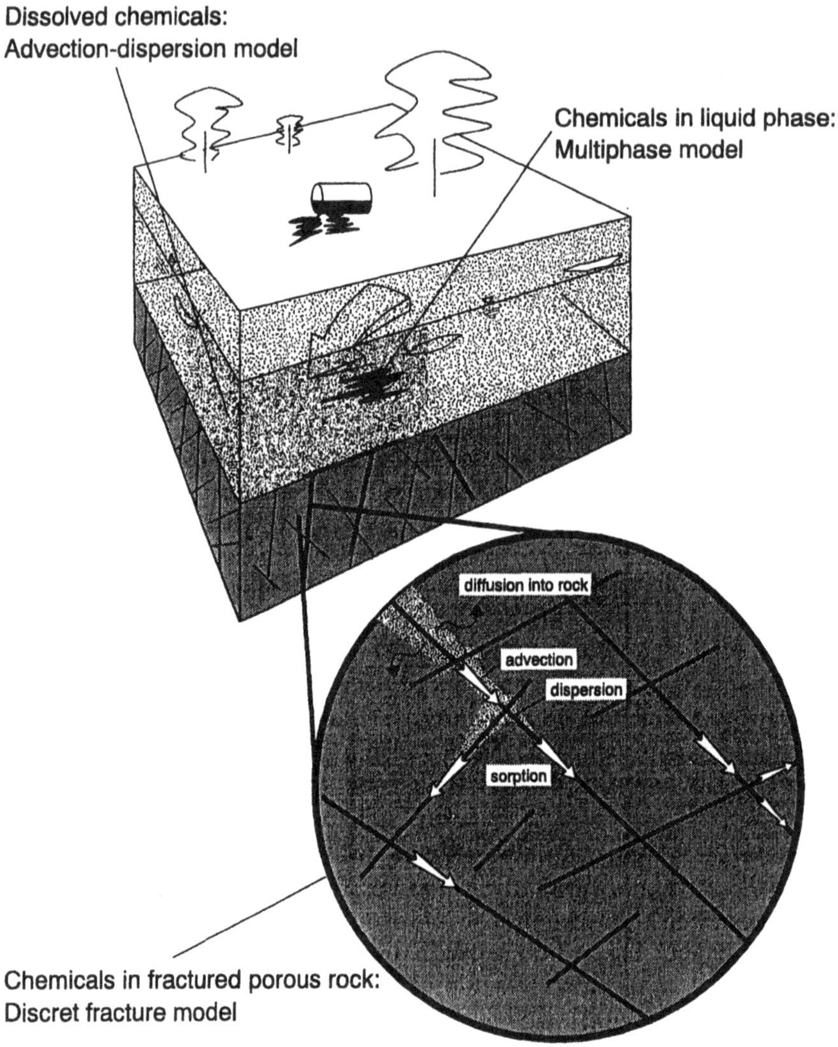

Figure 3.11 Spreading of dense hydrocarbons as a combination of various transport mechanisms.

port. Diffusive contaminant transport is small, but it can become important considering the high toxicity of many contaminants.

- Fractured-porous groundwater systems. At some landfill sites embedded in thick clay layers, contaminants have been detected at depths to which they could not migrate by diffusion only. The origin of the unexpected migration has been identified as fractures in the natural clay body. In fractured-porous materials the interrelationship of advection in the fractures and diffusion into the porous material becomes important in prediction of contamination migration. The effort to describe transport in fractured porous media is a recent development, and at present no standard models exist to simulate plume evolution in these groundwater systems.

- Unsaturated zone. Vertical migration by both advective and diffusive mechanisms are the dominant processes. This zone may act as a buffer (sorbing, dispersing, and volatilizing contaminants) or as a delayed source (gradually releasing stored contaminants). The problems of saturation-dependent coefficients (of hydraulic conductivity and dispersivity), hysteresis, and fissured soils make evaluations of unsaturated-zone transport complex.

Usually a groundwater system is built up of different types of aquifers. Contamination transport may therefore differ from aquifer to aquifer. As an example, the contamination propagation of dense hydrocarbons in a permeable aquifer overlying fractured bedrock, is schematically shown in Figure 3.11. The hydrocarbons move through the unsaturated and the saturated zone to the base of the sand/gravel aquifer. Dissolved hydrocarbons are transported with the groundwater and generate a contamination plume. At the same time the hydrocarbons enter the bedrock through the fracture network. Depending on the actual situation, the contamination of the fractured bedrock may be of more concern than the plume in the sand/gravel aquifer. Modeling contaminant spreading, the model user may be forced to describe with three different models transport at the source, transport with the groundwater, and transport in the fracture network.

4 Using Analytical Models for Idealized Flow and Transport Problems

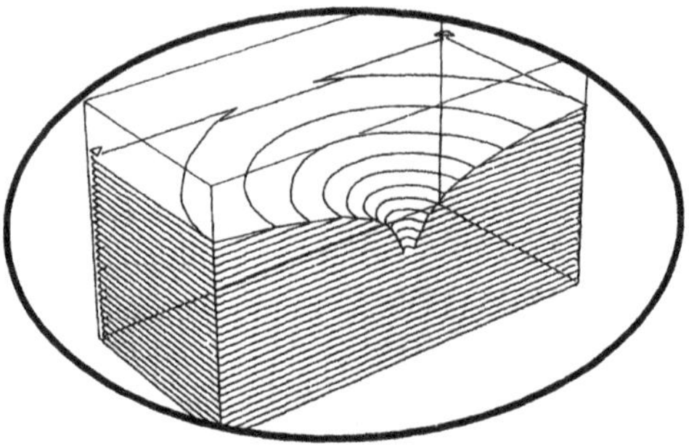

Analytical modeling uses the same approach as numerical modeling except for the solution of the mathematical model. The model user can benefit from familiarity with analytical solutions when dealing with numerical models. Chapter 4 provides a review of analytical predictive models for groundwater flow and transport as well as analytical models that analyze aquifer performance tests. This review should help the model user understand the more difficult numerical modeling described in the following chapters.

In view of the extensive literature on analytical solutions, such as Polubarinova-Kochina (1952) or Bear (1972), no detailed coverage is presented. The model user can benefit from the analogy of flow and diffusion to other relevant phenomena in physics by borrowing solutions established in these areas (Carslaw and Jaeger 1959; Crank 1956). Mathematical description of steady groundwater flow is well covered by Strack (1989). A prominent and comprehensive text on pumping test analysis is the book by Krusemann and de Ridder (1970). Walton (1970) compiled many illustrative cases studies of pumping test analysis. Practical aspects of pumping tests and well hydraulics are also summarized in Driscoll (1986). Ready-made computer programs that analyze pumping tests can be obtained from several sources, such as Boonstra (1989). Models based on analytical solutions are also listed in Figure 9.5. Note that analytical solutions for more complex flow or transport conditions often become so complex and unwieldy as to diminish the advantages of the analytical approach over a simple numerical solution. Analytical solutions that include infinite series may require significant programming and computational effort.

This chapter is organized in three sections. Section 4.1 is concerned with one-dimensional flow, either plane-symmetric or radially converging. The reader is also introduced to the method of superposition. Section 4.2 places emphasis on solutions of advective transport caused by wells and dispersive transport in uniform flow. Section 4.3 illustrates groundwater model application by a simple pumping test analysis that should be familiar to the model user, and briefly addresses analytical methods to analyze dispersion tests.

Whenever possible, the model user should give preference to analytical solutions over numerical modeling. A large number of groundwater problems can be greatly simplified and solved analytically without substantial loss of accuracy. There is no systematic approach for simplifying a given groundwater problem and for selecting the appropriate analytical solution. In fact it depends entirely on the capability of the model user to visualize the investigated problem and to judge professionally if a chosen analytical method is consistent with the hydrogeological controls. If the deviation between reality and the conceptual model is recognized and its impact properly estimated, analytical solutions can be beneficial.

4.1 APPLYING PREDICTIVE MODELS FOR GROUNDWATER FLOW

In nature, groundwater flow occurs under confined, semiconfined, or unconfined conditions. Semiconfined flow, also referred to as *leaky aquifer*, is more frequent than perfectly confined or unconfined flow. In this section we are interested in one-dimensional groundwater flow for all three conditions. General solutions are presented for plane-symmetric and radially converging steady groundwater flow. Transient solutions applicable to the infinite aquifer are given for radially converging flow (also referred to as *well flow*). We will end with a discussion on how to apply linear superposition of single-well solutions to describe multiple-well flow.

4.1.1 Describing One-Dimensional Plane-Symmetric Steady Flow

To understand one-dimensional plane-symmetric flow, consider the cross section shown in Figure 4.1. Assume that no groundwater flow normal to the shown cross section exists. Flow in the layer highlighted in Figure 4.1 experiences three different flow regimes:

- To the left, the aquifer is confined.
- In the middle section, leakage occurs. The aquifer is semiconfined.
- To the right, the aquifer is unconfined.

Assuming predominantly one-dimensional flow, the general solutions of groundwater flow in each flow regime are

Confined flow

$$h = Ax + B \tag{4.1}$$

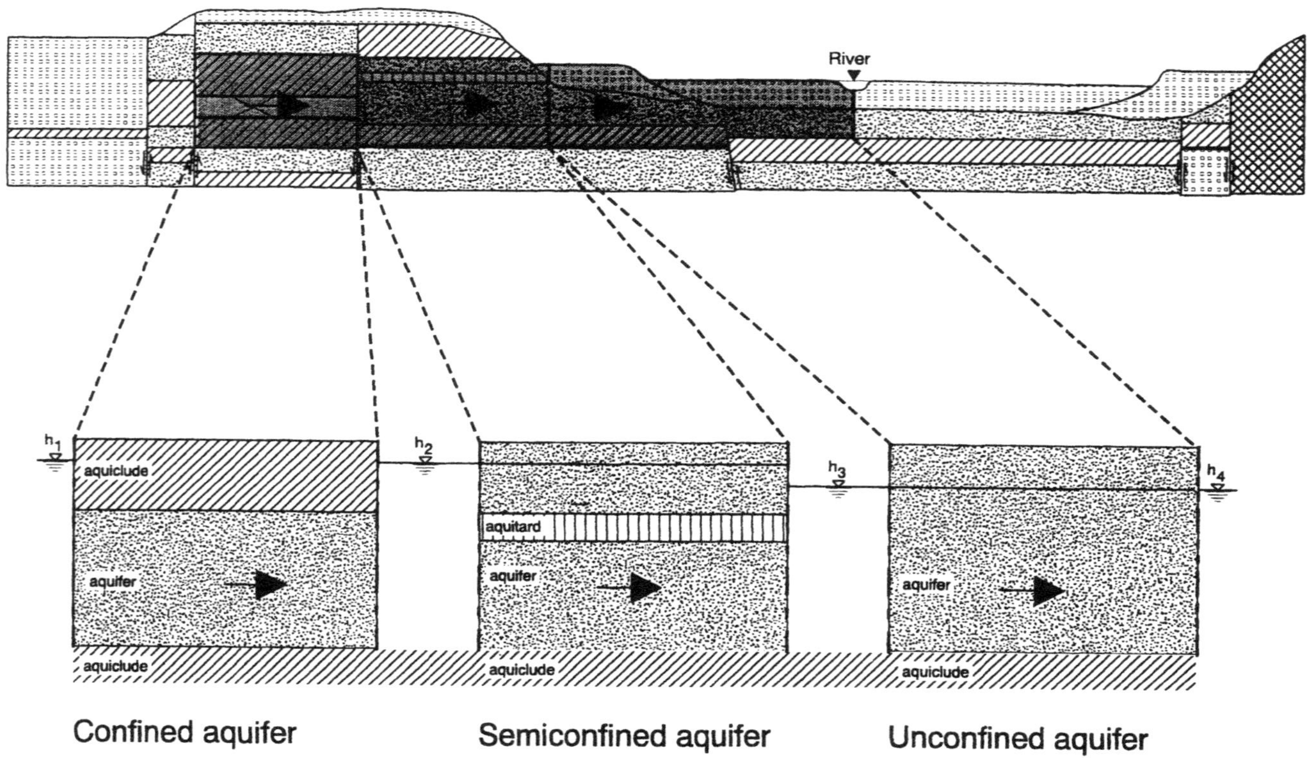

Figure 4.1 Application of method of fragments to a hypothetical flow problem.

Semi-confined flow

$$h = Ae^{x/\lambda} + Be^{-x/\lambda} \tag{4.2}$$

Unconfined flow

$$h^2 = -\frac{N}{K}x^2 + Ax + B \tag{4.3}$$

where

h = head at distance x in [L]
x = distance in [L]
K = hydraulic conductivity in [L/T]
λ = leakage factor in [L]
N = natural groundwater recharge in [L/T]
A, B = constants to be determined from boundary conditions

The leakage factor λ is defined as

$$\lambda = \sqrt{Tc} \tag{4.4}$$

where T is the aquifer transmissivity and c is the hydraulic leakance of the semipermeable layer as defined in Equation 2.6. The leakage factor is a measure of the amount of leakage into the aquifer. High values of λ indicate a great resistance of the semipermeable layer, compared with the resistance of the aquifer itself.

Applying different combinations of boundary conditions to Equations 4.1 to 4.3 yields head distribution for various one-dimensional plane-symmetric flow cases (Cohen and Miller 1983). Few flow problems conform with the restrictive assumptions underlying these flow cases. However, to solve a complex flow problem, try to subdivide the investigated flow domain into individual flow regimes as illustrated in Figure 4.1. Using Equations 4.1 to 4.3, groundwater levels and groundwater flow in each regime are found but with unknown groundwater level or unknown flux at the internal boundary between two regimes. The approach yields as many equations as unknowns. Hence more complex groundwater flow problems can be solved while using the basic solutions given in Equations 4.1 to 4.3.

This method, called the *method of fragments*, was furnished by Paulousky in 1935 (Paulousky 1956). It can be applied to vertical or horizontal two-dimensional flow assuming that equipotential lines at various critical parts of the flow can be approximated by straight lines. These lines may serve to subdivide the flow system into individual sections.

4.1.2 Describing One-Dimensional Radially Converging Flow

Thiem, in 1906, published the first formula to predict changes of groundwater head in the vicinity of a pumping well. Since then the description of flow toward wells has engaged countless scientists due to its prominent role in groundwater exploration and development. For flow at steady state the following general solutions for confined, semiconfined, and unconfined radially converging flow have been developed:

Confined flow,

$$h = \frac{Q}{2\pi T} \ln r + C \tag{4.5}$$

Semiconfined flow,

$$h - h_o = AI_o\left(\frac{r}{\lambda}\right) + BK_o\left(\frac{r}{\lambda}\right) \tag{4.6}$$

Unconfined flow,

$$h^2 = \frac{Q}{\pi K} \ln r + C \tag{4.7}$$

where

Q	=	constant well discharge or recharge in [L^3/T]
T	=	transmissivity in [L^2/T]
K	=	hydraulic conductivity in [L/T]
r	=	distance from well in [L]
h	=	head at distance r in [L]
h_o	=	constant head in adjacent aquifer in [L]
λ	=	leakage factor as defined in Equation 4.4 in [L]
I_o	=	Bessel function of order zero and of the first kind
K_o	=	Bessel function of order zero and of the second kind
A, B, C	=	constants to be determined from boundary conditions

Series expansions and tables for both Bessel functions are given in mathematical handbooks.

Common solutions for steady flow toward a well can be calculated as follows:

Confined flow,

$$h - h_1 = \frac{Q}{2\pi T} \ln\left(\frac{r}{r_1}\right) \tag{4.8}$$

Semiconfined flow,

$$h - h_o = \frac{Q}{2\pi T} K_o\left(\frac{r}{\lambda}\right) \tag{4.9}$$

Unconfined flow,

$$h^2 - h_1^2 = \frac{Q}{\pi K} \ln\left(\frac{r}{r_1}\right) \tag{4.10}$$

where h_1 is the known groundwater level at distance r_1; h_o in Equation 4.9 is the elevation of the groundwater head in the adjacent aquifer. At $r \rightarrow \infty$, the groundwater head in the pumped semiconfined aquifer equals h_o.

Equation 4.8 was published by Thiem (1906) and is also referred to as the *Thiem equation*. De Glee (1930, 1951) solved for the steady-state solution in the leaky aquifer (Equation 4.9). Equation 4.10 is known as the *Dupuit equation* (Dupuit 1863). The equation is derived assuming that flow is horizontal in each vertical cross section of the aquifer. This assumption is known as the *Dupuit assumption*.

Solutions for transient well flow are some of the most commonly used solutions for transient groundwater flow. They predict water level changes due to imposed pumping (or recharge), and they estimate transmissivities and storage coefficients when records of pumping and water level measurements are available. The latter application is found in the aquifer performance tests described in Section 4.3. For infinite aquifers, transient solutions for well flow are

Confined flow,

$$s = h - h_o = \frac{Q}{4\pi T} \int_u^\infty \frac{e^{-y} dy}{y} = \frac{Q}{4\pi T} W(u) \tag{4.11}$$

Semiconfined flow,

$$s = h - h_o = \frac{Q}{4\pi T} \int_u^\infty \frac{1}{y} \exp\left(-y - \frac{r^2}{4\lambda^2 y}\right) dy = \frac{Q}{4\pi T} W\left(u, \frac{r}{\lambda}\right) \tag{4.12}$$

In the above equations s is the drawdown, defined as the difference of measured groundwater level h at distance r at time t to water level h_o prior to pumping, S is storativity, and λ is leakage factor. In the case of the semiconfined aquifer, it is assumed that h_o equals the groundwater head in the adjacent aquifer and that the groundwater conditions in the adjacent aquifer are not affected by pumping.

In Equation 4.11 the function $W(u)$, called the *well function*, is introduced with

$$u = \frac{r^2 S}{4Tt} \tag{4.13}$$

$W(u, r/\lambda)$ in Equation 4.12 is called the *Hantush well function*. The function is similar to the well function but with two parameters in the integral. Tables of both mathematical functions are found in the literature related to well hydraulics.

To predict water level changes in an unconfined aquifer, Equation 4.11 is applicable as long as the drawdown is small compared to the saturated thickness and as long as water is released from storage without delay. Instantaneous release of water in response to the decline of the water table quite often differs from field observations in unconfined aquifers. The solution for unconfined aquifers incorporating delayed yield is a complicated differential equation explained by Boulton (1954).

Figure 4.2 graphically presents Equations 4.11 and 4.12 and the governing equation for transient well flow in an unconfined aquifer with delayed yield. The graph demonstrates the development of groundwater head at a given distance due to pumping. Time and drawdown are given at logarithmic scale as is common in pumping test analysis. In the case of confined and unconfined well flow, the drawdown will increase infinitely if no prescribed head boundary or natural groundwater replenishment is present. If the drawdown reaches a prescribed head boundary, decrease of water levels will eventually come to the steady state shown in Figure 4.2. An impermeable aquifer boundary will result in increased drawdown measurements. In a semiconfined aquifer the extent of the drawdown will be finite due to leakage from the adjacent aquifer.

Analytical solutions for well flow are applied mostly in analyzing pumping tests and in predicting drawdown due to planned well discharge. They are also helpful in approximating the effect of local groundwater recharge or discharge in more complex groundwater systems. This is illustrated in Figure 4.3. In this field study a two-dimensional numerical flow model was used to study the impact of infiltration pits on the elevation of the water table. Artificial infiltration was introduced to compensate for groundwater overabstraction in dry years, with little natural groundwater recharge. The solid lines in Figure 4.3 give the increase of the water table after 2 years of artificial infiltration calculated numerically. The infiltration rate was 12.7 million m^3/year. The

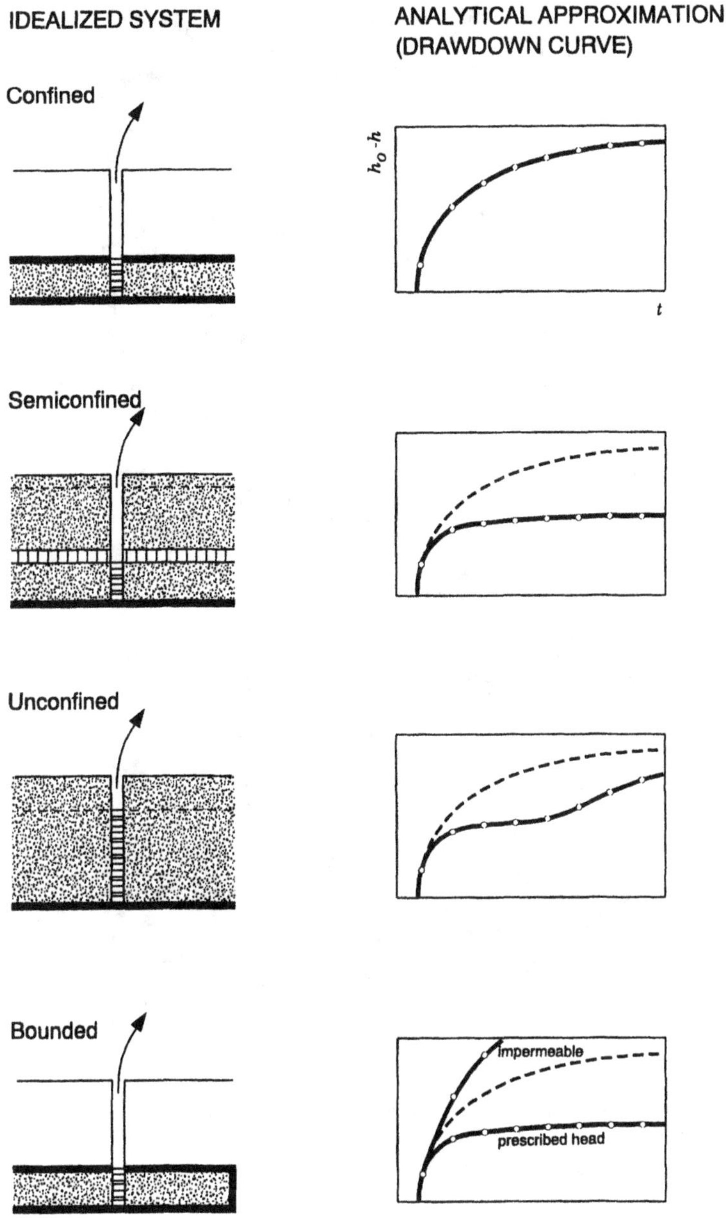

Figure 4.2 Log-log $h_0 - h$ versus log t data for confined, semiconfined, unconfined, and bounded systems (after Freeze and Cherry 1979).

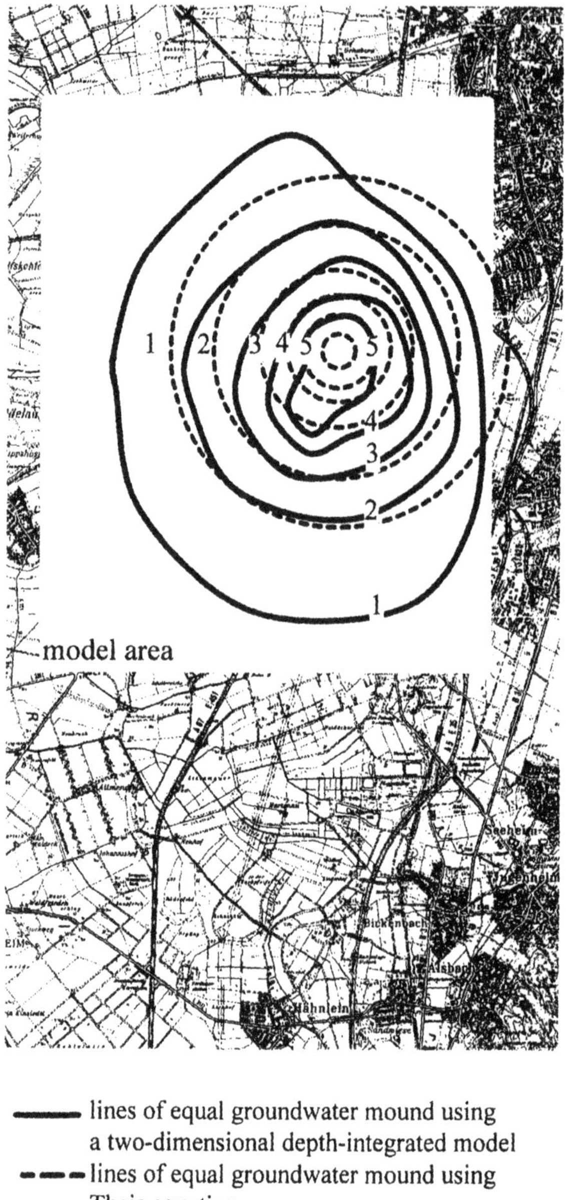

Figure 4.3 Analytical solution compared to numerical results.

dashed lines demonstrate the mound after recharge calculated analytically using Equation 4.11. For the analytical simulation, it was assumed that infiltration occurs locally by a well at the center of the infiltration pits. The comparison of solid and dashed lines, or respectively of numerical and analytical simulations, demonstrates that the simple analytical solution yields reasonable indications of the impact of artificial groundwater recharge. By applying an analytical solution to a complex field study, this example also underlines the importance of defining and simplifying the groundwater problem under investigation. Depending on the needs, cost, and complexity of the groundwater system, it may be possible to design a simple groundwater model to address key modeling aspects. The analytical model gives an approximate solution in a short time, while it takes weeks to months to design the numerical model.

4.1.3 Solving Multiple-Well Flow in Confined Aquifers

The differential equation that describes steady two-dimensional confined flow is Laplace's equation. Mathematics state that if two different functions are solutions of the Laplace equation, then the sum of both functions is also a solution of Laplace's equation. Since each single-well solution described in Equation 4.5 is a solution of the Laplace equation, the principle of linear superposition allows us to derive solutions for multiple-well flow in confined aquifers. In unconfined aquifers, superposition can serve only as an approximation as long as the drawdown is small compared to saturated thickness.

Combined use of the superposition principle and symmetry, also referred to as *method of images*, allows the introduction of various boundary conditions to the flow domain (e.g., Bear 1972). Streamlines can be interpreted as impermeable boundaries and equipotential lines for groundwater head equal to constant head boundaries. Walton (1970) presents numerous successful applications of the method of images to field cases. Two field applications are schematically presented in Figure 4.4. In the Arcola field study pumping wells are elongated along a fluvial deposit, with transmissivity values significantly higher than the ones of the adjacent groundwater system. In the conceptual model the aquifer was represented as a strip with two parallel impermeable boundaries allowing the determination of groundwater elevation by means of an analytical model. In the Pekin field study a buried valley forms the aquifer. At its edge the aquifer thickness decreases and the underlying bedrock forms an impermeable boundary. The Illinois River has been dredged into the alluvial materials. Pumping tests indicate a fair connection between the river and groundwater system. Again, a conceptual model could be defined that allows analytical approximation of groundwater elevation.

Method of images is mostly applied to straight-line boundaries, but it can be used to simulate circular boundaries. It also allows flow to be simulated in an aquifer with an abrupt change in transmissivity (Bear 1972). The following section demonstrates that using the basic solution for confined well flow

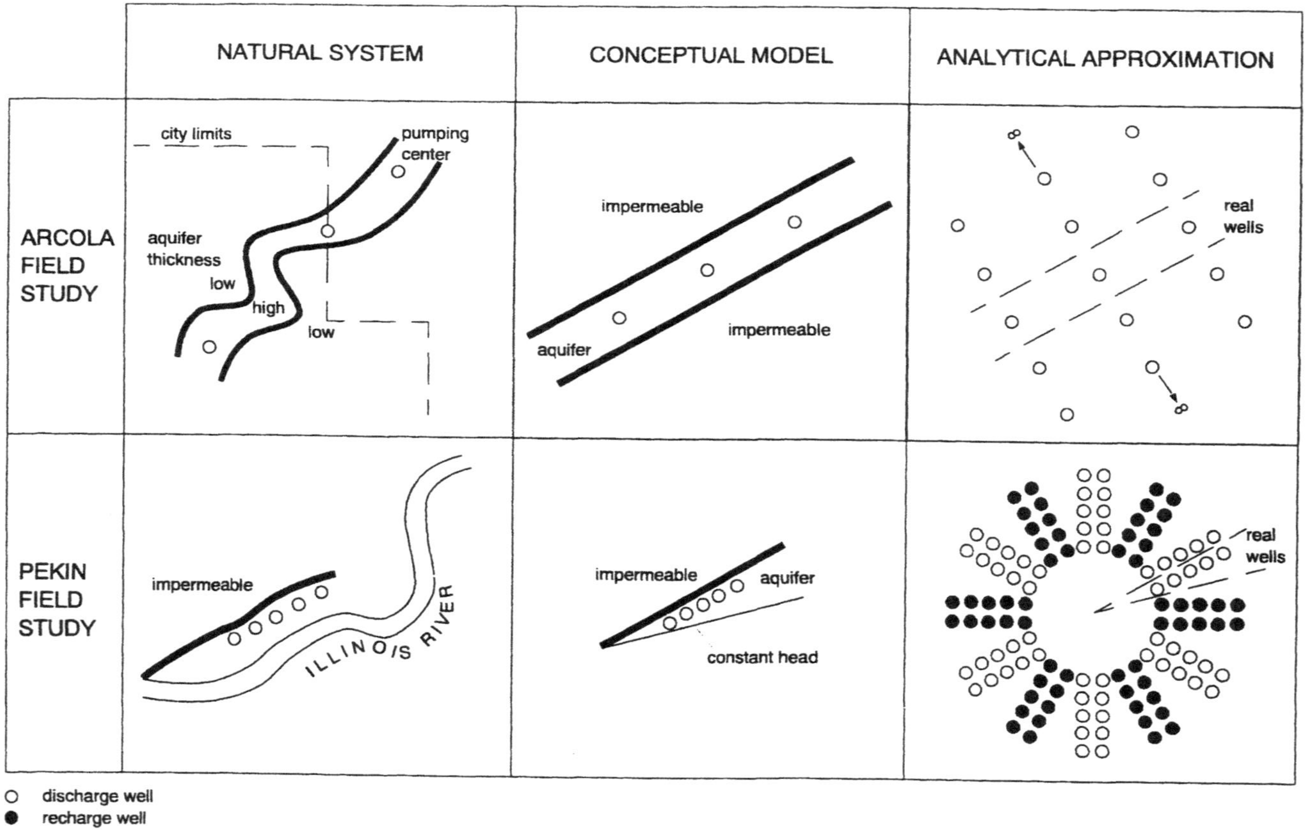

Figure 4.4 Application of method of superposition (after Walton 1970).

(Equation 4.5) also allows solution of practical transport problems involving multiple-well flow.

4.2 APPLYING PREDICTIVE MODELS FOR GROUNDWATER TRANSPORT

Analytical solutions for contaminant transport problems are not as common as flow solutions. The mean direction of contaminant migration and average travel times may be estimated by using analytical solutions that neglect the effect of dispersion. Practical problems that can be solved in this approximation are the design of groundwater protection zones and the design of decontamination wells in hydraulic remediation actions. A second class of transport solutions solves for dispersion but assumes uniform groundwater flow. Analytical solutions for space-dependent advective transport velocities are the most difficult to derive because of the nonlinearity of the governing transport equation.

4.2.1 Describing Advective Transport Caused by Wells

In the design of groundwater remediation actions particular attention must be given to the description of advective transport caused by wells. The model user is asked to position discharge wells in a way that will prevent the enlargement of the contamination plume and will eventually reduce groundwater contamination by pumping out contaminated groundwater. Simple analytical and semianalytical models have been developed to serve as design tools for groundwater remediation measures.

Groundwater flow toward a single well in an aquifer with no natural flow is radial, and streamlines are straight lines originating at the well. Transport velocities are obtained directly from continuity considerations. The amount of water that flows into a cylinder with radius r and thickness m around the well equals the well discharge (or, recharge) Q as shown in the following equation:

$$v(r) = \pm \frac{1}{n} \frac{Q}{2\pi m r} \tag{4.14}$$

Darcy flux is divided by porosity n to obtain average advective transport velocities. Lines of equal travel time toward/away from the discharge/recharge well are circles with the radius

$$r = \left(\frac{Q}{\pi m n} t \right)^{\frac{1}{2}} \tag{4.15}$$

The superposition principle may be employed to determine the flow and, subsequently, the travel times in multiple-well flow. Since uniform groundwater flow can be interpreted as flow due to a discharge and a recharge well at infinite distance and with infinite discharge and recharge rate, uniform flow can also be superimposed.

Figure 4.5 represents a typical flow field near a pumping well. A natural gradient and thus natural groundwater flow exist. Undisturbed flow varies little, and the system is idealized assuming uniform groundwater flow (q_o). Streamlines and travel times can be expressed analytically (Bear 1972) and are presented in dimensionless form. Hence the dimensionless graph given in Figure 4.5 can be applied to any flow problem involving one pumping well in uniform flow. Site-specific coordinates and travel times are obtained by inserting the actual values for the pumping rate, aquifer thickness, and so on, into the expression for dimensionless distance and time.

A simple method for determining the flow field in multiple-well flow is derived from the superposition principle. Flow velocities are calculated as the sum of the contributions of each well, given that all assumptions stated in Section 4.1 still apply. The flow velocity calculations are

$$v_x = \sum_{i=1}^{k} \frac{Q_i}{2\pi mn} \frac{\cos \beta}{\sqrt{(x_i - x)^2 + (y_i - y)^2}} + \frac{q_o}{n} \cos \alpha \tag{4.16}$$

$$v_y = \sum_{i=1}^{k} \frac{Q_i}{2\pi mn} \frac{\sin \beta}{\sqrt{(x_i - x)^2 + (y_i - y)^2}} + \frac{q_o}{n} \sin \alpha \tag{4.17}$$

where

β = angle between x-axis and connection line of well i with considered location (x, y) measured counterclockwise

α = angle between base flow direction and x-axis

Travel times along a streamline between two points (x_1, y_1) and (x_2, y_2) are found using the equation

$$t = \int_{x_1}^{x_2} \frac{dx}{v_x} = \int_{y_1}^{y_2} \frac{dy}{v_y} \tag{4.18}$$

In practice, closed-form solutions for streamlines and travel times are restricted to two-well flow (for more detailed discussions, see Muskat 1949; Bear 1972; Dacosta and Bennett 1960; Mehlhorn et al. 1981). Flow and travel times for simplified configurations based on two-well solutions are presented in

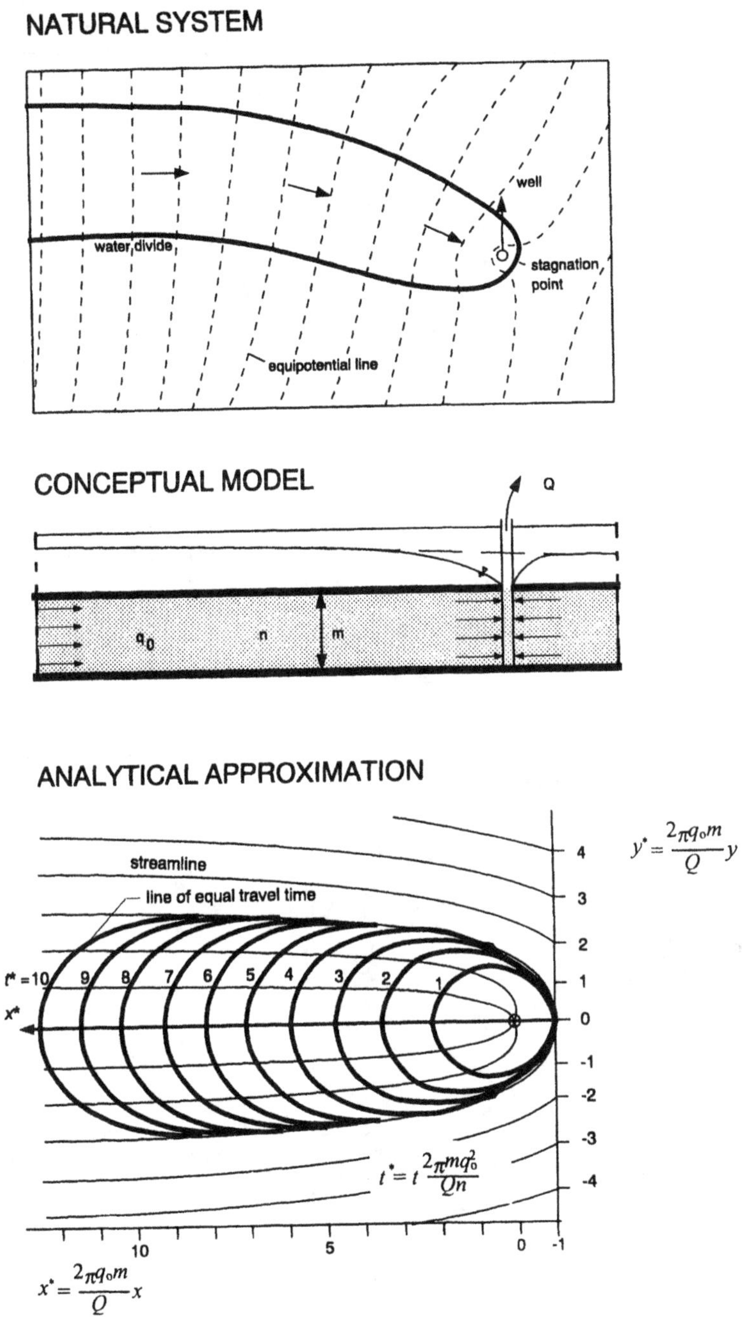

Figure 4.5 Well in uniform flow.

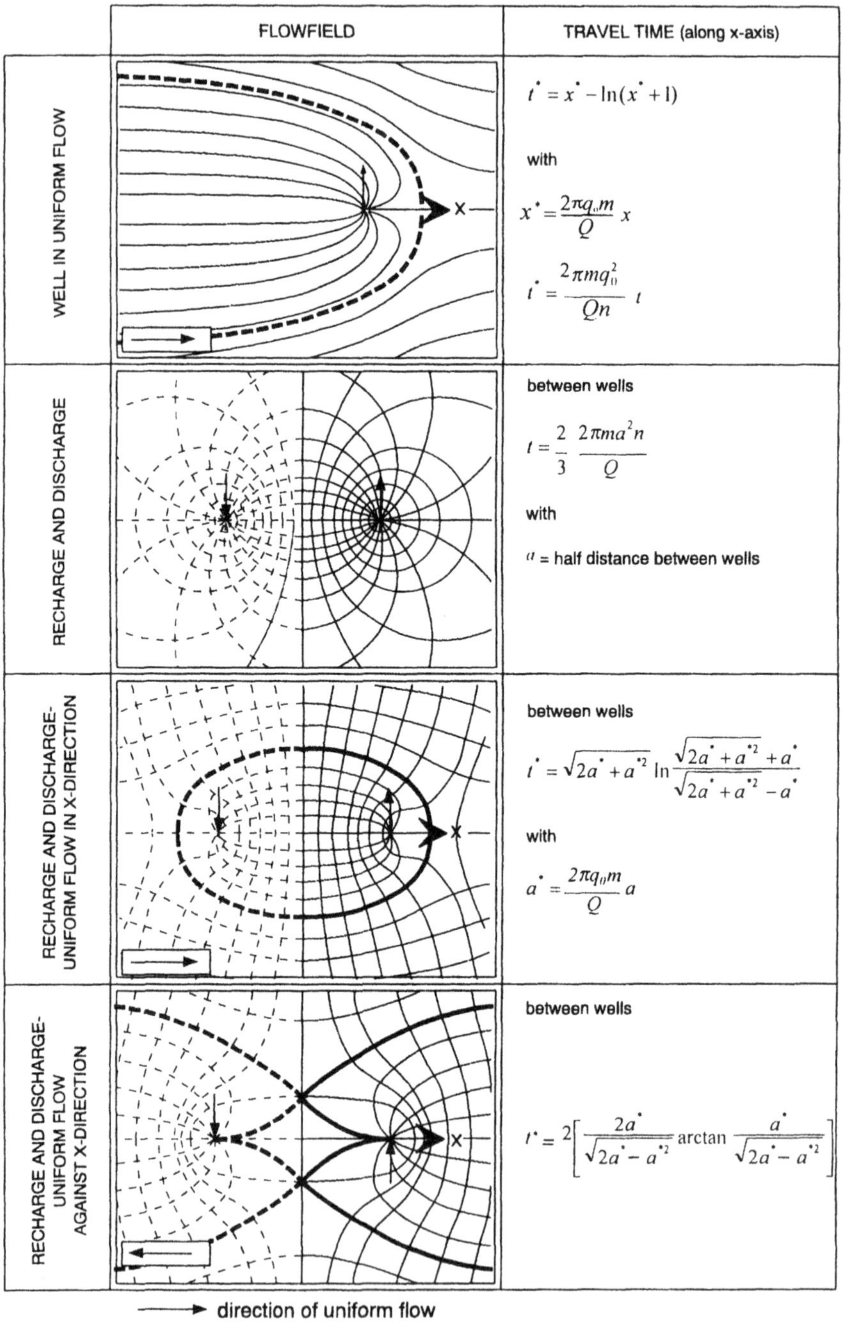

Figure 4.6 Two wells solution: Flow field and travel time along x-axis.

Figure 4.6. The solutions for two wells could also represent the flow of one well near the surface water by the method of images. This is indicated in Figure 4.6 by the dashed outline of the symmetric part of the flow net.

Closed-form solutions for multiple-well flow are especially helpful when designing remediation measures. They can also support prognosis of numerical models. This is illustrated in Figure 4.7. We consider the problem of intercepting flow between wells of equal discharge Q. The line connecting the wells is at right angles to the natural groundwater flow q_o. Closed-form solutions exist for the critical distances between 2, 3, 4, and 5 wells where the flow is on the verge of breaking through (Bear 1972; Kauch 1982). They are given in the upper part of Figure 4.7. The lower part of Figure 4.7 shows the application to a field study. Groundwater contamination had occurred due to a leaking storage basin. The model user is asked to determine the pumping rate of the two pumping wells such that no contaminated groundwater is transported downstream. Both pumping rates should be equal and minimized. The task was solved numerically and analytically. The model application for this engineering purpose is straightforward since undisturbed groundwater flow is basically uniform. Streamlines serve as no-flux boundaries; measured groundwater heads as prescribed head boundary. Clearly numerical model results must compare to the analytical model which imposes two restrictions on the pumping rate:

1. The combined catchment area of both wells must enclose the area of the storage basin. This leads to the first condition for Q, assuming that the asymptotic width of the catchment area equals the maximum width of the storage area.
2. The second condition for Q follows directly from the critical well distance between two wells.

Semianalytical models (Javandel et al. 1984) take advantage of the fact that with Equations 4.16 and 4.17 transport velocities are known everywhere. Streamlines and travel times are found numerically by tracking the movement of particles set into the flow field. The position of the particle after a time step dt calculates as

$$x_{\text{new}} = x_{\text{old}} + \int_{t_{\text{old}}}^{t_{\text{new}}} v_x \, dt \tag{4.19}$$

$$y_{\text{new}} = y_{\text{old}} + \int_{t_{\text{old}}}^{t_{\text{new}}} v_y \, dt \tag{4.20}$$

The error in the calculation of streamlines is small if time steps are small. The accuracy of the numerical integration can be improved by applying

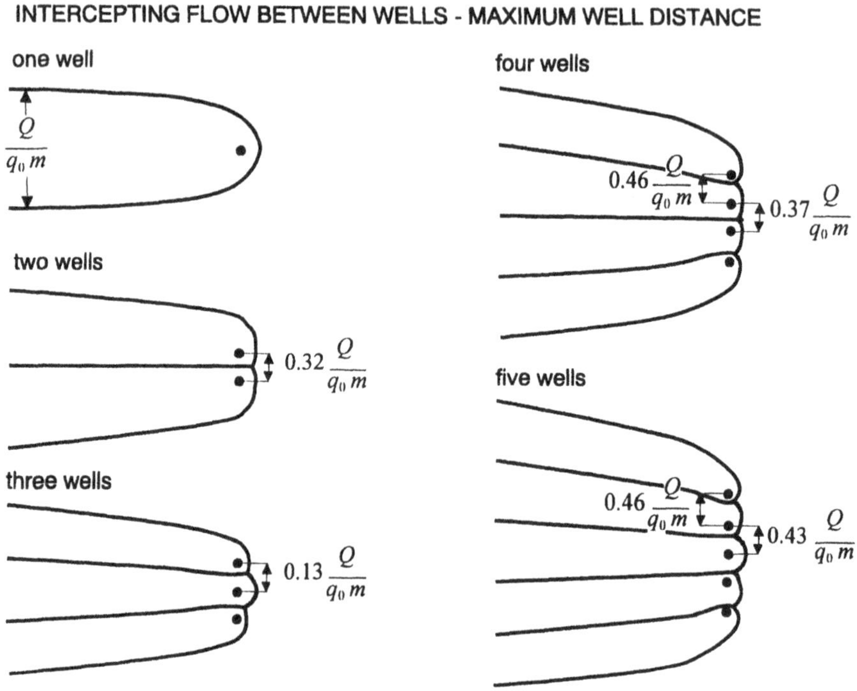

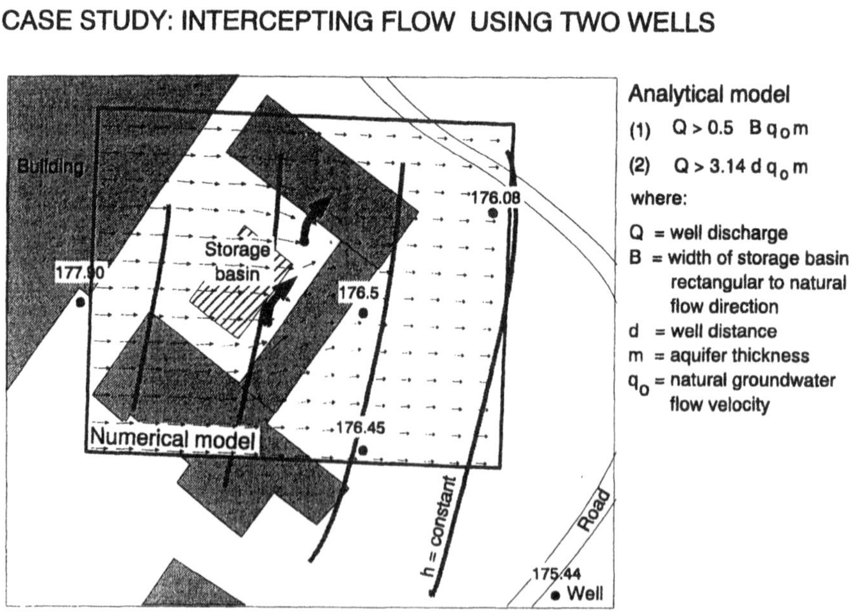

Figure 4.7 Intercepting flow between wells.

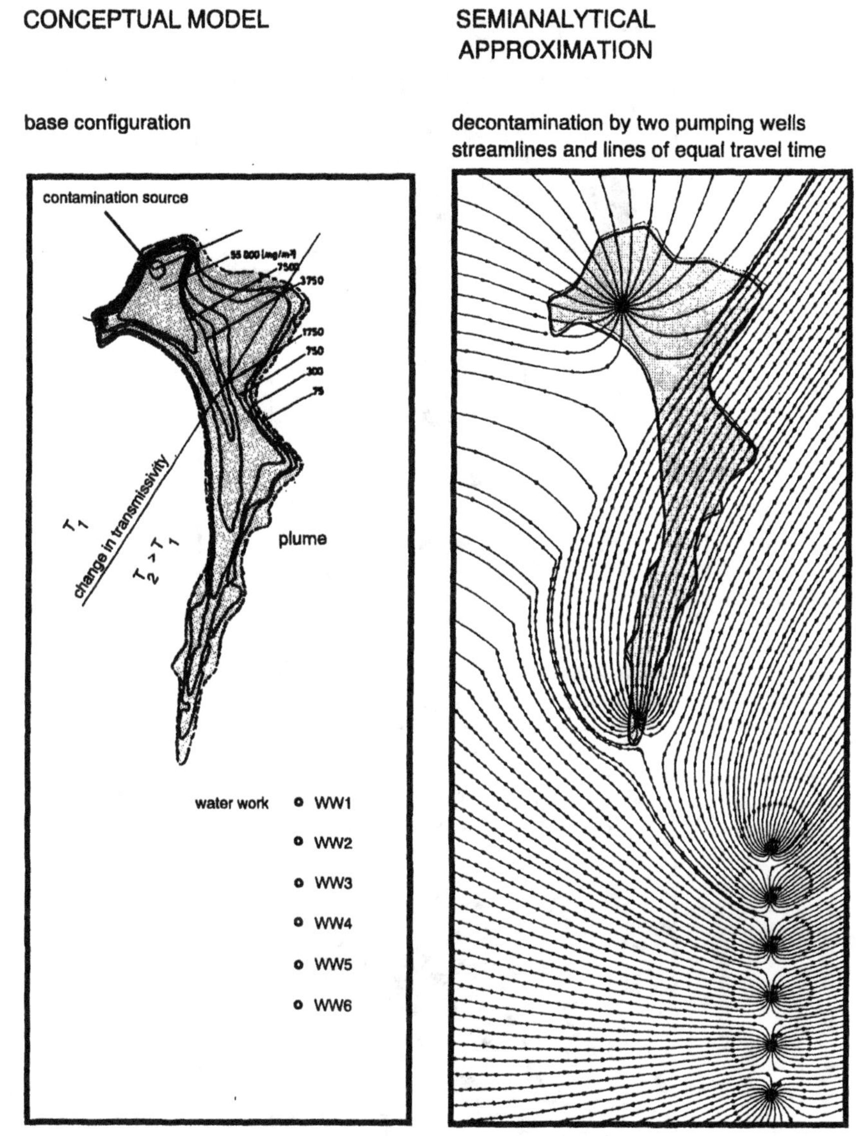

Figure 4.8 Dispersion-free transport simulation: Application of method of superposition to multiple well flow (after Herr 1985).

advanced integration methods such as Runge-Kutta-type formulas (Hornbeck 1975). Lines of equal travel times are easy to derive by keeping record of the time steps as particle movement begins.

Semianalytical models are a powerful method applicable to the design of decontamination schemes. Two field applications are shown in Figures 4.8 and 4.9. Figure 4.8 shows a decontamination scheme applied to a contamina-

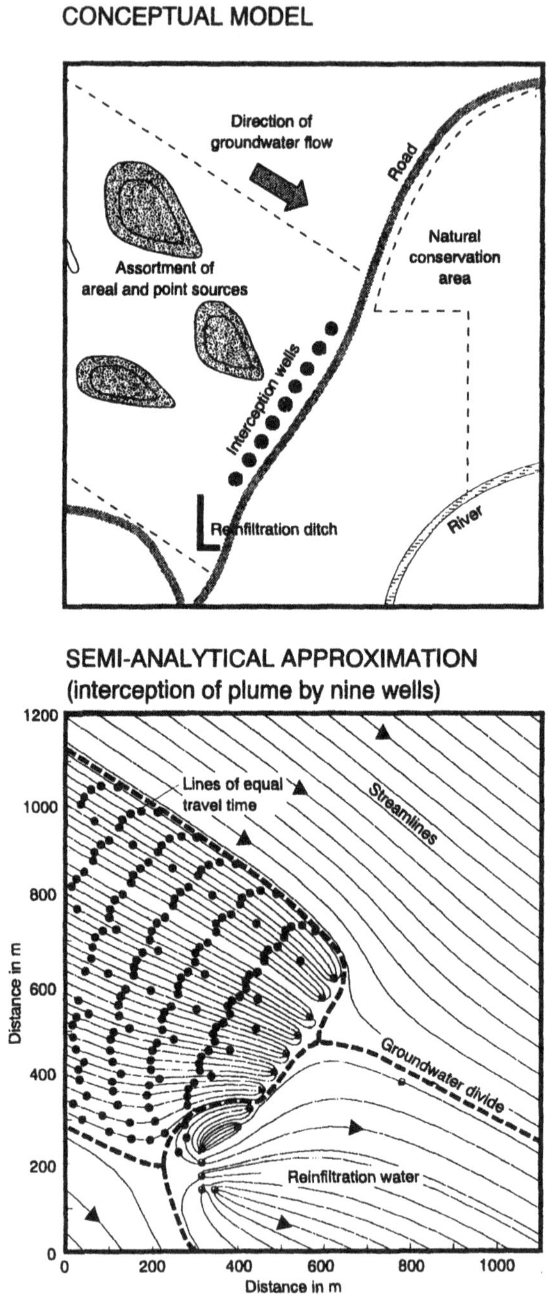

Figure 4.9 Dispersion-free transport simulation: Application of superposition to a decontamination scheme involving nine interception wells.

tion plume approaching the production wells of a water supply pumping station. Advective transport (streamlines and travel times) have been solved entirely semianalytically, using Equations 4.16 to 4.20. In the investigated aquifer the thickness and, respectively, the transmissivity changes abruptly due to a fault, causing the bend in the streamlines in the upper left-hand side of the illustration. The change in transmissivity was introduced by applying the method of images. To obtain an illustrative pattern, the release points of the streamline particles were equally distributed around the well and then were tracked upgradient.

Figure 4.9 shows a field application where the model user was asked to position remediation wells downstream from an industrial site to intercept the contamination plume. Several contamination sources were generating a relatively wide plume. Pumped groundwater was reinjected after water treatment. Detailed assessment of flow was carried out numerically. A semianalytical model provided, with little effort, reliable information on the expected flow and transport field.

4.2.2 Describing Dispersive Transport in Uniform Flow

The advection-dispersion equation can be solved analytically for only a few transport problems. Analytical models are limited to simple geometry and simple combinations of initial and boundary conditions. In the following cases four basic solutions of the advection-dispersion equation are briefly reviewed assuming uniform groundwater flow:

- Permanent point source
- Instantaneous point source
- One-dimensional movement of a tracer in a semiinfinite column
- One-dimensional transverse spreading of an initially sharp front in a plane

Case 1. Permanent Point Source

An experiment in statistics is helpful to visualize the typical form of analytical solutions for diffusive or, respectively, dispersive processes. In this experiment balls are dropped into a vertical nail board to illustrate random processes, as shown in Figure 4.10 (called Galton board after Galton, 1822–1911). At the base of the vertical nail board, the balls are collected in separate boxes. Assuming a large number of experiments and a large number of rows of nails to pass for each ball, the distribution of the balls would follow the Gaussian function, which is expressed as

$$f(y) = \frac{1}{\sigma\sqrt{2\pi}} \exp\left(-\frac{y^2}{2\sigma^2}\right) \tag{4.21}$$

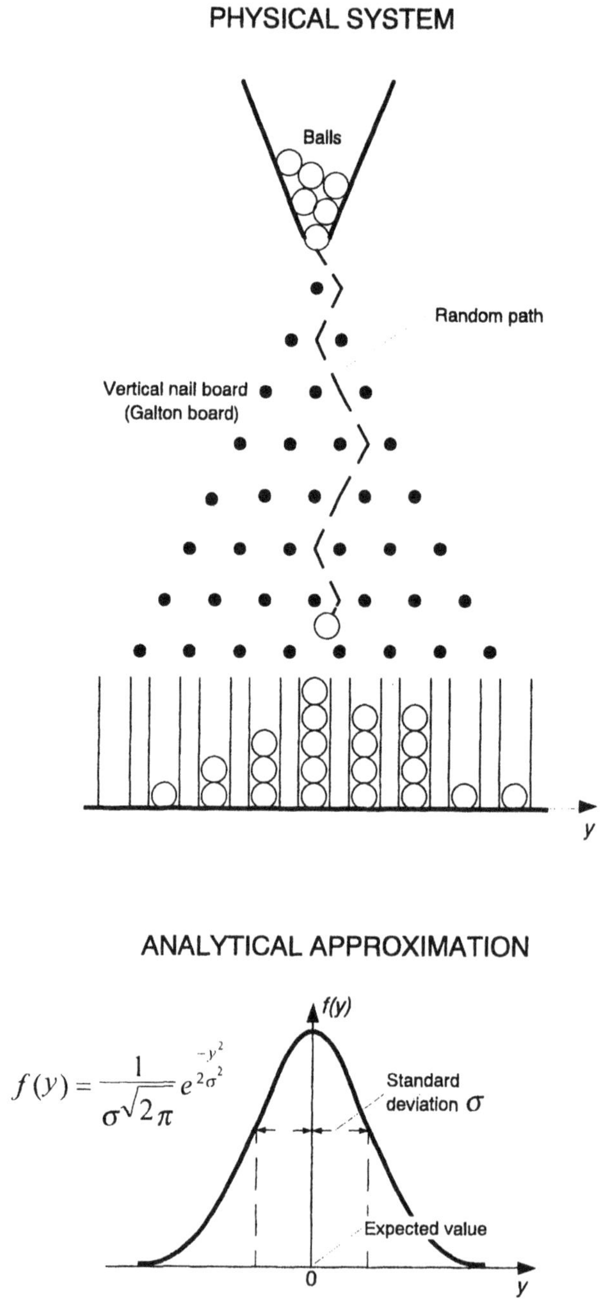

Figure 4.10 Using Galton's board to illustrate disperse lateral spreading.

where σ represents the effect of the nail placing at the nail board upon lateral spreading in y-direction and the number of rows of nail passed.

If we visualize the balls as contaminants, the nail board as the aquifer, and the nails as soil grains, we can expect a similar concentration distribution in the case of a continuous point source in uniform groundwater flow. Indeed the steady-state solution for this configuration may be approximated as (Sayre 1968)

$$\frac{c}{c_o} = \frac{1}{2\sqrt{\pi \alpha_T x}}\, e^{\left(\frac{-y^2}{4\alpha_T x}\right)} \tag{4.22}$$

where

x = distance in flow direction in [L]
y = distance transverse to flow direction in [L]
c = concentration in [M/L^3]
c_o = imposed source concentration in [M/L^3]
α_T = transverse dispersivity of aquifer in [L]

As expected, the concentration distribution is Gaussian $\left(\sqrt{2\alpha_T x}\right)$, since the standard deviation σ serves as a measure of dispersive spreading. Maximum concentrations occur along the axis of symmetry ($y = 0$) and can be calculated as

$$\frac{c_{\max}}{c_o} = \frac{1}{2\sqrt{\pi \alpha_T x}} \tag{4.23}$$

Equations 4.21 through 4.23 neglect dispersive spreading in the flow direction. The error introduced is small since concentration gradients in the flow direction are small. For the complete time-dependent solution of the dispersive spreading of a permanent point source in uniform flow, incorporating longitudinal and transverse spreading, see Kinzelbach (1986).

Analytical solutions of the advection-dispersion equation often include the Gaussian function. It is often hidden in the error function, defined as

$$erf(x) = \frac{2}{\sqrt{\pi}} \int_0^x e^{-\zeta^2}\, d\zeta = 1 - erfc(x) \tag{4.24}$$

The $erfc(x)$ in the above equation is called the *complementary error function*. The Gaussian function as well as the error function and complementary error function are tabulated in mathematical books (Abramowitz and Stegun 1970).

Case 2. Instantaneous Point Source

A solution that combines longitudinal and transverse spreading, as shown in Figures 4.11 and 4.12*a*, can be derived for an instantaneous point source in uniform flow. The solution approximates contamination spreading caused by an accidental spill. Contaminated loading where mass M is added to the system at one point and where concentrations will still be zero at the end of the system, no matter how large time gets, is expressed in three dimensions as

$$c(x, y, z) = \frac{M}{n}\,\delta(x)\,\delta(y)\,\delta(z), \qquad t = 0 \tag{4.25}$$

$$c(\pm\infty, \pm\infty, \pm\infty) = 0, \qquad t \geq 0 \tag{4.26}$$

where $\delta(x)$, $\delta(y)$, and $\delta(z)$ are the Dirac function (unit step function). The solutions for the concentration distribution are

One-dimensional (Bear 1972),

$$c(x, t) = \frac{M/n}{\sqrt{4\pi\alpha_L vt}} \exp\left[-\frac{(x - vt)^2}{4\,\alpha_L vt}\right] \tag{4.27}$$

Two-dimensional (De Josselin de Jong 1958),

$$c(x, y, t) = \frac{M/n}{4\pi m v_x t\sqrt{\alpha_L\alpha_T}} \exp\left[-\frac{(x - v_x t)^2}{4\,\alpha_L v_x t} - \frac{y^2}{4\,\alpha_T v_x t}\right] \tag{4.28}$$

Three-dimensional (Baetsle 1969),

$$c(x, y, z, t) = \frac{M}{8(\pi v_x t)^{\frac{3}{2}}\sqrt{\alpha_x\alpha_y\alpha_z}} \exp\left[\frac{(x - v_x t)^2}{4\,\alpha_x v_x t} - \frac{y^2}{4\,\alpha_y z_x t} - \frac{z^2}{4\,\alpha_z v_x t}\right] \tag{4.29}$$

The two- and three-dimensional solutions are given as the product of one-dimensional solutions similar to Equations 4.22 and 4.27. The concentration distribution is Gaussian with greater spreading in the flow direction. Again the standard deviations are a measure for spreading, and are calculated in two-dimensions as

$$\sigma_L = \sqrt{2\alpha_L v_x t} \tag{4.30}$$

$$\sigma_T = \sqrt{2\alpha_T v_x t} \tag{4.31}$$

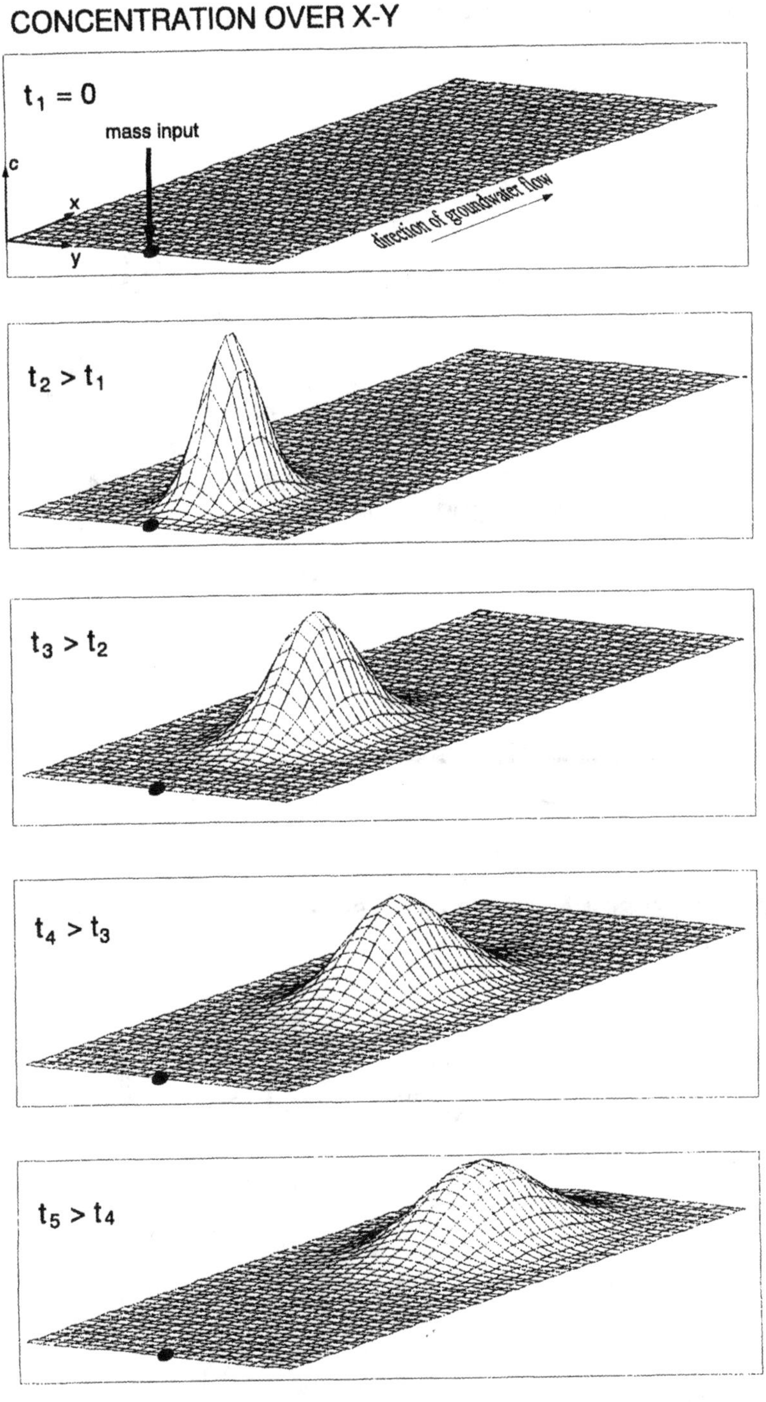

Figure 4.11 Point source in uniform groundwater flow.

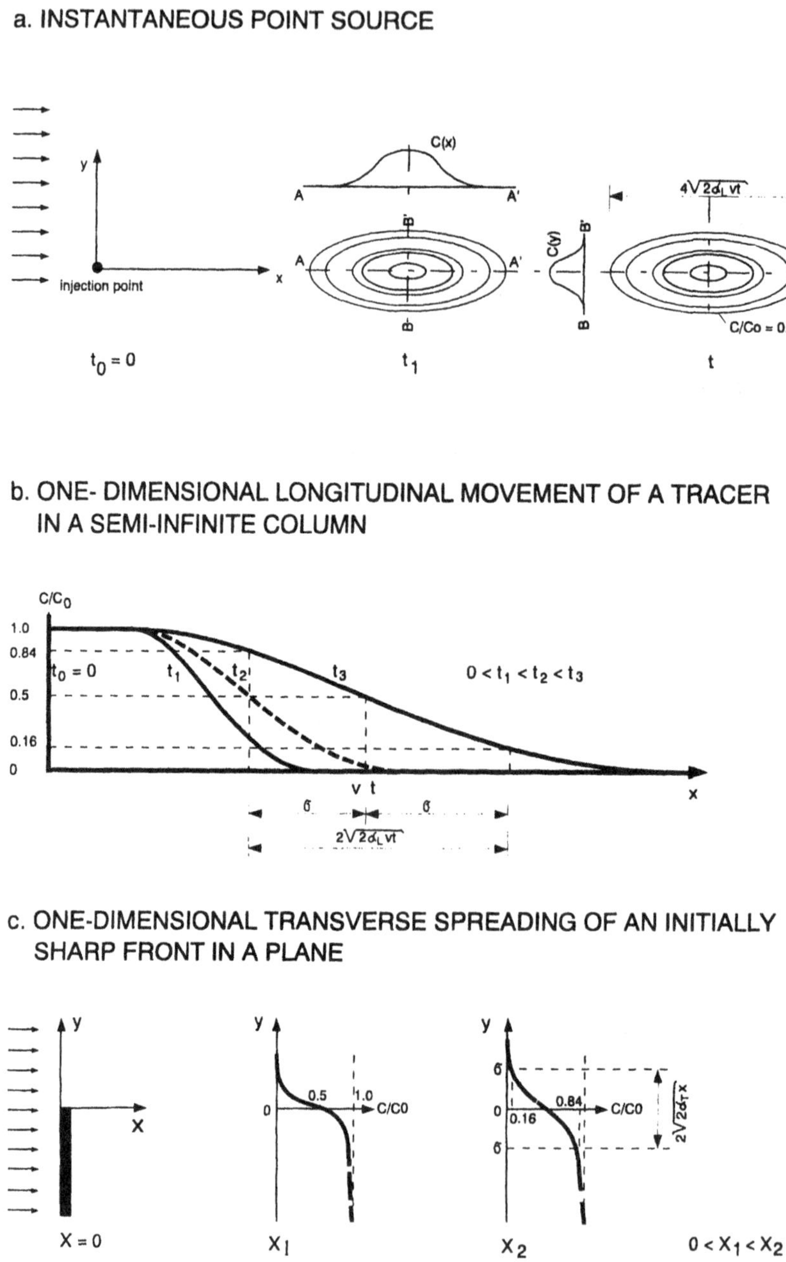

Figure 4.12 Disperse spreading in uniform flow.

Maximum concentration occurs at the center of the plume at $x = v_x t$, and it is calculated in two-dimensions as

$$c_{\max} = \frac{\Delta M}{4\pi n m v_x t \sqrt{\alpha_L \alpha_T}} \tag{4.32}$$

In Equations 4.27 to 4.29 contaminant input is given as mass. A wide range of solutions exists for which the contaminant concentration is specified as the source term. The boundary condition of this type is a prescribed concentration boundary.

Cases 3 and 4 give two examples with prescribed concentration boundaries for longitudinal or, respectively, transverse dispersion in uniform flow.

Case 3. One-Dimensional Movement of a Tracer in a Semi-Infinite Column

Ogata and Banks (1961) present the one-dimensional solution of the movement of a tracer in a semi-infinite column. The initial and boundary conditions are

$$c = 0, \quad t \leq 0,\ x \geq 0 \tag{4.33}$$
$$c = c_o, \quad t > 0,\ x = 0 \tag{4.34}$$
$$c = 0, \quad t > 0,\ x \rightarrow \infty \tag{4.35}$$

which yield the concentration distribution:

$$\frac{c}{c_o} = \frac{1}{2}\left[erfc\left(\frac{x - vt}{2\sqrt{\alpha_L vt}}\right) + \exp\left(\frac{x}{\alpha_L}\right) erfc\left(\frac{x + vt}{2\sqrt{\alpha_L vt}}\right)\right] \tag{4.36}$$

The second term may be neglected when the ratio of x/α_L becomes large ($x/\alpha_L > 300$). The approximate form of the solution is then

$$\frac{c}{c_o} = \frac{1}{2}\, erfc\, \frac{x - vt}{2\sqrt{\alpha_L vt}} \tag{4.37}$$

The boundary conditions (Equations 4.33 to 4.35) are first-kind boundaries and indicate a step change in concentration. In nature, a sudden change of concentration is less likely than continuous leakage of contaminated water into the groundwater. Sauty (1980) derives the solution for a second-kind boundary where concentration is injected continuously. The solution is identical to Equation 4.36 except that the second term is subtracted rather than added. Since the second term is much smaller than the first term, at large distances from the source, the solution (Equation 4.37) applies to both boundary conditions in practical applications.

Equations 4.36 and 4.37 of one-dimensional dispersion prove helpful in estimating longitudinal spreading. Defining the contaminant front as the c/c_o line of 16% (one standard deviation of the underlying Gaussian function), the width of dispersive spreading may be approximated as

$$\sigma_L = \sqrt{2\alpha_L vt} \tag{4.38}$$

The distance σ_L provides general information about the impact of dispersion assuming that the aquifer dispersivity is known. The length σ_L of the mixing zone does not give the actual front of measurable concentrations. Contaminants will be dispersed over larger distances but at lower concentrations. This is shown in Figure 4.12*b*, which gives at three time increments the concentration distribution in flow direction.

Case 4. One-Dimensional Transverse Spreading of an Initially Sharp Front in a Plane

A solution for transverse spreading is derived by Harleman and Rumer (1963). Figure 4.12*c* illustrates the investigated transport configuration. With

$$c(0, y) = c_o, \qquad -\infty < y \leq 0 \tag{4.39}$$

$$c(0, y) = 0, \qquad 0 < y < +\infty \tag{4.40}$$

$$\frac{\partial c}{\partial y} = 0, \qquad y \rightarrow \pm\infty \quad \text{for all } x \tag{4.41}$$

the steady-state solution for dispersion in uniform flow in a plane becomes

$$\frac{c}{c_o} = \frac{1}{2}\, erfc\, \frac{y}{2\sqrt{\alpha_T x}} \tag{4.42}$$

The effect of longitudinal dispersion can be neglected due to small concentration gradients in the flow direction. The distance between the 16% and 50% concentration lines measures dispersive spreading:

$$\sigma_T = \sqrt{2\alpha_T x} \tag{4.43}$$

4.3 ANALYZING AQUIFER PERFORMANCE TESTS

The scope of model applications presented in the introduction of this book is important in discussing the analysis of aquifer performance tests. One purpose of modeling is the identification of missing aquifer parameters, most often storativity and transmissivity. This problem is generally referred to as the *inverse problem* where known aquifer responses (e.g., drawdowns) to a known groundwater stress situation (e.g., pumping) are used to identify missing model input data. The difference to predictive model application is emphasized where aquifer parameters are known and where a solution, or response, of the aquifer to a specific operation is sought.

Regarding analytical solutions as a specific type of groundwater model, aquifer performance test analyses are equivalent to the calibration procedure in numerical modeling. As a consequence aquifer performance test analyses rely on most phases in modeling design as summarized in Section 1.2, although this is not immediately obvious. This is illustrated in the following subsections.

4.3.1 Solving the Inverse Flow Problem of Pumping Tests

Pumping test analysis using analytical solutions, or analytical models, were introduced to the hydrogeologic community long before numerical models. They have become a standard method in groundwater exploration. The steps of modeling follow:

Developing the Conceptual Model

To develop the conceptual model, certain conditions of the hydrogeological setting must be understood. As in numerical modeling, pumping test analysis also requires an appreciation and understanding of the hydrogeological setting of the test site. The relevant conditions are:

- Aquifer. Location, depth, thickness, stratigraphy.
- Regional features. Surface water, natural aquifer boundaries.
- Pumping and observation wells. Construction details.
- Field measurements. Water table over time, pumping rate, and depth.

Although information about the actual aquifer geometry and stratigraphy may be available, the model user can still apply analytical solutions to dramatically simplify the natural aquifer conditions. Both the limited pumping duration and the decreasing speed of the drawdown development with time are justification for this. Nevertheless, applying analytical solutions to groundwater problems always implies a rigorous simplification of natural conditions.

Selecting the Model

The selection of the appropriate model, or formula, for a specific pumping test is difficult. Using the most common formulas prevents flexibility in simplifying nature, and directly or indirectly we use the solutions available to us. Most often Theis's equation (Equation 4.11) is applied. Unlike numerical modeling the Theis equation allows us to predict the aquifer's response, or the change in hydraulic head due to pumping, at any time and at any location.

Calibrating the Model

The observed hydraulic heads during the pumping event is the hydrological record used in the model calibration. But applying Theis's equation, we can obtain the missing aquifer parameters, which are transmissivity and storativity. Unlike model studies of complex hydrogeological environments, where the model user relies on tedious trial-and-error adjustment of simulated and observed values, analytical solutions permit direct parameter identification. Since Equation 4.11 does not allow for the explicit solution of T and S, several well-known and easily applied methods have been developed to circumvent the problem (Theis 1935; Cooper and Jacob 1946; Jacob 1950; Chow 1952).

This direct method of parameter identification yields aquifer parameters in the vicinity of the well. Theoretically it appears relatively easy to apply the method and to obtain good aquifer estimates. In practice the analysis often requires sound hydrogeologic understanding. The reasons for this are manifold, but all relate to common problems experienced in modeling. They are explained in the following step.

Analyzing Model Errors

In numerical models there are errors inherent in the numerical approximation itself, in the limited accuracy of the computer used, and from simplification of the natural system. In analytical solutions, model errors are solely caused by our rigorous simplification of the real world.

In reviewing all the assumptions made, one might wonder that analytical models often produce good results, since there is no real aquifer that conforms fully to the rigorous simplifications required. To address this concern, the model user must relate deviations between theory and field observation to deviations between simplification and nature. The following two observations on pumping test analysis, or model calibration, are important:

- Matching pumping test data by a specific theoretical curve does not ensure that the solution is unique. This is a basic difficulty in solving the inverse problem, which is not found in forecasting problems. Among others, the main reason for nonunique solutions is due to incompletely representing reality. Unknown factors may be embedded in field observations. Based on our aquifer simplification, for example, we may interpret a limited drawdown as the result of leakage that in reality is due to the influence of an unknown recharge boundary or due to unconfined groundwater flow.
- The pumping test normally focuses on the vicinity of the well. Most analytical solutions applied to pumping test analyses assume that drawdown develops with radial symmetry. Geologic anomalies, however, may exist within the radius of influence of the pumping test, and they can

cause a significant deviation from radially converging flow. This occurs when the cone of depression reaches a boundary. If the conceptual model of radially converging flow is wrong, a reasonable match between observed and calculated drawdowns for all points cannot be achieved. The model calibration cannot make up for errors in the conceptual model.

Figure 4.13 gives a field example. Pumping test data, for a multilayer aquifer, are given for an observation interval of 80 hours. Approximately 700 m away from the test site is the sea. Applying the Theis's equation for the confined and infinite aquifer yields strong discrepancies between simulated and observed drawdowns for large time. By assuming instead a semiconfined condition, a good match for most observations can be achieved. Similar results yield the assumption of a bounded aquifer. The observed periodical oscillations in the drawdowns at larger time, not clearly seen in Figure 4.12, definitively indicate the influence of the tidal fluctuation on the groundwater system.

4.3.2 Solving the Inverse Transport Problem of Dispersion Tests

Field experiments on dispersion have increased our understanding about large-scale dispersion. The basic idea underlying the dispersion test is as simple as in the pumping test. In a controlled experiment a tracer is introduced into the flow field. Subsequently over time tracer concentrations at one or more observation locations are recorded. This information is inserted into an analytical model consistent with the flow conditions and with the boundary conditions for the transport problem. Dispersivities in the analytical model are varied until observations and simulations match within a prescribed tolerance. Hence the same modeling steps as in the aquifer pumping test are required. An explanation of these steps follows.

Developing the Conceptual Model

To develop the conceptual model, an understanding of the hydrogeological setting of the test site and a record of any event during the experiment that may affect the tracer spreading are mandatory. Accurate collection and interpretation of field data form the basis of an accurate analysis of test results. The description of the hydrogeological setting should account for aquifer geometry, temporal variation of the flow field, and the way the tracer is added into the flow regime. The number of monitoring locations must be sufficient to define accurately the tracer plume under investigation.

Selecting the Model

Useful solutions for tracer experiments in sandbox models are the solution of Ogata and Banks (1961) to determine longitudinal dispersivity (Equation 4.37) and the solution of Harleman and Rumer (1963) to study transverse spreading (Equation 4.41). Field experiments are often designed according to Equation 4.31 for the instantaneous point source in uniform flow and yield information on both the transverse and the longitudinal dispersivities.

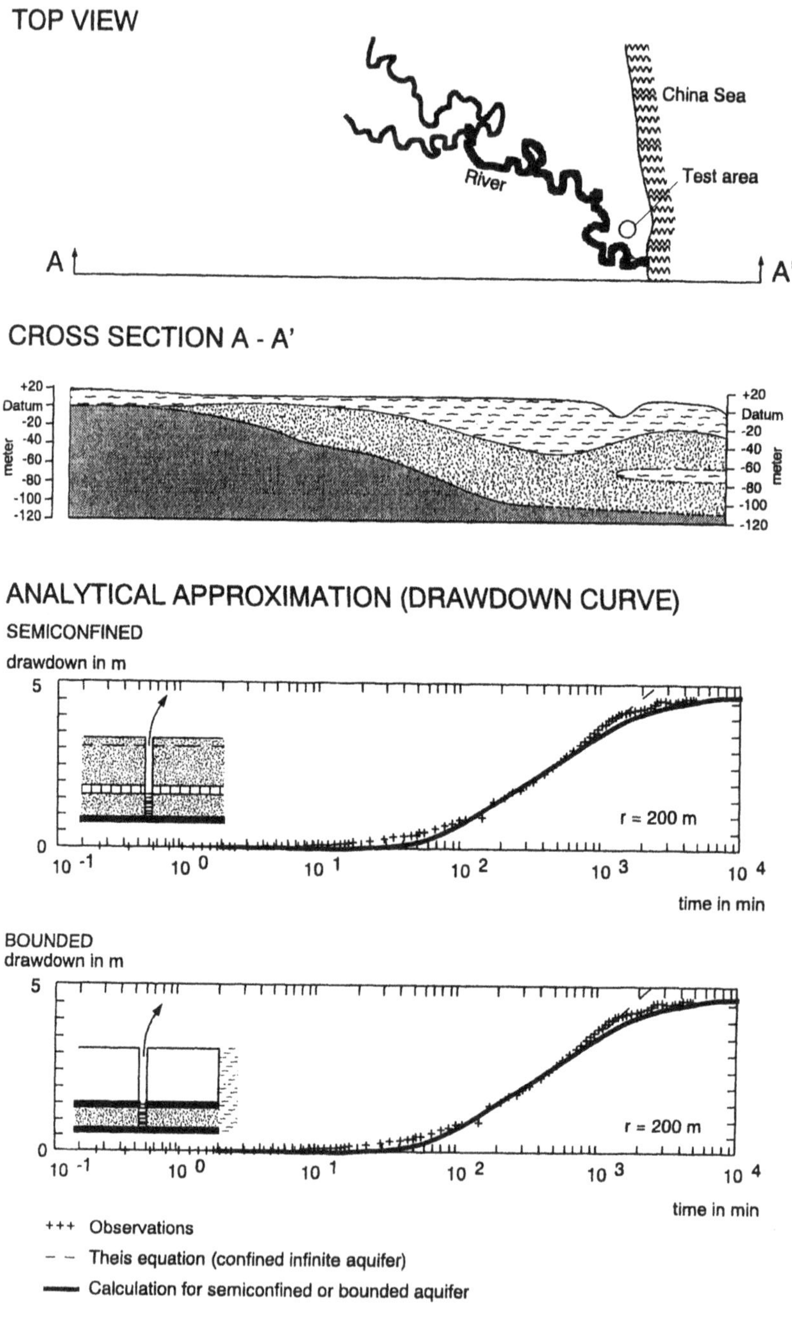

Figure 4.13 Analyzing pumping test data using analytical models.

Calibrating the Model

Analytical solutions enable the model user to predict tracer concentrations at any time and at any location, assuming that the tracer test is consistent with the underlying assumptions. Hence the record of field observations enables the model user to modify the unknown dispersivities such that differences between simulation and observation diminish. There is, however, a drawback in the model application, which illustrates a fundamental difficulty in transport modeling. Assume that during the tracer experiment longitudinal concentration distributions have been measured at two different times. Suppose that we successfully applied the analytical model to both data sets. Surprisingly we could calibrate different dispersivity values for each data set or different aquifer properties for the same aquifer. This phenomenon leads to an interesting source of modeling error.

Analyzing Model Errors

To perform model analysis for dispersion tests, recall the comments on dispersion in subsection 3.2.3. While in the description of fluxes a REV for flow may be derived, an equivalent REV to approximate dispersion as a Fickian process cannot necessarily be assumed. It is observed that the approximation of dispersive spreading in a real aquifer by Fick's law fails at early times. Any efforts in the model calibration using Fickian assumptions is bound to fail. Instead, time-, or travel distance-dependent dispersivities must be introduced to compensate for the inadequate mathematical representation of dispersive processes.

The deviation between theory and field observation can be traced back to a deviation between the mathematical model and reality. Now, however, the problem is fundamental. The correct way to circumvent the problem is to apply a model in which the underlying theory conforms with the actual transport mechanisms. Since modeling efforts increase exponentially as soon as Darcy's law or Fick's law fail to describe flow or transport respectively, the model user must often rely on classical models and address introduced errors in an intensive model sensitivity and error analysis.

This chapter has presented a comprehensive view of model application, and model calibration, from which we can benefit in numerical modeling. As we have seen, model calibration that proceeds with too little input data is not generally successful; a unique solution cannot be assured. The selection of an analytical formula, or of a conceptual model that fits the investigated problem best, is crucial in the interpretation because no systematic methodology for model selection is available. Model users rely on experience to recognize hydrogeologic controls appropriately. The task of the model user, who in the course of pumping test analysis must be able to visualize the physical nature of the aquifer and how it deviates from the basic assumptions (Driscoll 1986), is magnified in a complex model study. If the paucity of data does not suggest abandoning the simplified local cause-and-effect analysis by means of well hydraulic equations, the same logic holds for regional cause-and-effect evaluation using numerical groundwater models.

5 Understanding Numerical Flow Models

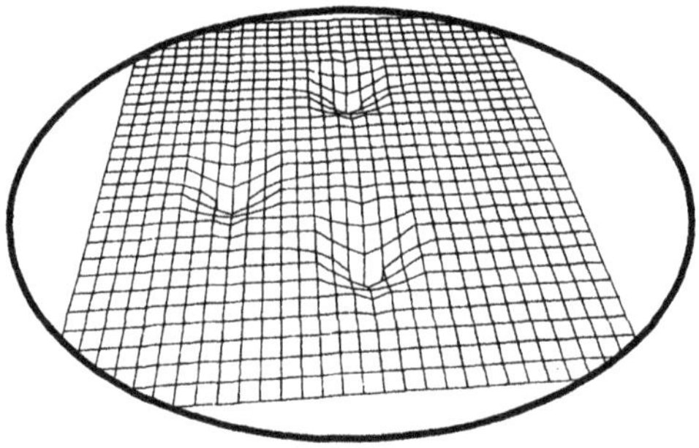

Numerical models allow analysis of flow or transport solutions if the complexity of the mathematical model prevents an analytical analysis. In flow modeling, as in transport modeling, the simplest numerical model considers the groundwater system as a single control volume. Changes in groundwater heads or concentrations are expressed as averages over the groundwater system and are calculated by a global mass balance. Single-cell or black box models are helpful when little field data are available and when the model user is interested in global changes of the groundwater quantity and quality. If information about local groundwater heads or concentrations is required, the model user applies more sophisticated models such as multiple-cell or numerical models.

The resolution of results provided by the model relates directly to the degree of discretization of the groundwater system under investigation, as shown in Figure 5.1. The single-cell model yields information on one investigated parameter, the average over the entire model area. A multiple-cell model provides more detailed information. For the three-cell model shown in Figure 5.1, three values of the investigated parameter averaged over each cell are calculated. The numerical model discretizes the model area in a great number of cells. For example, Frind et al. (1987) uses about 1,000,000 cells in the study of dispersion in heterogeneous media. Hence information on the investigated parameter is obtained at 1,000,000 locations.

Starting from the global water balance, this chapter presents the multiple-cell model, the finite-difference (FD) model, and the finite-element (FE) model for groundwater flow. Also some insight is provided into the numerics,

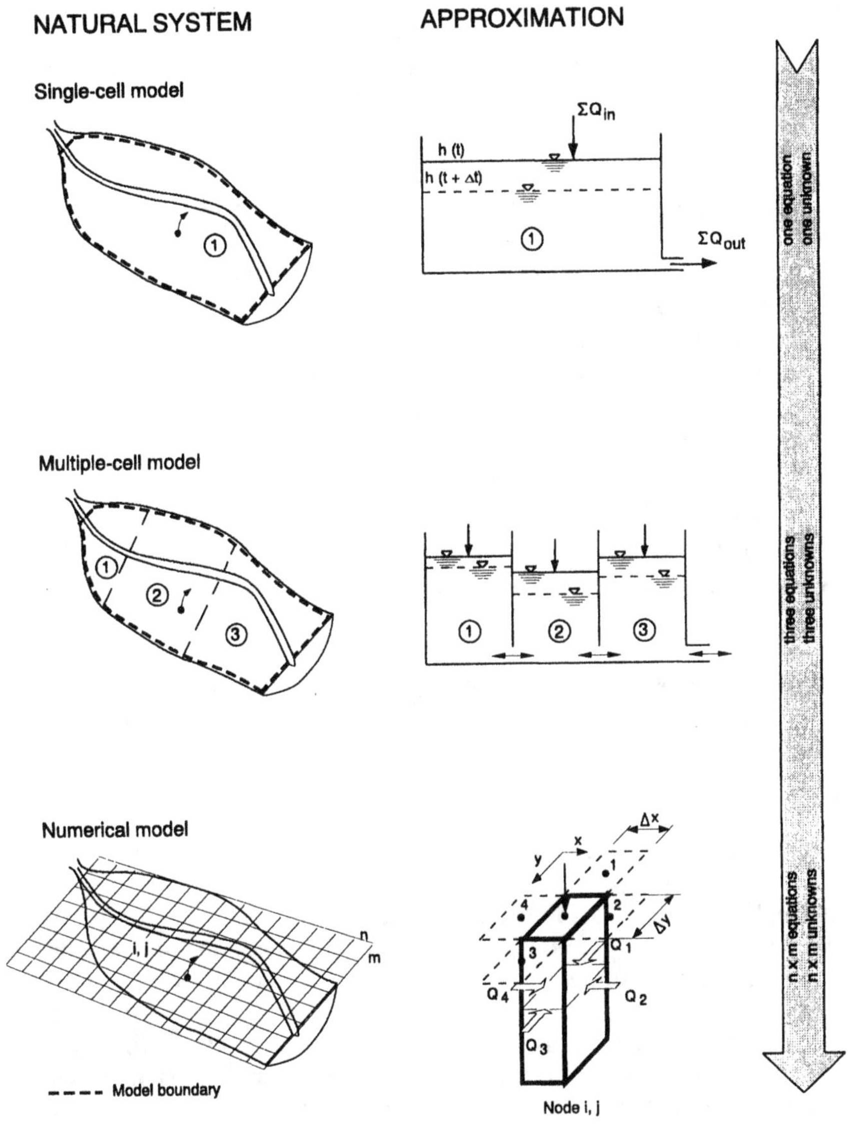

Figure 5.1 Discretization of model area using a single-cell model, a multiple-cell model, and a numerical model.

but without burdening the text with theory. Illustration of mathematical concepts behind numerical models has been mostly restricted to FD models. Once the mathematical approach for this model type is understood, it can be applied to the other model types without any complications.

5.1 DEVELOPING A SINGLE-CELL MODEL FOR FLOW

The model user should try to gain an appreciation of the global water balance before starting, for this ensures accurate modeling. To gain an appreciation, one begins by developing a single-cell model for flow by identifying and quantifying the following elements of the global water balance:

- Change in storage
- Natural groundwater re/discharge
- Exchange with surface water
- Subsurface flow over boundary
- Leakage component coming from adjacent aquifers
- Total local ex/infiltration

Without any exception these elements must be known for any complex numerical model study. See Appendix A.1 for a detailed case study of a single-cell model.

5.1.1 Identifying Elements of the Single-Cell Model

The single-cell model, also referred to as a *hydrologic budget*, *mass balance*, or *black box model*, is a quantitative evaluation of water gained or lost from the groundwater system under investigation during a specific period of time. The entire groundwater system is visualized as a single-cell, as shown in Figure 5.2. Since the change of groundwater stored in the system relates directly to the change of the average groundwater water table within the system, it is also a numerical groundwater model in its simplest form: a single-cell model. Hence the basic ideas underlying the evaluation of the global mass balance are also true for numerical (multiple-cell) models. In any modeling study it is important to establish the global mass balance in parallel with modeling efforts as a check on modeling accuracy.

To establish the global mass balance, the investigated groundwater system is represented as a black box where the quantities of water passing through in a given period of time are summarized in the continuity equation:

$$\text{change in storage} = \text{inflow} - \text{outflow} \tag{5.1}$$

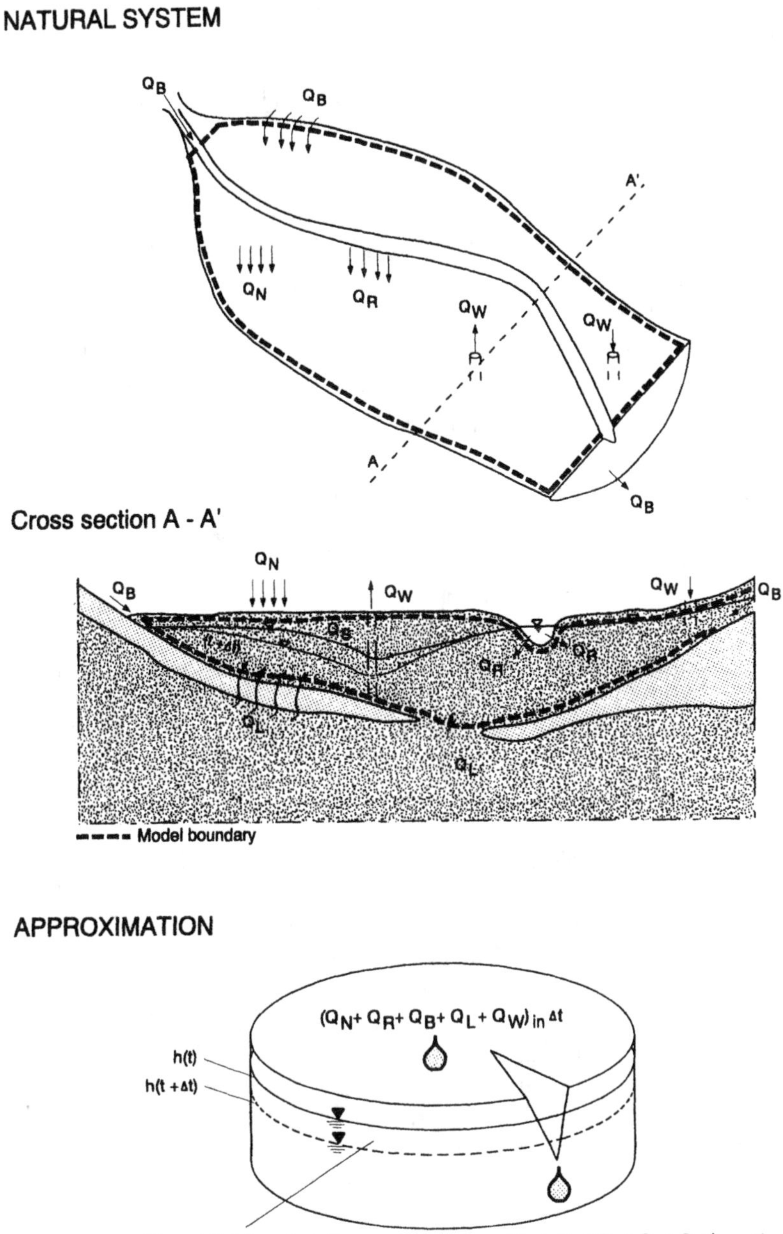

Figure 5.2 Single-cell model for groundwater flow.

The elements of the equation are shown in Figure 5.2 and may be written for a specific groundwater system

$$Q_s\,(t + \Delta t) - Q_s\,(t) = \Delta Q_s = (\pm\, Q_N \pm Q_R \pm Q_B \pm Q_L \pm Q_W)\,\Delta t \qquad (5.2)$$

where

ΔQ_s = change in storage Q_s in $[L^3]$ (see subsection 5.1.2)
$\pm Q_N$ = natural groundwater recharge or discharge in $[L^3/T]$ (see subsection 5.1.3)
$\pm Q_R$ = exchange with surface water in $[L^3/T]$ (see subsection 5.1.4)
$\pm Q_B$ = subsurface flow over boundary in $[L^3/T]$ (see subsection 5.1.5)
$\pm Q_L$ = leakage component leaving adjacent aquifers in $[L^3/T]$ (see subsection 5.1.6)
$\pm Q_W$ = local exfiltration or infiltration such as wells in $[L^3/T]$ (see subsection 5.1.7)
t = time in [T]
Δt = time increment in [T]

Steady-state groundwater flow, where the storage term is zero, is a special case of Equation 5.2. Most often the model user seeks the average fluctuation of the water table in the groundwater system resulting from changes of inflows and outflows.

5.1.2 Calculating the Storage Term

In the study of transient groundwater flow, we are directly concerned with gain or loss of water from storage. The amount of water released from storage per unit surface area of an aquifer, per unit change in head is referred to as the *specific storage coefficient*. As discussed in Chapter 2, the specific storage coefficient in an unconfined aquifer is equivalent to the specific yield or effective porosity n_e. It may range from 1% to 30%, as shown in Figure 2.2. In a confined aquifer, storage is attributed to compression of both the aquifer and the water, and storage coefficient is comparatively small. In both cases the amount of the water (ΔQ_s) added or released from storage is equal to the product of the volume of rock (V_{rock}) through which the change in water level occurs, the storage coefficient (S_s) and the head difference ($h(t + \Delta t) - h(t)$), which is written

$$\Delta Q_s = V_{\text{rock}}\, S_s\, [h(t + \Delta t) - h(t)] \qquad (5.3)$$

Although the specific storage coefficient for a confined aquifer is relatively small, the amount of stored or released water becomes significant in field studies where large aquifer systems are investigated. For unconfined aquifers Equation 5.3 simplifies to

$$\Delta Q_s = A n_e \left[h(t + \Delta t) - h(t)\right] \tag{5.4}$$

where A is the area of the investigated aquifer. Equation 5.3 or Equation 5.4 can be introduced into Equation 5.2. Hence change in the water table can be calculated, assuming that all inflow and outflow elements of Equation 5.2 are known. How to estimate these elements is presented in the following subsections.

5.1.3 Estimating Natural Groundwater Recharge and Discharge

Groundwater recharge due to precipitation, or groundwater discharge due to evaporation, is usually the most prominent element in any mass balance evaluation. Both recharge and discharge depend on numerous factors such as precipitation, evaporation, surface runoff, and vegetation or climatic conditions, making an accurate calculation or estimation difficult.

In modeling transient flow, we have to understand the delay between precipitation and percolation events, which can range from a few hours to several weeks. The actual processes of natural groundwater recharge require assessment by means of sophisticated unsaturated groundwater flow models and a formidable database. A crude approximation of the percolation behavior is suggested by Kinzelbach (1985), using the single linear reservoir model:

$$\begin{aligned} Q(\tau) &= 0, && \tau < t \\ Q(\tau) &= Q_N(1 - \exp(-\alpha(\tau - t))), && t \le \tau < t + \Delta t \\ Q(\tau) &= Q_N(1 - \exp(\alpha \Delta t))\exp(-\alpha(\tau - (t + \Delta t))), && \tau > t + \Delta t \end{aligned} \tag{5.5}$$

where

Q_N = natural groundwater recharge (% of a precipitation event eventually contributing to groundwater recharge) in [L^3/T]

α = coefficient in [1/T] depending on factors such as size and properties of unsaturated zone

Groundwater inflow due to precipitation is approximated to decline exponentially, as shown in Figure 5.3. Consecutive precipitation events are linearly superimposed.

If delay of groundwater recharge needs to be considered in the mass balance or in numerical modeling, the approximation in Equation 5.5 may prove to be sufficient. The dominating coefficient α may be chosen in accordance with field observations to allow a percentage of the total percolation to occur, for example, within one month of the precipitation event.

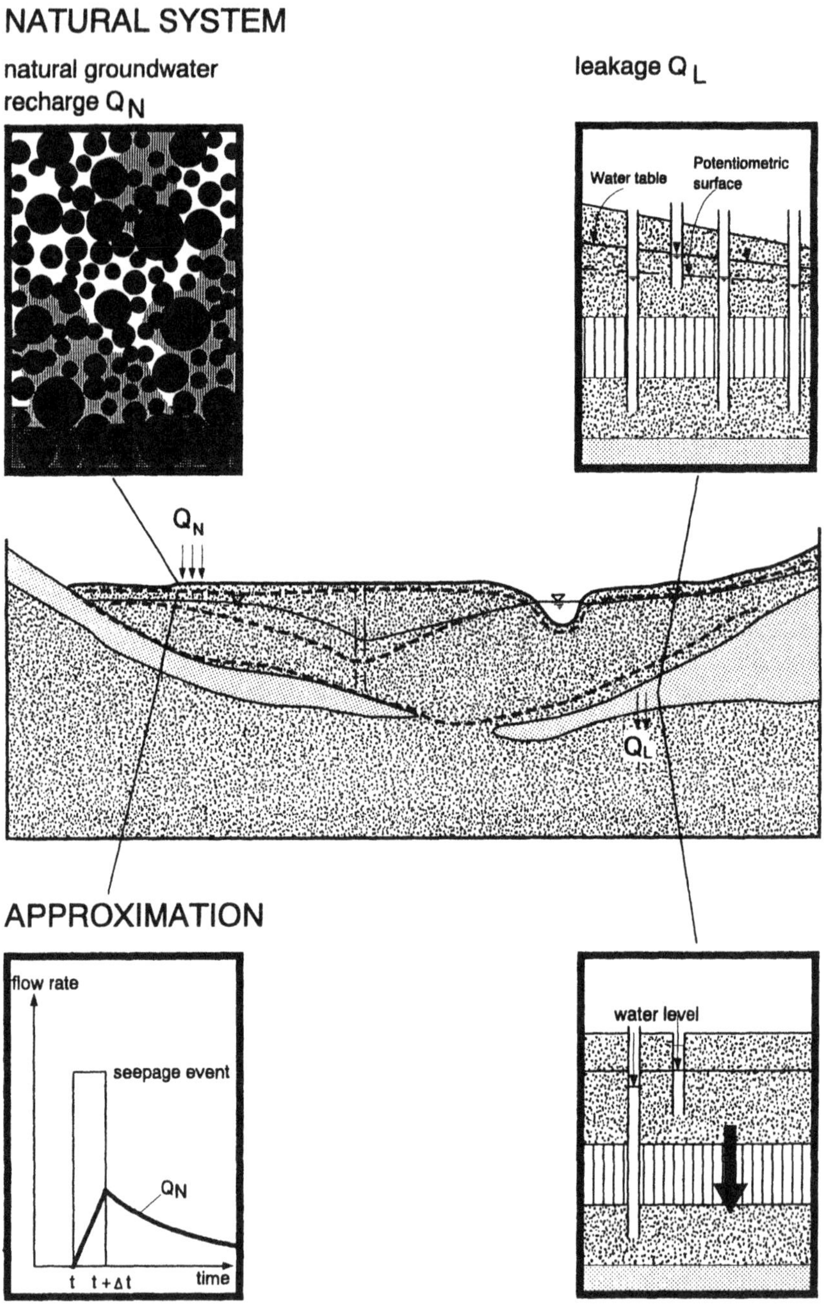

Figure 5.3 Estimation of model inflow and outflow.

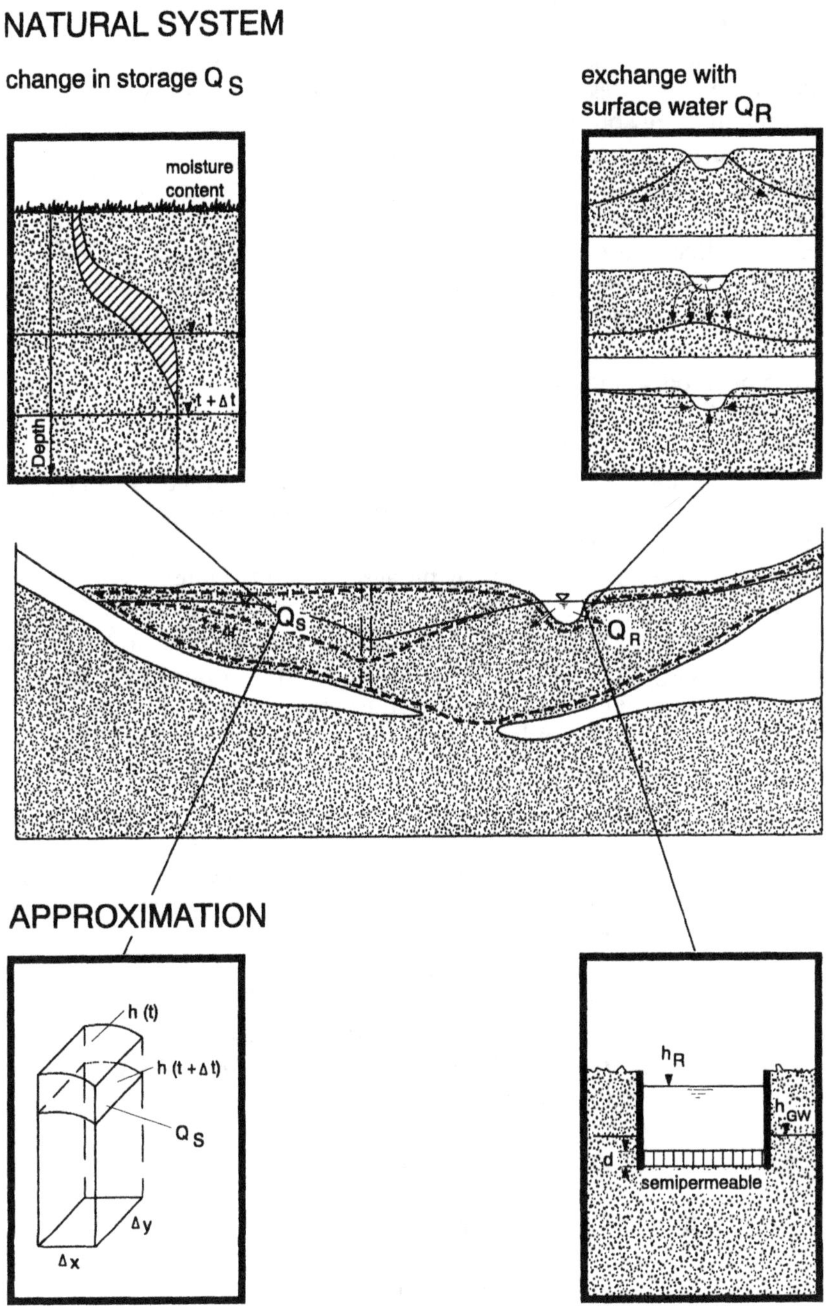

Figure 5.3 (*Continued*)

5.1.4 Approximating Exchange with Surface Water

Section 2.4 gave a qualitative account of groundwater exchange with surface water bodies when describing hydraulic flow boundaries. In the quantitative assessment of flow problems, water fluxes need to be realistically estimated. For this purpose measured groundwater levels can be used to advantage. Assuming a dense network of observation wells in the vicinity of surface water bodies, recharge areas from, or discharge areas to, the surface water can be identified. Lee and Cherry (1978) apply minipiezometer and flow-meter measurements to obtain quantitative field information. Direct measurement using a lysimeter may be the more accurate but also the more expensive approach.

In most model studies, however, model users have to rely on their own estimates. A good initial attack at the problem is provided by records of groundwater level measurements at monitoring wells close to the surface water. Although they do not yield quantitative information on groundwater/surface water interactions, they indicate whether or not the aquifer responds to changes in the surface water elevations.

The conclusions in the following paragraphs, for the interaction of a river and the groundwater body, will be the same for any other surface water bodies such as fish ponds, lakes, irrigation channels, drains, ditches, or the sea. The degree to which groundwater responds to river levels is controlled by the water levels of surface water and groundwater, the permeability of both, the riverbed and the adjacent aquifer, and the size of contact area. The model user differentiates infiltration of surface water into one of the following two groups:

- Free seepage with groundwater levels below the riverbed.
- Direct infiltration with the riverbed penetrating the groundwater body. Depending on the elevation of the water table, flux will be reversed and the aquifer actually loses water to the river.

Numerous formulas are available to estimate the infiltration rate for configurations such as these (Vassilios 1986). In numerical models the leakage principle is commonly used to introduce infiltration from or to surface water bodies, as shown in Figure 5.3. Free seepage is estimated as

$$q_R = c(h_R - h_B) \tag{5.6}$$

and direct infiltration estimated as

$$q_R = c(h_R - h) \tag{5.7}$$

where

q_R = infiltration or exfiltration rate per unit square meter in [L/T]

h_R = elevation of surface water table in [L]
h_B = elevation of surface water bed in [L]
h = elevation of groundwater table in [L]
c = resistance of semipermeable riverbed in $[T^{-1}]$

The resistance factor can be visualized as representing the flow resistance due to a thin semipermeable layer at the riverbed with thickness d and hydraulic conductivity K:

$$c = \frac{K}{d} \tag{5.8}$$

Equation 5.8 is identical to Equation 2.6, defining the resistance of the leaky layer in the case of flow in a semiconfined aquifer.

Both parameters in Equation 5.8 are generally unknown. In areas with permanent infiltration of river water, the hydraulic conductivity of the riverbed may even decrease with time, due to fine particles suspended in the infiltrating water. Eventually these particles close pores in the riverbed, building up a layer with high flow resistance.

The model user is not normally concerned about the thickness of the riverbed and its hydraulic conductivity but summarizes them into one estimate for the resistance factor, as defined in Equation 5.8. The resistance factor is then subject to the model calibration (see Section 8.5). A large resistance factor implies that the river is in close communication with the aquifer and will act as a prescribed head boundary. Since the water flux between the river and groundwater must remain finite, the head difference between them must diminish with increasing leakage factor. In more complex hydrogeologic environments, the model user should not be tempted to simulate every river as a model boundary with prescribed heads.

Equations 5.6 and 5.7 show that the infiltration rate is only unaffected by the groundwater heads in the case of free infiltration. Otherwise, this term contributes a nonlinearity to the flow equation. In particular, flux may reverse if the water table rises above the river surface, as shown in Figure 5.3.

5.1.5 Quantifying Subsurface Inflow and Outflow

No method exists to quantify subsurface inflow and outflow directly. The model user must rely on approximations; the most common approximations follow:

- An estimate of subsurface flow is obtained by measuring the hydraulic gradient along the boundary and calculating flow by applying Darcy's law

$$q_B = -Ti \tag{5.9}$$

where

q_B = boundary flux per meter boundary in [L²/T]
T = transmissivity in [L²/T]
i = hydraulic gradient perpendicular to boundary in [1]

- A water balance of the neighboring areas can provide useful estimates of boundary fluxes. The single linear reservoir model presented in subsection 5.1.3 will permit approximation of the delay of lateral inflow due to precipitation events.
- A descriptive rather than a quantitative evaluation of the inflow rate can be made by the interpretation of chemical and isotopic groundwater hydrology (Desaulniers et al. 1981).

Quite often, the model calibration includes quantifying the flow rate at boundaries.

5.1.6 Estimating Interaction with Adjacent Aquifers

Figure 5.3 illustrates that vertical groundwater flow may exist between aquifers separated by an aquitard. The exchange rate is estimated, using the leakage principle represented in the following equation:

$$q_L = \frac{-K}{d}(h_1 - h_2) = c(h_l - h_2) \tag{5.10}$$

where

q_L = leakage flux per unit square meter in [L/T]
K = hydraulic conductivity of aquitard in [L/T]
d = thickness of aquitard in [L]
h_1 = head in aquifer 1 in [L]
h_2 = head in aquifer 2 in [L]
c = resistance factor of aquitard in [T^{-1}]

A representative resistance factor is estimated in areas where the aquitard is missing. Flow at such locations is concentrated, and a minimum or maximum in the water table may occur depending on whether the aquifer is losing water to or receiving water from the adjacent aquifer.

The interaction between numerous aquifers may force one to apply a multilayer model to describe groundwater flow or transport. A multilayer model can be interpreted as several individual two-dimensional flow models, one for each layer, connected to each other through a source and sink term, according to Equation 5.10 (see also Section 7.5).

In a layered aquifer the potentiometric surface normally varies with depth. Depending on the construction of monitoring wells, one may measure potentiometric surface in different layers, but this will make it difficult to compare values with each other.

5.1.7 Summarizing Local Exfiltration or Infiltration

Local exfiltration or infiltration includes groundwater abstraction by wells. In most field studies local exfiltration or infiltration are most accurately determined as an element of the total water balance. In areas with intensive groundwater use, compilation of field data may be fairly demanding.

5.2 UNDERSTANDING THE FINITE-DIFFERENCE MODEL

One of the first model studies using the finite-difference (FD) method was conducted in 1968 by the California Department of Water Resources to study the Los Angeles coastal plain groundwater basin. In FD models the model user establishes a regular grid over the model area that subdivides the total model area into rectangular subdomains. Constant system parameters are then assigned to each cell as cell value. Depending on the FD model, groundwater heads or concentrations are calculated as discrete values at the grid nodes, or at the center points of the cells.

The mathematical approach to develop the FD model is to expand the governing differential equation term by term into a Taylor series. This section introduces the finite difference approximation by extending the single-cell approach of Section 5.1 to a three-cell model. The three-cell model illustrates how to:

- Discretize the model domain.
- Assign field data to each cell.
- Establish the continuity equation for each cell.
- Determine inflow and outflow for each cell.
- Express the continuity equation for the unknown heads.
- Solve the set of equations.

This section also demonstrates that a multiple-cell model ultimately leads to a numerical model if the model user further pursues the discretization.

5.2.1 Developing the Finite-Difference Approximation

Figure 5.4 illustrates the finite-difference approximation to a hypothetical flow regime. Consider a valley with an approximate width of ($B = 2500$ m) and

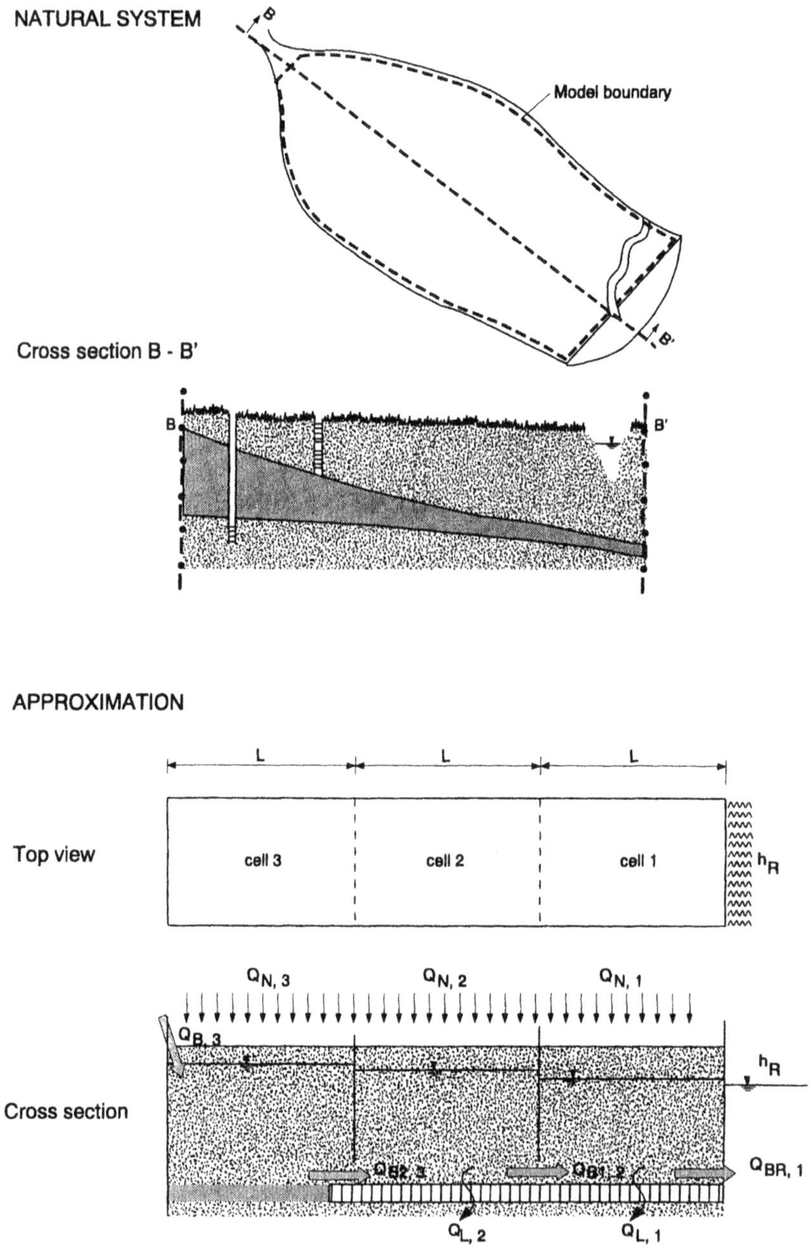

Figure 5.4 Demonstration of a three-cell model.

a length of (L = 12,000 m). Leakage occurs on the right-hand side through an aquitard with an average thickness of (d = 1 m) and vertical hydraulic conductivity ($K_v = 10^{-5}$ m/d). Thus the resistance factor is calculated as (10^{-5} 1/d) using Equation 5.8. The head in the adjacent aquifer is assumed to be constant (h_o = 149 m). The only significant subsurface inflow exists to the left (q_B = 0.5 m³/day/m). To the right, the valley ends at a river controlling the groundwater level at this end (h_R = 150 m).

Approximation by finite differences proceeds as follows:

- Discretize the flow domain. To estimate the groundwater levels, subdivide the valley into three balance volumes of equal length, as shown in Figure 5.4.
- Assign field data to each cell. Assume as constant, transmissivity, natural groundwater recharge, and all other model input data within each cell. These model input data are:

Cell	Width in [m]	Length in [m]	Trans-missivity in [m²/d]	Leakage in [1/d]	Natural Ground-water Recharge in [mm/y]	External Boundary Flux in [m³/d/m]
1	2500	4000	2000	10^{-5}	400	—
2	2500	4000	1400	10^{-5}	300	—
3	2500	4000	1000	10^{-5}	300	0.5

- Establish the continuity equation for each cell using Equation 5.2.

$$-Q_{BR,1} + Q_{B1,2} + Q_{N,1} - Q_{L,1} = 0$$
$$-Q_{B1,2} + Q_{B2,3} + Q_{N,2} - Q_{L,2} = 0$$
$$-Q_{B,3} + Q_{B2,3} + Q_{N,3} = 0$$

- Determine inflow and outflow for each cell. Natural groundwater recharge in [m³/d] is calculated as unit recharge times total cell area:

$$Q_{N,1} = 10{,}954 \quad \text{(cell 1)}$$
$$Q_{N,2} = 8219 \quad \text{(cell 2)}$$
$$Q_{N,3} = 8219 \quad \text{(cell 3)}$$

Subsurface flow entering or leaving the total control volume is flow per unit width times length of boundary. The boundary flux through the common boundary of two adjacent cells, i and $i + 1$, is calculated as:

$$Q_{Bi,i+1} = T_{i,i+1} B \frac{h_i - h_{i+1}}{0.5(L_i + L_{i+1})} \tag{5.11}$$

where h_i and h_{i+1} are the potentiometric surface in the cells and L_i and L_{i+1} are the length of each cell. Normally the harmonic instead of the arithmetic mean of the transmissivities is chosen as the representative transmissivity because it ensures flux continuity across transmissivity discontinuities. The transmissivity between both cells is calculated as

$$T_{i,i+1} = \frac{L_i + L_{i+1}}{\dfrac{L_i}{T_i} + \dfrac{L_{i+1}}{T_{i+1}}} \tag{5.12}$$

For example, the relevant transmissivity for the subsurface flow between cells 1 and 2 is found as ($T_{1,2} = 1647$ m^2/day). Hence subsurface inflow or, respectively, outflow for each cell in [m^3/d] calculates as

$$\begin{aligned} Q_{BR,1} &= -2.500(h_r - h_1) \quad &\text{(cell 1)} \\ Q_{B1,2} &= -1.029(h_1 - h_2) \quad &\text{(cell 2)} \\ Q_{B2,3} &= -0.729(h_2 - h_3) \quad &\text{(cell 3)} \end{aligned}$$

Leakage flux in [m^3/d] is calculated using Equation 5.10. We find that

$$\begin{aligned} Q_{L,1} &= 100(h_1 - h_0) \quad &\text{(cell 1)} \\ Q_{L,2} &= 100(h_2 - h_0) \quad &\text{(cell 2)} \\ Q_{L,3} &= 0 \quad &\text{(cell 3)} \end{aligned}$$

- Express the continuity equations for the unknown groundwater heads. The set of equations for the unknowns h_1, h_2, h_3, is found by inserting the expressions for the different fluxes into the continuity equation. Three equations for three unknowns remain:

$$\begin{aligned} -3629h_1 + 1029h_2 \qquad\qquad &= -400{,}859 \quad &\text{(cell 1)} \\ 1029h_1 - 1858h_2 + 729h_3 &= -23{,}119 \quad &\text{(cell 2)} \\ 729h_2 + 729h_3 &= -9469 \quad &\text{(cell 3)} \end{aligned}$$

- Solve the set of equations. The set of equations can be solved fairly easily by hand. For completeness, results are given to two decimal places, which infers an accuracy that does not exist:

$$h_1 = 160.00 \text{ m} \quad h_2 = 174.68 \text{ m} \quad h_3 = 187.67 \text{ m}$$

In numerical models with a larger number of equations and unknowns, alternative solution techniques must be applied.

5.2.2 Implementing Finite-Difference Approximation into a Numerical Scheme

In Section 5.1 and subsection 5.2.1, we established water balances for one- and three-cell models. More than one cell chosen results in the numerical FD model shown in Figure 5.1. In modeling practice one does not develop a numerical scheme in each model study. Existing numerical codes are general enough to apply to different scenarios. The tasks of the model user are to:

- Subdivide the model area into regular cells.
- Simplify the boundaries as a series of interconnected vertical and horizontal lines.
- Assign constant system parameters to each cell.
- Assign fluxes to each cell except for the subsurface flow between adjacent cells.
- Assign initial conditions.

These four tasks generate input data. The numerical code then automatically:

- Establishes the continuity equation for each cell, yielding one equation containing the unknown head in the considered cell and in its neighboring cells.
- Arranges the equations in the form of a set of n linear equations for the n unknown heads.
- Obtains the unknown discrete head distribution by solving the entire set of equations.

For illustration, we extend the three-cell model of subsection 5.2.1 to a general one-dimensional finite difference scheme. The governing flow equation solved is

$$\frac{\partial}{\partial x}\left(T\,\frac{\partial h}{\partial x}\right) + Q = S\,\frac{\partial h}{\partial t} \tag{5.13}$$

where Q, in $[L^3/T]$, stands for the external sinks and sources. For simplification we assume that the aquifer is confined so that the transmissivity becomes independent of h.

Now subdivide the flow system into n cells, each with equal length of Δx. The center of a cell is called a *node* and represents the cell. Consequently calculated heads are assigned to nodes. Discretize time into time levels t_1 to t_n separated by time intervals Δt. Establish a water balance for each time level, starting with the initial condition at time t_o. This approach is called *time stepping.*

The water balance for the node i for a given time increment depends on the fluxes received from, or lost to, the neighboring nodes $i - 1$ and $i + 1$. To calculate inflow, or respectively, outflow into cell i, use the following equations derived from Darcy's law:

$$q_{i,i-1} = -\frac{T_{i,i-1}}{\Delta x}(h_i - h_{i-1})\Delta t \tag{5.14}$$

$$q_{i+1,i} = -\frac{T_{i+1,i}}{\Delta x}(h_{i+1} - h_i)\Delta t \tag{5.15}$$

In Equations 5.14 and 5.15 the harmonic mean of the cell transmissivities represents the average transmissivity between nodes.

A constant flux over the time step Δt is assumed. This is acceptable if the time step is short and flow rates vary slightly. Under these assumptions flow rates can be expressed as averages. To express flow between nodes, choose the hydraulic heads at any time t' within the time increment Δt. For example, choosing $t' = t$ in Equation 5.14 leads to

$$q_{i,i-1} = -\frac{T_{i,i-1}}{\Delta x}(h_i(t) - h_{i-1}(t))\Delta t \tag{5.16}$$

If $t' = t + \Delta t$ is chosen, Equation 5.14 is expressed as

$$q_{i,i-1} = -\frac{T_{i,i-1}}{\Delta x}(h_i(t + \Delta t) - h_{i-1}(t + \Delta t))\Delta t \tag{5.17}$$

In addition to fluxes between nodes, a local source or sink with strength Q_i per unit width may contribute to the water balance of node i. The amount of groundwater Q_{Si} stored or released in cell i over Δt is calculated as

$$Q_{Si} = S_i \Delta x(h_i(t + \Delta t) - h_i(t)) \tag{5.18}$$

Hence the water balance for cell i, expressed per unit width, yields

$$T_{i,i-1} h_{i-1}(t') - [\mathrm{T}_{\mathrm{i,i-1}} + T_{i+1,i}]h_i(t') + T_{i+1,i} h_{i+1}(t') = \frac{\Delta x^2}{\Delta t} S_i (h_i (\mathrm{t} + \Delta t) - h_i (t)) - Q_i \Delta x \tag{5.19}$$

The equation is identical to the equations for unknown groundwater heads found in the three-cell model in subsection 5.2.1, except for the missing leakage component.

For simpler notation, consider the case where the transmissivity T and the storativity S are constant over the model area. Thus Equation 5.19 simplifies to

$$h_{i-1}(t') - 2h_i(t') + h_{i+1}(t') = \frac{\Delta x^2}{\Delta t}\frac{S}{T}(h_i(t+\Delta t) - h_i(t)) - \frac{Q_i}{T}\Delta x \tag{5.20}$$

Since i is arbitrary ($i = 1$ to n where n is the total number of nodes), Equation 5.20 represents n equations for the n unknown heads. To solve for h we must now specify what intermediate time level t' contained in the time interval $[t, t + \Delta t]$ at which we will choose to express flow rates between nodes.

To avoid solving for a set of simultaneous equations, explicit finite difference models make use of the known hydraulic heads at the beginning of the time step (forward differences). Approximating t' by t, we can rearrange Equation 5.20 as

$$h_i(t+\Delta t) = h_i(t) + \frac{\Delta t}{\Delta x^2}\frac{T}{S}\left[\frac{\Delta x}{T}Q_i + h_{i-1}(t) - 2h_i(t) + h_{i+1}(t)\right] \tag{5.21}$$

At time t_o all values of h_i are known (initial condition). Thus we can solve explicitly for the potentiometric surface at the new time level $t_o + \Delta t$. These results provide the known conditions for the next time level, and so on. The method is straightforward and easy to program on a computer, but the big disadvantage is that the solution becomes unstable for large Δt. For one-dimensional flow we must ensure that

$$\frac{T}{S}\frac{\Delta t}{\Delta x^2} \leq \frac{1}{2} \tag{5.22}$$

In two dimensions, the stability criterion is

$$\frac{T}{S}\left(\frac{\Delta t}{\Delta x^2} + \frac{\Delta t}{\Delta y^2}\right) \leq \frac{1}{2} \tag{5.23}$$

In the model application to field situations, this rigid constraint on the time step calls for small time steps or longer calculation times.

Implicit solution schemes are unconditionally stable. They express the water balance for each cell using the unknown heads at the end of the time step (backward differences). Unlike the explicit scheme the model user now has to solve simultaneously a set of equations. Approximating t' by $t + \Delta t$, we rearrange Equation 5.20 as

$$h_{i-1}(t+\Delta t) - \left(2 + \frac{\Delta x^2}{\Delta t}\frac{S}{T}\right) h_i(t+\Delta t) + h_{i+1}(t+\Delta t) = \frac{\Delta x^2}{\Delta t}\frac{S}{T}\left(h_i(t+\Delta t) - h_i(t)\right) - \frac{Q_i}{T}\Delta x \tag{5.24}$$

The right-side of the equation is known. For $n = 6$, and in implementing given heads at node 1 and 6 as boundary conditions, we find that

$$\underbrace{\begin{bmatrix} -(2+a) & 1 & 0 & 0 \\ 1 & -(2+a) & 1 & 0 \\ 0 & 1 & -(2+a) & 1 \\ 0 & 0 & 1 & -(2+a) \end{bmatrix}}_{\text{coefficient matrix}} \underbrace{\begin{Bmatrix} h_2(t+\Delta t) \\ h_3(t+\Delta t) \\ h_4(t+\Delta t) \\ h_5(t+\Delta t) \end{Bmatrix}}_{\text{unknowns}} = \underbrace{\begin{Bmatrix} -a\,h_2(t) - b_2 - h_1(t) \\ -a\,h_3(t) - b_3 \\ -a\,h_4(t) - b_4 \\ -a\,h_5(t) - b_5 - h_6(t) \end{Bmatrix}}_{\text{knowns}} \tag{5.25}$$

where

$$a = \frac{\Delta x^2}{\Delta t}\frac{S}{T} \tag{5.26}$$

$$b_i = Q_i \frac{\Delta x}{T} \tag{5.27}$$

The coefficient matrix is tridiagonal and is always symmetric when solving flow problems. This feature can be used to check the numerical formulation and to save storage. The coefficient matrix reflects the characteristics of the aquifer as well as the chosen time and space discretization. The vector for the knowns represents known fluxes and heads. It is obvious that in model studies based on more than three or four cells, the solution of the matrix equation becomes somewhat demanding. A model study with 1000 cells yields a set of 1000 equations with 1000 unknowns. Thus the equations have to be solved numerically, applying either direct or iterative matrix solvers. Tridiagonal matrices are best solved by the Thomas algorithm (Hornbeck 1975). The program subroutine that solves the matrix equation forms the heart of each numerical model and controls the calculation time. This solution procedure is repeated at each time level.

Besides explicit and implicit schemes, general time weighting is also applied to the difference Equation 5.20. Then $h(t')$ is expressed as

$$h(t') = (1 - \theta)h(t) + \theta h(t + \Delta t) \tag{5.28}$$

Note that $\theta = 1$ yields the implicit solution and $\theta = 0$ the explicit one. The choice of $\theta = 0.5$ results in the widely used Crank-Nicholson scheme.

FD models for two- or three-dimensional flow are equivalent to the FD model for one dimension. Two- and three-dimensional models require more bookkeeping. The set of equations to be solved is more complicated. For the model user the increasing complexity is not only noticeable in more complex data input files, or in increasing calculation times, but also in increasing efforts when interpreting numerical results.

There is a similarity between writing a water balance for a cell or for three cells, as done previously, and writing an FD model for the general two or three-dimensional case. This is illustrated in Figure 5.5, which summarizes an example of the two-dimensional modeling of a simplified aquifer system. Again, the model area is subdivided into several cells, and the model user assigns field data to each cell. The numerical code automatically sets up the corresponding matrix equation, which then is solved using matrix solvers such as Gauss-Seidel-Iteration, ADI, Jacobi, L-R-decomposition, preconjugate-gradient, and so on (for an overview of these solvers, see Remson et al. 1971). Figure 5.5 also presents calculated heads, assuming steady-state flow.

An example of grid discretization in a field study is shown in Figure 5.6. To adjust the numerical scheme to site-specific characteristics of a flow regime, variable grid spacing is necessary. The grid comprises 2500 nodes.

5.3 UNDERSTANDING THE FINITE-ELEMENT MODEL

FD models and FE models are very popular in flow modeling, and they yield accurate results. Both model types are equivalent in view of their accuracy, but the FE approach is more versatile.

The FE model differs from the FD model by approximating the flow equation by integration rather than differentiation. As in the FD model, the model area is subdivided into subareas, called *elements.* One normally chooses triangular elements as subareas. Since there are basically no restrictions on the shape of the elements, the model user is more flexible in the model discretization than when using the finite difference scheme.

Section 5.3 introduces the basic concept of FE models. The mathematical approach to solve for the unknown heads is briefly presented. A mathematical coverage of the theory is not intended.

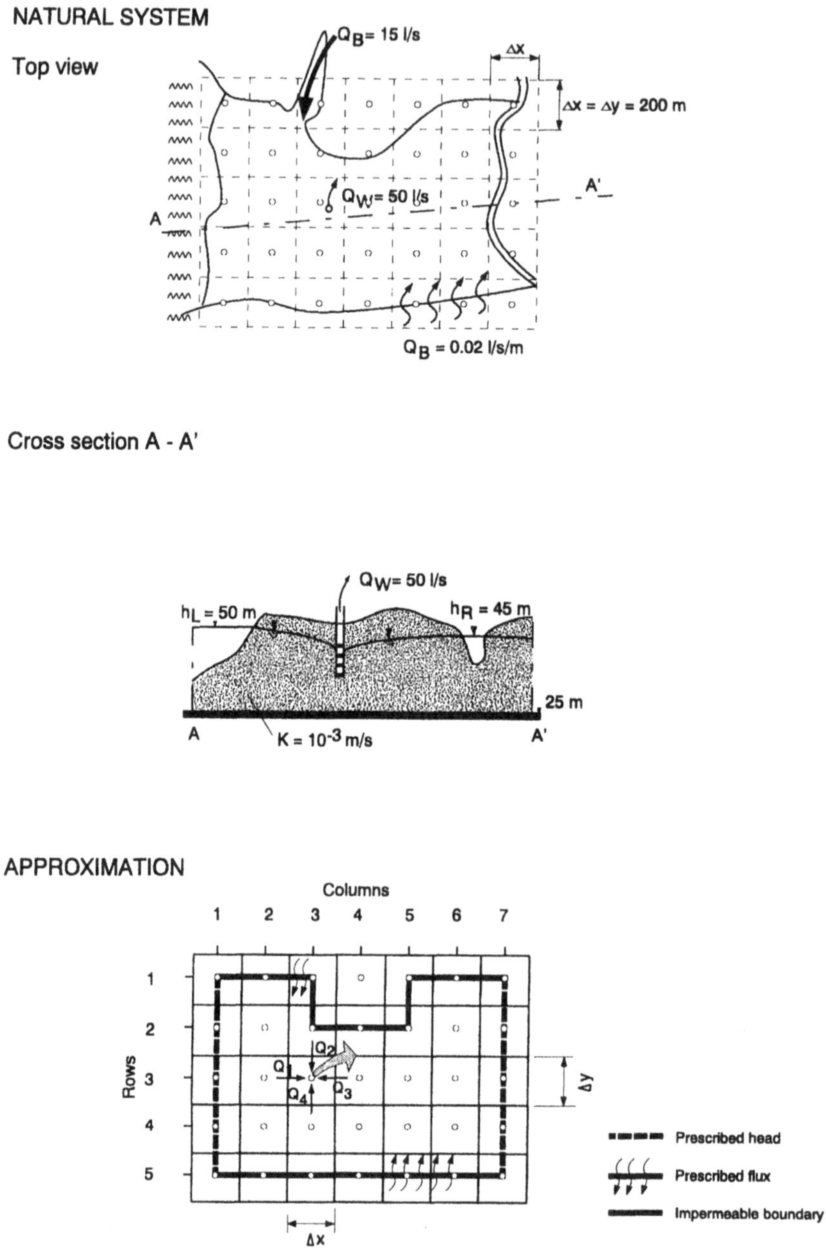

Figure 5.5 Demonstration model for two-dimensional FD model.

MODEL INPUT DATA

Geological parameters

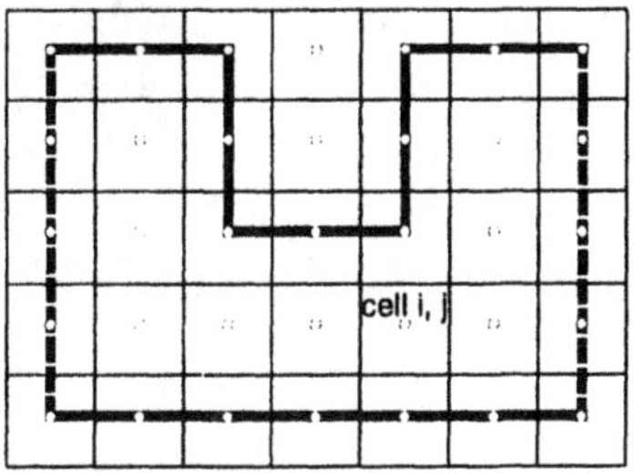

Hydraulic conductivity:

$K_{i,j} = 10^{-3}$ m/s

Aquifer base:

$h_{i,j} = 25$ m   i = 1 to 7
j = 1 to 5

Boundary conditions

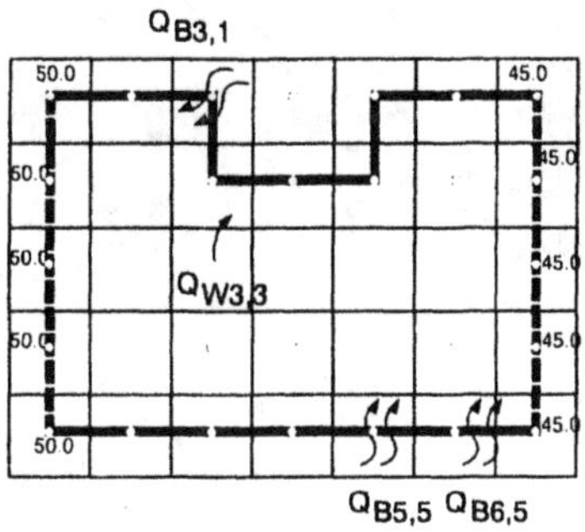

Prescribed head
$h_R = 45$ m
$h_L = 50$ m

Boundary flux:
$Q_{B3,1} = 15$ l/s
$Q_{B5,5} = 4$ l/s
$Q_{B6,5} = 4$ l/s

Internal sources/sinks:
$Q_{W3,3} = 50$ l/s

NUMERICAL RESULTS

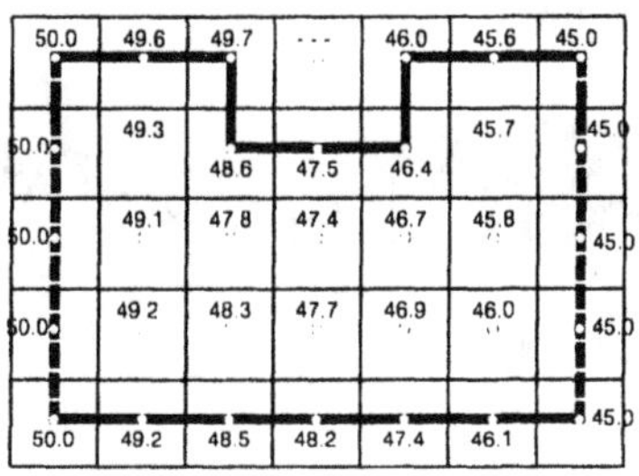

Heads in m

Figure 5.5 (*Continued*)

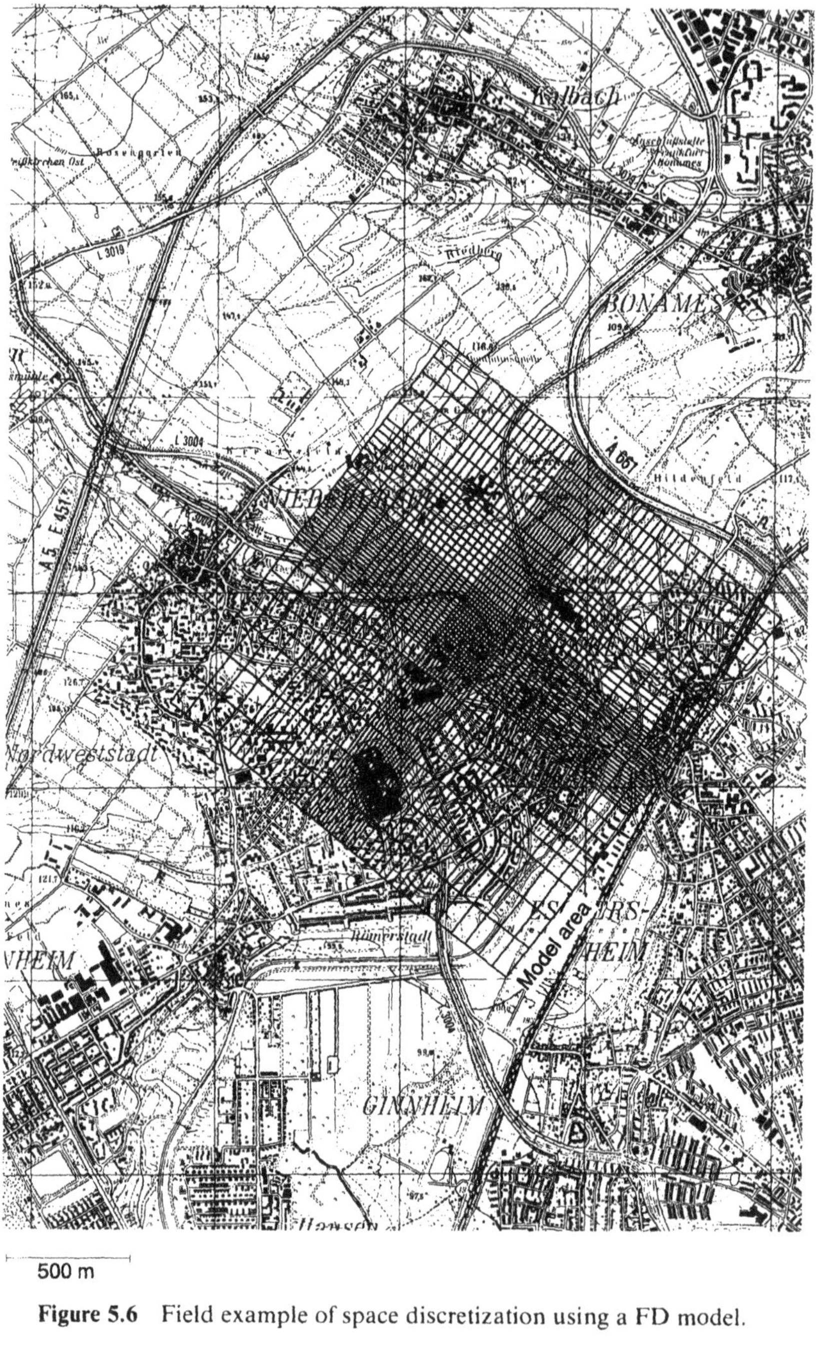

Figure 5.6 Field example of space discretization using a FD model.

5.3.1 Developing the Finite-Element Approximation

Numerical programs are an approximation of the partial differential equation describing groundwater flow (Equation 2.4). In the case of two-dimensional flow in a confined aquifer, this equation is

$$\frac{\partial}{\partial x}\left(T\frac{\partial h}{\partial x}\right) + \frac{\partial}{\partial y}\left(T\frac{\partial h}{\partial y}\right) + Q - S\frac{\partial h}{\partial t} = 0 \tag{5.29}$$

Expressing boundary conditions in terms of the hydraulic head or of flux, the model user must search for the head distribution h that satisfies Equation 5.29 everywhere inside the model area at each time. Equation 5.29 is fulfilled everywhere in the case of an exact solution.

The exact head distribution cannot be determined in complex systems. Most FE models approximate the solution of h by piecewise linear functions, which will be described next. Higher-order functions can also be considered (Pinder and Gray 1977). The model area is divided into elements. Then the head distribution is approximated for each element by a linear function.

To illustrate the finite element approximation, consider the aquifer with irregular thickness as shown in Figure 5.7. The governing flow equation is

$$\frac{\partial}{\partial x}\left(T\frac{\partial h}{\partial x}\right) - S\frac{\partial h}{\partial t} = 0 \tag{5.30}$$

Divide the groundwater system into four elements. In one dimension for a given time, straight lines as elementwise approximation for the head distribution, for a given time, are found:

$$\hat{h} = a + bx \tag{5.31}$$

where $\hat{h}$ represents the approximate head.

For example, consider element 3 with the end points (x_3, h_3) and (x_4, h_4). The head distribution approximates as

$$\hat{h} = \begin{cases} h_3 + \dfrac{h_4 - h_3}{x_4 - x_3}(x - x_3) & \text{for } x_3 \leq x \leq x_4 \\ 0 & \text{elsewhere} \end{cases} \tag{5.32}$$

An alternative way to express $\hat{h}$ is by using interpolation functions, also referred to as *basis functions*, ϕ defined in Figure 5.7 and by formulating $\hat{h}$ as a weighted average of the nodal values h_3 and h_4. The approximation for the head distribution in element 3 is

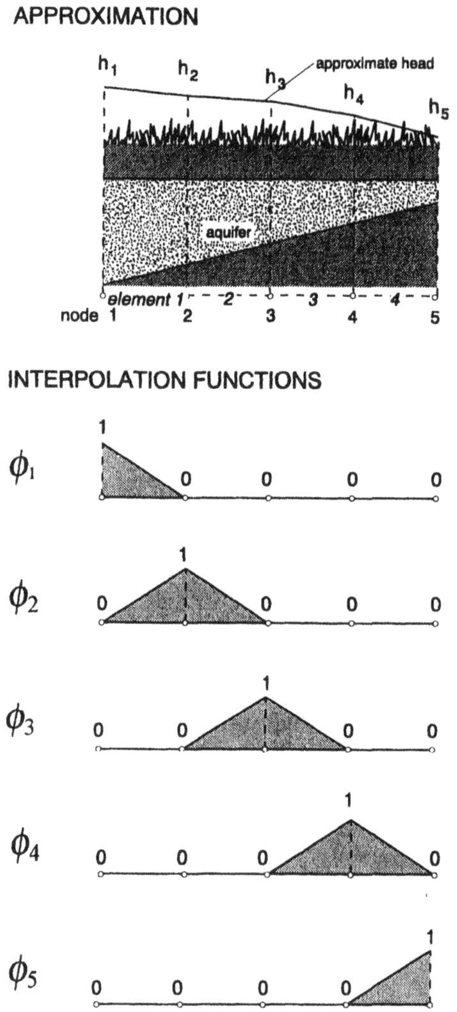

$$\phi_i = \begin{cases} 0 & |x| > x_{i+1} \\ & |x| < x_{i-1} \\ \dfrac{x - x_{i-1}}{x_i - x_{i-1}} & x_{i-1} \le x \le x_i \\ \dfrac{x_{i+1} - x}{x_{i+1} - x_i} & x_i < x \le x_{i+1} \end{cases}$$

Figure 5.7 Basis functions in one dimension (example).

$$\hat{h} = \begin{cases} h_3 + \dfrac{x_4 - x}{x_4 - x_3} + h_4 \dfrac{x - x_3}{x_4 - x_3} & \text{for } x_3 \leq x \leq x_4 \\ 0 & \text{elsewhere} \end{cases} \tag{5.33}$$

The approximation for the entire head distribution results from the summation of the contribution of each element, and we find that

$$\hat{h} = h_1\phi_1 + h_2\phi_2 + h_3\phi_3 + h_4\phi_4 + h_5\phi_5 \tag{5.34}$$

Equation 5.34 represents an approximate solution to the flow problem, but the nodal head values h_i are still unknown. The solution becomes exact for an infinitesimal discretization:

$$h = \lim_{n\to\infty} \sum_{i=1}^{n} h_i\phi_i \tag{5.35}$$

where n represents the number of nodes. For finite n, the solution remains an estimate:

$$h \approx \hat{h} = \sum_{i=1}^{n} h_i\phi_i \tag{5.36}$$

Consequently we find for the one-dimensional example in Figure 5.7:

$$\frac{\partial}{\partial x}\left(T \frac{\partial}{\partial x} \hat{h}\right) - S \frac{\partial \hat{h}}{\partial t} = \in \neq 0 \tag{5.37}$$

where $\in$ is called the *residual*. The nodal heads h_i are still undetermined.

5.3.2 Determining Unknown Head Values

We wish to keep the residual $\in$ as small as possible since $\in = 0$ for the exact solution. Although the governing flow equation cannot be satisfied everywhere, we determine h_i such that the residual on average is forced to zero over the solution domain as expressed in

$$\int_x \in dx = 0 \tag{5.38}$$

Locally $\in$ will not generally be zero. To force the residual on average to zero, the FE model applies the method of weighted residuals. Using this method, the residual is weighted by an arbitrary function $f(x)$ demanding that the area integral still be zero:

$$\int_x \in f(x) dx = 0 \tag{5.39}$$

If one could choose the unknown values of h_i such that Equation 5.39 remains true for every possible arbitrary weighting function $f(x)$, then $\in$ would have to be identical to zero. Since there are only n unknowns (see Equation 5.36), only n conditions or n functions can be applied to Equation 5.39. In other words, we determine h_i so that Equation 5.39 holds for n functions. For $n \rightarrow \infty$, in the case of infinitesimal discretization, Equation 5.39 can be satisfied for each function and the solution will be exact.

The popular Galerkin method is a special form of weighting method in which the n interpolation functions ϕ_i are chosen as the n weighting functions ω_i:

$$f(x) = \omega_i = \phi_i \tag{5.40}$$

Equation 5.39 now becomes

$$\int_x \in \omega_i \, dx = 0 \qquad \text{for } i = 1 \text{ to } n \tag{5.41}$$

Combining Equations 5.36, 5.37, and 5.41 yields

$$\int_x \left[\frac{\partial}{\partial x} \left(T \frac{\partial}{\partial x} \sum_{i=1}^{n} h_i \omega_i \right) - S \frac{\partial}{\partial t} \sum_{i=1}^{n} h_i \omega_i \right] \omega_j \, dx = 0$$

$$\text{for } j = 1 \text{ to } n \tag{5.42}$$

Developing Equation 5.42, step by step, a set of algebraic equations is formed for the unknown groundwater heads h_i. Because geometric irregularities are lacking in the time domain, finite differences are commonly chosen in time. Identical schemes with similar advantages and disadvantages as in FD models are possible. Depending on the chosen FD approximation for time, the following numerical schemes exist:

- Explicit (forward difference) eliminating the need for matrix solution but having stability constraints.
- Implicit (backward difference) being unconditionally stable.

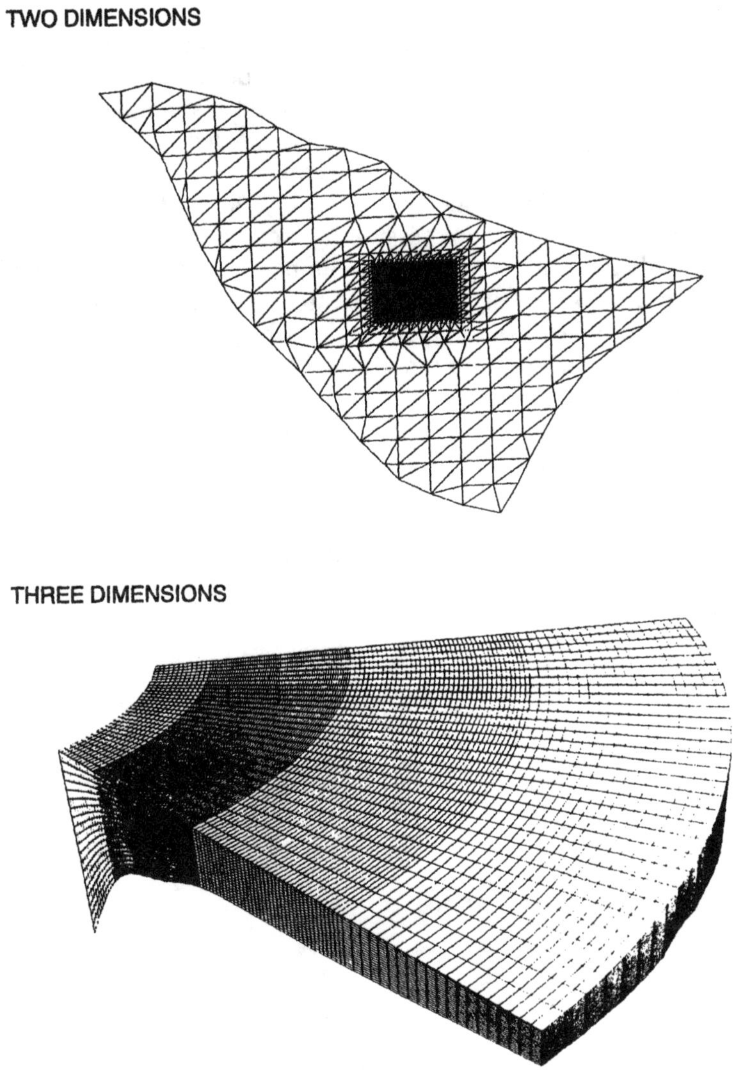

Figure 5.8 Space discretization using FE models in two field studies (two-dimensional: Kinzelbach et al. 1991; three-dimensional: Molson 1988).

- Centered (Crank-Nicholson).
- General time weighted.

For more rigorous mathematics on finite element techniques, refer to Pinder and Gray (1977).

Contrary to the FD model, the head distribution in FE models is given as a surface. In the case of a triangular subdivision of the model area and of linear interpolation functions, the surface representing the simulated head distribution is formed by triangular plane segments. Space discretization for two field studies are given in Figure 5.8, illustrating the flexibility of grid design in FE models. In the two-dimensional grid shown, a model area with fine discretization is embedded. The fine discretization was used to simulate transport. The three-dimensional grid shown in Figure 5.8 is elongated along the aquifer base and along the estimated average flow direction.

Applying a specific flow model to a field study, the model user may treat the numerical model as a black box (assuming that the numerical model is sufficiently tested and that the model user can be assured that no errors in the numerical formulation exist). The model user provides the grid information and all required system parameters such as hydraulic conductivities and external fluxes for each subdomain. Based on this information, called *model input data*, the numerical model will calculate the head distribution on discrete points (FD model) or as an approximate surface (FE model). If the flow problem is reasonable and the model-specific restrictions on the time and space discretization are fulfilled, the model can provide reasonable approximations of the solution. The realism of this solution depends on the approach used in developing the input data, to be discussed in Chapter 8.

6 Understanding Numerical Transport Models

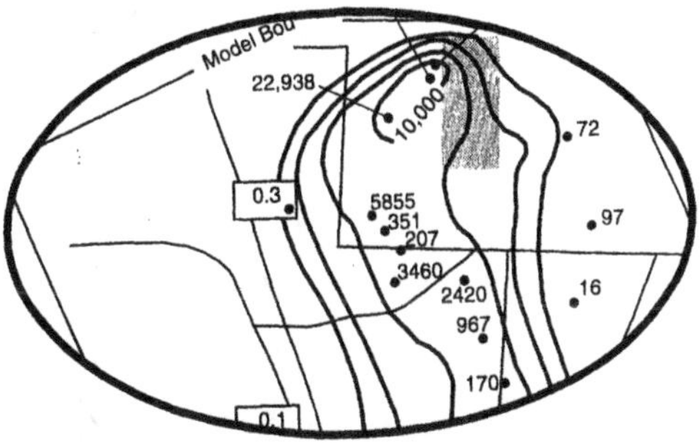

Chapter 3 discussed groundwater contamination involving a number of processes and mechanisms. Therefore it is not surprising that different numerical models have been developed to solve the complex mathematical model and to assess advective, purely diffusive, advective-dispersive, density-dependent, multiphase, or other transport problems. Most numerical models are based on one of the following four model types illustrated in Figure 6.1:

- Finite-difference (FD) model
- Finite-element (FE) model
- Method of characteristics (MOC)
- Random walk (RW) model

In this book the model types are differentiated by the underlying theory of model development, and not by their capability to handle various transport mechanisms. Transport models for advection and dispersion are at a mature stage of development and are applied with success to two and three-dimensional transport problems. Transport modeling of chemically reacting contaminants, however, achieves only a fair success if adsorption is linear. Besides large demands on computing power, uncertainties in describing complex reactions and difficulties in obtaining reliable field data limit modeling success. Transport predictions are more sensitive to a lack of field data than flow prognoses. Although the final objective of a model analysis is predictive, transport predictions in an absolute sense are impossible to achieve, even with the best models.

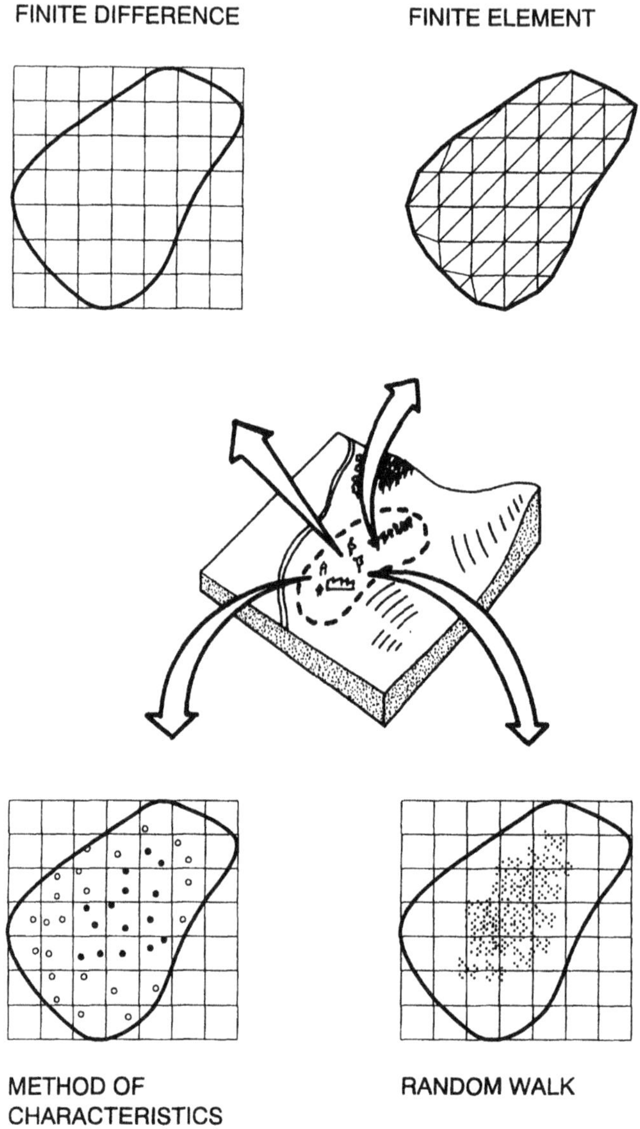

Figure 6.1 Types of transport models.

Why do several types of transport model exist? Transport models are fundamentally different from flow models in that the objective is to determine the travel paths that the contaminant is expected to migrate. Consequently the governing equations for transport are different from those for groundwater flow. Only when advection is negligible and diffusion is the controlling transport factor do both equations have identical forms. If advection becomes a significant transport mechanism, as it does in most transport problems, the numerical solution of the advection term in the transport equation invokes numerical errors in the FD and FE models. These errors appear in the simulations as dispersive transport. This phenomena is referred to as *numerical dispersion.*

Careless use of FD or FE models may cause numerical dispersion that outweighs the actual physical dispersion. Hence numerical errors may render a solution meaningless. MOC and RW models have been developed to circumvent these problems.

This chapter reviews the four model types starting with the most simple FD model, the single-cell model. Because FD and FE models were explained in Chapter 5, illustrations of the FD and FE models are minimal and the mathematical presentation is not complete. Section 6.6 introduces stochastic transport modeling. Practical aspects of model application are presented in detail in Chapter 8. Chapter 8 also provides criteria for the grid design that minimizes numerical dispersion.

6.1 DEVELOPING A SINGLE-CELL MODEL FOR TRANSPORT

Single-cell models, which are used to estimate average concentration in an aquifer, are most appropriate when contaminants enter the aquifer equally distributed as is the case when contamination originates from agricultural pesticide input. The global mass balance over the aquifer for a water compound is established as in flow modeling, as illustrated in Figure 6.2. Assuming for simplification steady groundwater flow, the mass balance equation for the contaminant is

$$
\begin{aligned}
Q_s \frac{\Delta c}{\Delta t} &= \sum_i Q_{\text{in},i}\, c_i - c \sum Q_{\text{out},i} \\
&= (Q_{N,\text{in}}\, c_N + Q_{R,\text{in}}\, c_R + Q_{B,\text{in}}\, c_B + Q_{L,\text{in}}\, c_L + Q_{W,\text{in}}\, c_W) \\
&\quad - c(Q_{N,\text{out}} + Q_{R,\text{out}} + Q_{B,\text{out}} + Q_{L,\text{out}} + Q_{W,\text{out}})
\end{aligned}
\tag{6.1}
$$

with

$$Q_s = V n_e \tag{6.2}$$

where

V = balance volume in [L^3]
n_e = effective porosity in [1]
Q_s = groundwater stored in balance volume in [L^3]

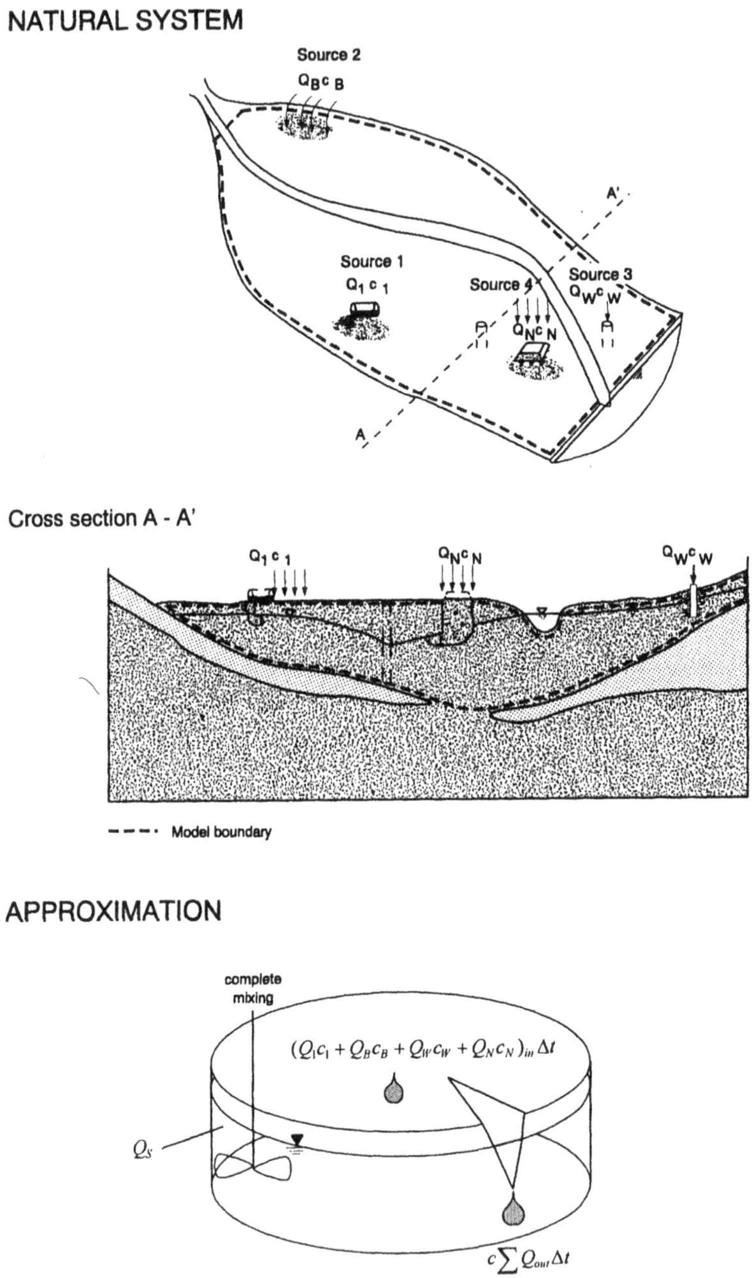

Figure 6.2 Single-cell model for transport.

Q_N = natural groundwater recharge or discharge in [L^3/T]
Q_R = exchange with surface water in [L^3/T]
Q_B = subsurface flow over boundary in [L^3/T]
Q_L = leakage component coming from or leaving adjacent aquifers in [L^3/T]
Q_W = local exfiltration or infiltration such as wells in [L^3/T]
c = concentration averaged over balance volume in [M/L^3]
c_N = concentration of natural recharge in [M/L^3]
c_R = concentration of inflow from surface water in [M/L^3]
c_B = concentration of subsurface inflow over boundary in [M/L^3]
c_L = concentration of leakage inflow in [M/L^3]
c_W = concentration of local infiltration in [M/L^3]
Δc = change in average concentration over time Δt in [M/L^3]
Δt = time increment in [T]

For many chemical contaminants inflow concentrations for recharge c_R, leakage c_L, and so on, become zero and the relevant inflow term drops out of the equation. Outflow, however, is always affected by the actual concentration c averaged over the total balance volume V.

The typical behavior of the single-cell model is best illustrated by the case of a single-source input. With

$$t = 0: \quad c = 0 \tag{6.3}$$

$$t > 0: \quad c_{\text{in}} = c_o \tag{6.4}$$

the change in concentrations averaged over the aquifer becomes

$$\frac{c}{c_o} = 1 - \exp\left(\frac{-t}{\bar{t}}\right) \tag{6.5}$$

with

$$\bar{t} = \frac{Q_s}{Q} \tag{6.6}$$

where

c = concentration averaged over balance volume in [M/L^3]
c_o = concentration of inflow component Q in [M/L^3]
Q_s = groundwater stored in balance volume in [L^3]
Q = inflow component in [L^3/T]

t = time in [T]
$\bar{t}$ = average replacement time of contaminants in balance volume in [T]

Since groundwater flow is at steady state, the inflow component Q equals groundwater outflow and is identical to the total groundwater flux through the balance volume. Stored groundwater Q is constant over time.

The calculations can be expanded relatively easily to the case where several independent contamination sources exist. If the mass input and the water fluxes are constant over time so that the sum of groundwater inflow is equal to the sum of groundwater outflow, we find that

$$c(t) = a\left[1 - \exp\left(\frac{-t}{\bar{t}}\right)\right] \tag{6.7}$$

with

$$a = \frac{\sum_i Q_{in,i}\, c_{in,i}}{\sum_i Q_{in,i}} \tag{6.8}$$

The average replacement time calculates as

$$\bar{t} = \frac{Q_s}{\sum_i Q_{in,i}} = \frac{Q_s}{\sum_i Q_{out,i}} \tag{6.9}$$

To apply Equation 6.5 to time-dependent mass input, the total water flux, as well as the contaminant input, have to be evaluated first as a function of time. Contaminant input rates (e.g., input rate of pesticides) are then expressed as step functions over time. The mass-balance equation is solved either by time-stepping (finite differences in time) or simply by superposition of individual solutions for each mass input.

Time-dependent mass input solved by superposition is shown in the following equation for a step-function type single contaminant source. With

$$t = 0: \qquad c = 0 \tag{6.10}$$

$$0 < t < t_1: \qquad c_{in} = c_o \tag{6.11}$$

$$t \geq t_1: \qquad c_{in} = 0 \tag{6.12}$$

average concentrations calculate

$$c = 0 \qquad \text{for } t \leq 0 \tag{6.13}$$

$$\frac{c}{c_o} = 1 - \exp\left(\frac{-t}{\bar{t}}\right) \qquad \text{for } 0 < t < t_1 \tag{6.14}$$

$$\frac{c}{c_o} = \exp\left[-\frac{(t - t_1)}{\bar{t}}\right] - \exp\left(\frac{-t}{\bar{t}}\right) \qquad \text{for } t \geq t_1 \tag{6.15}$$

6.2 UNDERSTANDING NUMERICAL ERRORS IN FINITE-DIFFERENCE MODELS

As in flow modeling, the finite-difference approach to solving the transport equation is the oldest and the simplest. Difference equations are substituted for the partial differential equations for transport. Difference equations can be derived either by a mass balance as presented in Section 5.2 for flow modeling, or by a Taylor series development of each term of the transport equation. Both approaches yield identical finite-difference approximations of the governing transport equation. The FD model discretizes space and time in intervals. All values such as concentrations are represented as nodal values. Discretization, averaging, and solution procedures invoke numerical errors. Since FD models were explained in Chapter 5, this section discusses the types, causes, and controls of numerical errors in FD models for transport.

6.2.1 Types of Numerical Errors

Model application is subject to various model errors. Errors in data collection and interpretation, errors in conceptualizing nature, and errors in presentation of model results are discussed in Chapter 8. This section discusses numerical errors, including the following:

- Truncation error. Truncation errors are created when expressing a complicated function such as the potentiometric head distribution as a series of piecewise simpler functions. The error is embedded in numerical theory.
- Roundoff error. Roundoff error is always present and is due to the limitation of the computer to represent digits. Roundoff errors may invoke increasing numerical errors in time. Cumulative roundoff errors will eventually cause numerical instability. Most models apply double precision to increase accuracy.

- Distance-related numerical dispersion. Distance-related numerical dispersion originates in FD models because grid concentrations are known only at the nodes which represents the entire cell. This means that a node downstream of the contaminant front will show concentrations even when the actual front has not actually reached the midpoint of the considered downstream cell. This is schematically shown in Figure 6.3.
- Angular-numerical dispersion. To understand angular-numerical dispersion, consider a pulse that moves under 45 degrees relative to the chosen discretization. Since internodal advective flow exists only in the direction of the discretization (internodal velocities), mass transport is also only approximated between direct neighbor cells. The true direction of the pulse cannot be represented on the discrete grid, resulting in a gradually dispersive broadening of the pulse, as illustrated in Figure 6.3.
- Oscillation. Oscillation is a common numerical problem visible in simulated concentration higher than the source concentration (overshoot) or concentration less than zero (undershoot). These errors are normally experienced when coarse discretization in time and space are used to simulate transport in highly variable concentration fields. Such conditions are encountered close to prescribed concentration boundaries or in the vicinity of internal contamination sources.
- Instability. The stability or, respectively, instability of a numerical model is of considerable practical importance. Instability is the result of a feedback process in which errors grow as each succeeding step is taken to obtain the solution. The growth of errors is often exponential and oscillatory. The numerical solution fails to converge to the exact one. Stability is a necessary and sufficient condition for convergence. Convergence implies stability. In almost all situations, an unstable solution is easy to distinguish from a stable one. The model user must understand however that stability or, respectively, convergence does not automatically imply convergence to the correct solution. Hence consistency must be tested before an approximate solution is accepted.
- Iteration residual error. The iteration residual error is involved in iterative solution procedures. When iteration is involved, the model user specifies a convergence criterion to determine iteration. Iteration is stopped if the numerical solution is within a specified error tolerance. The choice of the convergence criterion determines the size of the iteration residual error. Absolute and relative convergence criteria are common. The absolute convergence criterion specifies a limit of changes between two iteration steps for each node:

$$\left| c_i^{k+1} - c_i^k \right| \leq \varepsilon \tag{6.16}$$

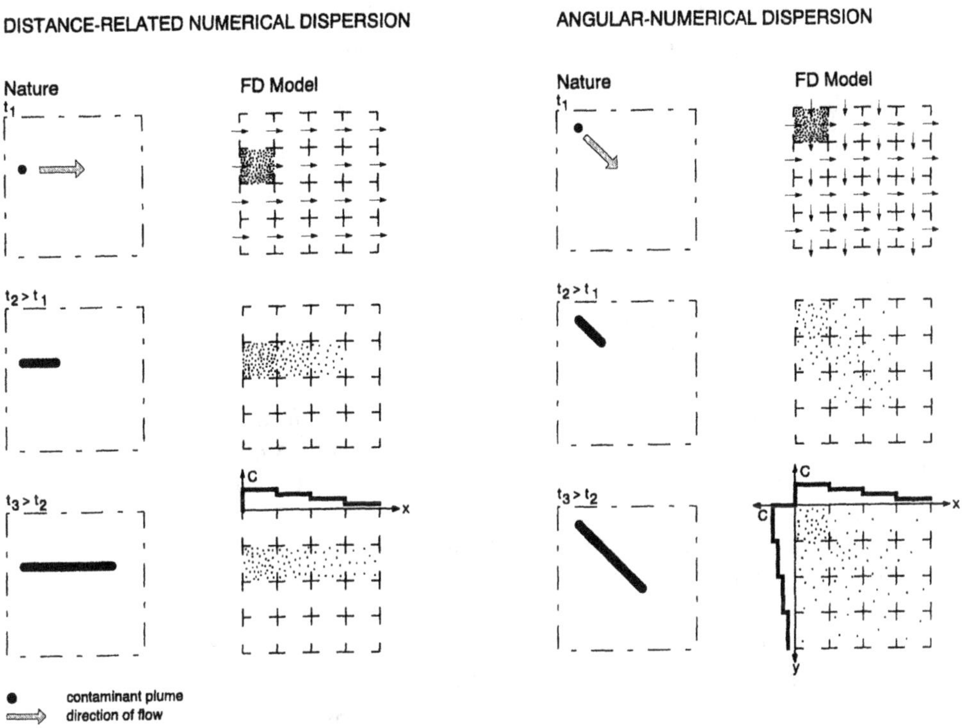

Figure 6.3 Origin of numerical dispersion in FD models.

where

c = unknown concentration in $[M/L^3]$
i = node number
k = number of iterations
ε = convergence criterion

The absolute criterion is used when the magnitude of the results and the desired accuracy are known beforehand. The relative convergence criterion tests the percentage change in each unknown concentration:

$$\left| \frac{c_i^{k+1} - c_i^k}{c_i^{k+1}} \right| \leq \varepsilon \tag{6.17}$$

Equation 6.17 indicates that the relative convergence criterion is problematic with concentrations close to zero. Hence, in almost all numerical models, absolute convergence criterion are used. Sometimes the absolute convergence criterion is applied to the sum of all changes rather than to nodal changes. Occasionally the model user specifies a convergence criterion too strict to be achieved by the computer due to roundoff errors. For such numerical models involving iterative solutions, the model user should specify the maximum number of iterations to avoid endless iteration loops.

6.2.2 Causes of Numerical Errors

To reduce numerical errors, the model user must understand the parameters in model application that can cause numerical errors. Some simple "cause and error" relationships can be established. In practice, however, most of these relationships are more complex. Numerical errors are closely related to each other. When a scheme is designed to minimize oscillation, numerical dispersion occurs, but when numerical dispersion is controlled, it is generally at the expense of increased overshoot.

The following observations help to identify causes of numerical errors in model application:

- Large grid spacing leads to numerical dispersion.
- Large time steps stimulate oscillation.
- Backward differences in time (implicit scheme) tend to numerical dispersion.
- Forward differences in time (explicit scheme) are subject to instability.
- High nonlinearity in flow or transport processes may cause instability.
- Large length-to-width ratio (cell aspect ratio) can also introduce roundoff errors.
- Abrupt changes in time or space discretization affect stability.
- Initial sharp concentration fronts at an inflow boundary, caused by prescribed concentration boundaries, lead to overshoot.
- Coarse space discretization in areas where large responses are expected, in areas where aquifer properties change abruptly, or in areas where boundary conditions change rapidly cause numerical dispersion and may result in unstable solutions.

6.2.3 Controlling Errors

Numerical errors can be influenced by time and space discretization. The following controls exist to handle numerical errors and to ensure accurate results:

- Peclet criterion. The Peclet criterion controls the spatial discretization in each of the coordinate directions x_i and is represented as

$$Pe_{xi} = \frac{v_{xi} \Delta x_i}{D_{xxi}} < 1 \tag{6.18}$$

where

v_{xi} = particle velocity in the x_i direction in [L/T]
Δx_i = cell size in the x_i direction in [L]
D_{xxi} = dispersion coefficient in x_i direction in [L^2/T]

- Courant criterion. The Courant criterion controls the time step Δt. It is written for two-dimensional transport as

$$Co_x = \left| \frac{\Delta t \, v_x}{\Delta x} \right| \leq 1 \tag{6.19}$$

$$Co_y = \left| \frac{\Delta t \, v_y}{\Delta y} \right| \leq 1 \tag{6.20}$$

The Courant criterion states that the contaminants released by advection from each cell within a time step cannot exceed those contained in the cell at the beginning of the time step. It applies to each direction as indicated by Equations 6.19 and 6.20. The combination of the Courant criterion in the x- and y-direction applies to upwind differences (see Kinzelbach 1985):

$$Co_x + Co_y \leq 1 \tag{6.21}$$

In implicit solution of the matrix equations the fulfilment of the Courant criterion helps to minimize numerical errors. Implicit solutions use the unknown concentrations at the end of each time step to calculate advective and dispersive transport. In explicit solutions it is essential to satisfy the criterion to obtain a stable solution. Similar to flow simulations, the explicit solution scheme urges one to work with small time steps that often yield unreasonably long calculation times. Explicit solutions are numerical solutions in which the known concentrations at time t are used to calculate the unknown concentrations at time $t + \Delta t$. The advantage of explicit schemes is that numerical dispersion can be minimized in some cases. The disadvantage is instability when not all stability criteria can be satisfied.

- Neumann criterion. The Neumann criterion restricts the permissible dispersive flux within a time step and is represented as

$$\frac{D_{xx}\Delta t}{\Delta x^2} + \frac{D_{yy}\Delta t}{\Delta y^2} \leq 0.5 \tag{6.22}$$

Sometimes it is expressed as

$$\frac{D_{xx}}{\Delta x^2} \approx \frac{D_{yy}}{\Delta y^2} \tag{6.23}$$

If one term is significantly larger than the other, which is physically plausible, it will dominate the other.

- Cell aspect ratio. The cell aspect ratio is important because computers can only handle a finite number of significant figures. Ideally the model user wants to fulfill the following criterion:

$$K_x \frac{\Delta y}{\Delta x} = K_y \frac{\Delta x}{\Delta y} \tag{6.24}$$

Other criteria may become important when chemical transport is simulated. Unfortunately, the model user is not always able to satisfy the Courant and the Peclet criteria everywhere within the investigated flow field because of restricted computer capacity. Meaningful numerical solutions can still be achieved, but model results have to be reviewed carefully so that the effect of the numerical errors can be understood.

The causes of numerical errors listed in subsection 6.2.2 suggest that additional practical controls should be used to reduce numerical errors. Errors can be minimized by:

- Orienting the discretization grid along main flow direction.
- Ensuring gradual variation in time and space discretization.
- Choosing third-kind boundaries rather than first-kind boundaries in transport modeling.
- Keeping prescribed potentiometric surface numbers as small as possible to reduce roundoff errors. This can be done by subtracting the datum. Decreasing real numbers in numerical operations results in reducing loss of significant digits.
- Keeping the cell aspect ratio within at least one order of magnitude.
- Decreasing the time step to improve stability. Unstable numerical effects depend on the time step. If two solutions obtained with two significantly different and reasonably small time steps are essentially the same, then the numerical solution is probably stable.
- Concentrating fine discretization on areas with large changes in aquifer properties or expected aquifer responses.

It is probably a fair summary to state that model users reduce numerical errors best by reducing time and space discretization. Discretization has to be sufficiently small to give accurate results. There is no other technique to circumvent these constraints imposed by finite difference theory.

6.3 IDENTIFYING THE ADVANTAGES OF FINITE-ELEMENT MODELS

FE models are very popular in transport modeling. The foremost reason is the flexibility in the spatial discretization. The FE grid can be adjusted to match the requirements originating in the geometry or in the specific characteristic of the transport problem as shown in Figure 6.4:

- As in flow modeling, elements can be made flexible to adjust to the geometry of the problem.
- Elements can be oriented along flow, which practically eliminates numerical dispersion transverse to flow, referred to as the *principle direction technique* (Frind and Pinder 1982; Frind and Matanga 1985).
- In layered heterogeneous aquifers, the grid can be oriented along the principle direction of dispersion, which allows one to represent the tensorial character of dispersion in an elegant way.
- The method allows one to choose coarse discretization in areas that are not affected by the plume but are important for the flow submodel. Fine discretization can be concentrated in the area of the plume.

Identical sources of errors for FE models exist as in FD models. The flexibility of the space discretization gives the model user better control on numerical errors. An additional advantage, at least from the programmer's point of view, is that in the finite element formulation boundary conditions follow directly from the FE theory.

The principal direction technique takes advantage of a major difference between flow and transport problems that often resides in the time scale. The time scale is generally larger for transport problems (years to decades) allowing one to average out transient flow conditions. As a consequence the model user often applies steady-state flow conditions for transient contaminant transport. If flow, however, is at steady state and the flow system is linear (i.e., flow is unaffected by contaminants), it is advantageous to formulate the transport problem in terms of the principal direction of contaminant migration by orientating the discretization along the direction of flow (Frind and Pinder 1982). To define the flow system, the dual theory of potentials and stream functions (Frind and Matanga 1985) is applied. Once the flow net, or the grid, is defined, estimates of advective transport can easily be made manually. Transport velocity is inversely proportional to the distance between adjacent streamlines. Although the principal direction technique bears distinct

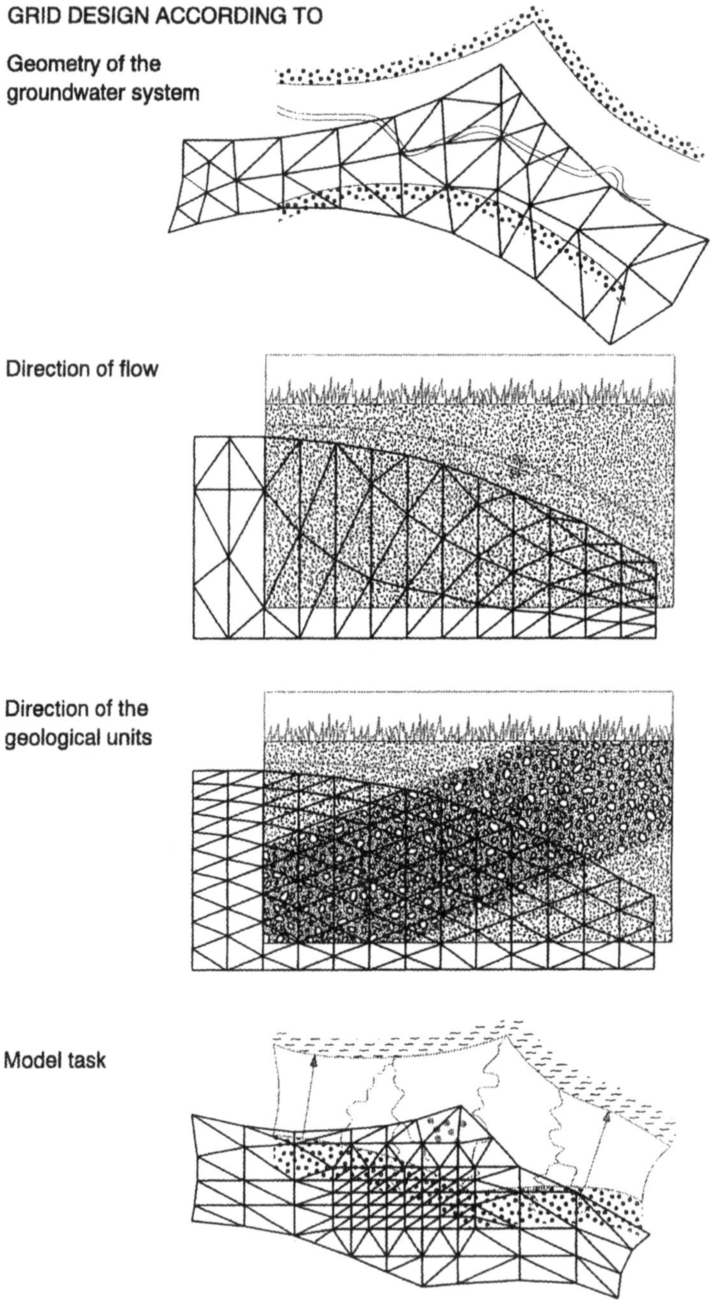

Figure 6.4 Choice of spatial discretization in FE models depending on the transport problem.

numerical advantages, its application is limited due to difficulties encountered to determine streamlines, or advective transport. This is particularly true if internal sources and sinks are present in the model area.

From a practical point of view, the ability of the FE model to refine space discretization in areas where contaminant transport occurs and to maintain coarse characterization in areas without transport is possibly the dominant advantage over the FD model. In modeling transport, the model area is generally much larger than the actual contamination plume to ensure flow condition in the area of interest that is unaffected by insufficiently defined boundary conditions. For example, investigating a plume of the size of several hundreds of meters, one may have to consider a flow regime of a few kilometers. Now compare the size of the entire model area with the degree of discretization required for transport modeling. The Peclet criterion states that space discretization should be less than the value of the relevant dispersivity to keep numerical smearing of concentration fronts small. Hence our discretization of the contamination plume is most likely to be done in few meter intervals. Data handling and computational effort become unreasonable if we extend the fine discretization required for transport over the entire model area. The flexibility in space discretization provided by the FE model does not constrain the model user. The grid can be designed according to the geometry of the boundaries as well as to the size of transport area, as illustrated in Figure 6.4.

A carefully designed grid will decrease numerical errors, but the following constraints on the spatial and temporal discretization exist:

- The time step should be chosen according to the Courant criterion to avoid overshooting (Equations 6.19 and 6.20).
- The element size should not exceed the value for the longitudinal dispersivity (Peclet criterion, Equation 6.18).
- The element shape should not be too distorted (Equation 6.24).

Grid generation can be a tedious if done manually. Design the grid simply but accurately. Information is related to the FE grid by numbering elements and nodes. Keep in mind that the grid numbering influences the bandwidth of the coefficient matrix and hence computational effort. The bandwidth of the coefficient matrix is calculated as the largest difference between node numbers for an element plus one. Hence, choose node numbering that keeps the difference between node numbers small for each element. Subsection 8.4.2 gives an example how the nodal numbering scheme controls the bandwidth.

6.4 USING THE METHOD OF CHARACTERISTICS

The MOC has been introduced to groundwater transport modeling to reduce numerical errors. Both FD and FE models approximate the solution for groundwater flow and for contaminant transport with an identical theory. In models based on the MOC or the RW method, the numerical approach for the flow and transport part differs.

In the widely used MOC model of Konikow and Bredehoeft (1978), flow is simulated by a standard finite-difference scheme. The method of characteristics is then applied to the transport equation. One might visualize the method as tracing equally distributed particles through the flow field and observing changes in concentration of each particle as it moves. Hence advection and dispersion are uncoupled. As in finite-difference models, concentrations are represented by grid nodes. These grid nodes are generally identical to the ones used in the flow submodel.

Advective and dispersive transport are simulated stepwise in MOC. Advective transport is simulated by particles distributed in a geometrically uniform pattern over the entire model area. Each particle is assigned an initial concentration associated with the concentration in the cell containing the particle. In areas without contamination, the assigned concentration is zero. For each time step all particles are moved according to the local average transport velocities.

Since velocities in finite-difference models are only internodal velocities (velocities along the direct connection line between two nodes), velocities at an arbitrary location must be interpolated (Prickett and Lonnquist 1971; Kinzelbach 1985). The new positions of the particles in the grid are calculated from

$$x_{\text{new}} = x_{\text{old}} + v_x \Delta t \tag{6.25}$$

$$y_{\text{new}} = y_{\text{old}} + v_y \Delta t \tag{6.26}$$

After each time step and after all points have been moved, concentration at each cell node is temporarily assigned the average concentration of all particles located in that cell. Consequently concentration changes due to advection are simulated.

Dispersive transport, or the change of concentrations due to dispersion, is obtained by a finite-difference calculation on the rectangular grid. After the calculation of dispersion, particle concentrations are updated according to the changes in the grid concentrations, and advective transport in the next time step is calculated.

The MOC requires a significant amount of bookkeeping of the particle positions and back and forward switching from grid to particle concentrations. The popularity of the MOC is justified because numerical dispersion can be kept relatively small. To achieve accurate results with the MOC, consider the following aspects:

- Accuracy depends on the spatial discretization used in the finite-difference calculation. Decreasing grid distance used in flow simulation results in an increase in MOC accuracy.

- Accuracy increases with an increase in particles per cell. Konikow and Bredehoeft (1978), for example, use nine particles.
- The size of the time step is limited to reproduce the correct pathline of each particle. As a rule of thumb, each particle needs at least four time steps in order to pass one cell. This directly corresponds to the Courant criterion.

6.5 VISUALIZING RANDOM WALK MODELS

The RW model is the most illustrative of all transport model types and is easy to understand. The model is recommended when the presentation of numerical simulations is a major concern.

The RW model has various advantages. Transport simulation can be visualized by the model user as in Figure 6.5, which illustrates the actual transport mechanisms of advection and dispersion. Hence it is an excellent model type to use for presenting model results to anyone unfamiliar with contamination transport modeling.

Where FD and FE transport models are plagued with numerical dispersion, RW models do not admit numerical dispersion in the classical sense. When large ratios of longitudinal to transverse dispersivities are involved, the RW model outclasses other model types.

Without much effort, RW particle transport can be added to any flow model. Three-dimensional transport simulation is easily achieved. In addition, by setting dispersion coefficients to zero, streamlines are generated.

RW models are robust and provide accurate results, but model constraints do exist. The main constraint is the limitation in time discretization. One has to allow each particle to move over several time steps within the cells used in the flow submodel. This directly corresponds to fulfilling the Courant criterion. Five to 10 steps per cell have proved to be appropriate (Prickett et al. 1981).

Finally, a high number of particles (approximately 5000) are required to simulate plume evolution realistically. Hence RW models tend to be time-consuming.

The main disadvantage of RW models is that the method yields particle distribution rather than concentrations. As far as the estimation of local concentrations is concerned, the RW model is a fairly crude method. Chemical transport simulations are difficult to realize in RW models, since chemical reactions depend on concentrations. Also RW models are not used where flow and transport is coupled.

As in other transport model types, RW models combine submodels for flow and transport. In most applications simulation for flow is based on an FD model (Prickett et al. 1981). All steps required for flow modeling are required. The submodel for flow yields discrete nodal head values or discrete internodal velocities when Darcy's law is applied. As in MOC, local velocities within the entire model area are then found by interpolation.

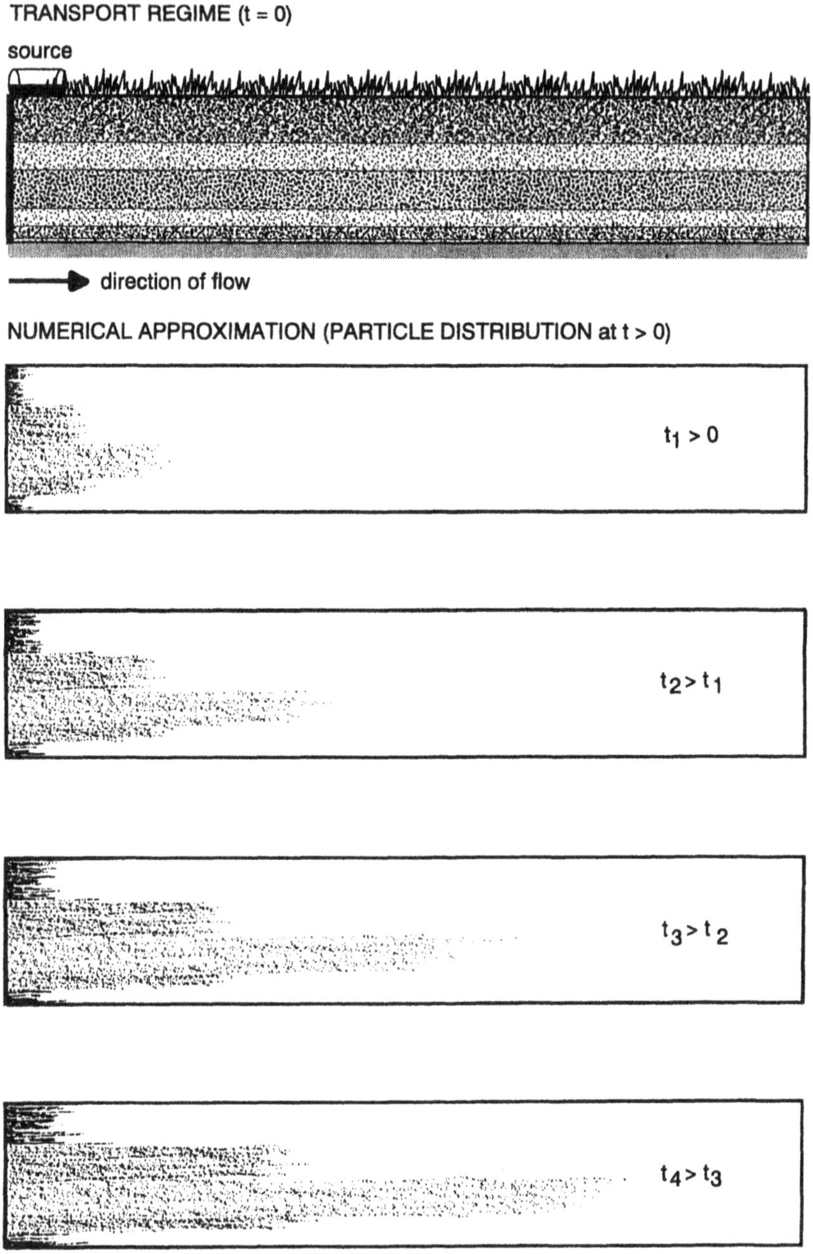

Figure 6.5 Simulation of contaminant transport in a layered aquifer using a RW model.

Based on velocities found in the flow submodel, the transport submodel approximates contaminant transport using random walk. The RW model is based on the theory of the Brownian molecular motion, which is summarized in Chandrasekhar (1943). Scheidegger (1954) and De Josselin de Jong (1958) were the first to apply the model to simulate mass transport in porous media. They stated that transport can be approximated with particles that are set into the flow field and that have both the advective and a random component of transport. The random step in and transverse to the flow direction representing dispersive transport is allowed after each advective step. In two-dimensions we find the following two steps (Figure 6. 6):

- Advective transport step Δx and Δy.

$$\Delta x = v_x \Delta t \tag{6.27}$$

$$\Delta y = v_y \Delta t \tag{6.28}$$

- Dispersive transport step P_L in flow direction.

$$P_L = \pm Z_1 \sqrt{2D_L \Delta t} \tag{6.29}$$

- Dispersive transport step P_T transverse to the flow direction.

$$P_T = \pm Z_2 \sqrt{2D_T \Delta t} \tag{6.30}$$

where

P_L = dispersive transport step in flow direction in [L]
P_T = dispersive transport step transverse to flow direction in [L]
D_L = longitudinal dispersion coefficient in [L^2/T]
D_T = transverse dispersion coefficient in [L^2/T]
Δt = time step in [T]
$Z_{1,2}$ = independent random variables

Random variables that are Gaussian distributed are obtained from standard routines available in each computer language (e.g., RANF( ) in FORTRAN). Each transport step is independent of the previous one. The summation of consecutive advective and dispersive transport steps yields the trail of particles, as shown in Figure 6.6. Path lines are generated based solely on advective transport.

The accuracy of the method increases with increasing numbers of particles representing the contamination source or plume. Each particle is assigned a certain mass calculated by dividing the total contaminant mass by the number of particles released in the flow field. For example, if one simulates the trans-

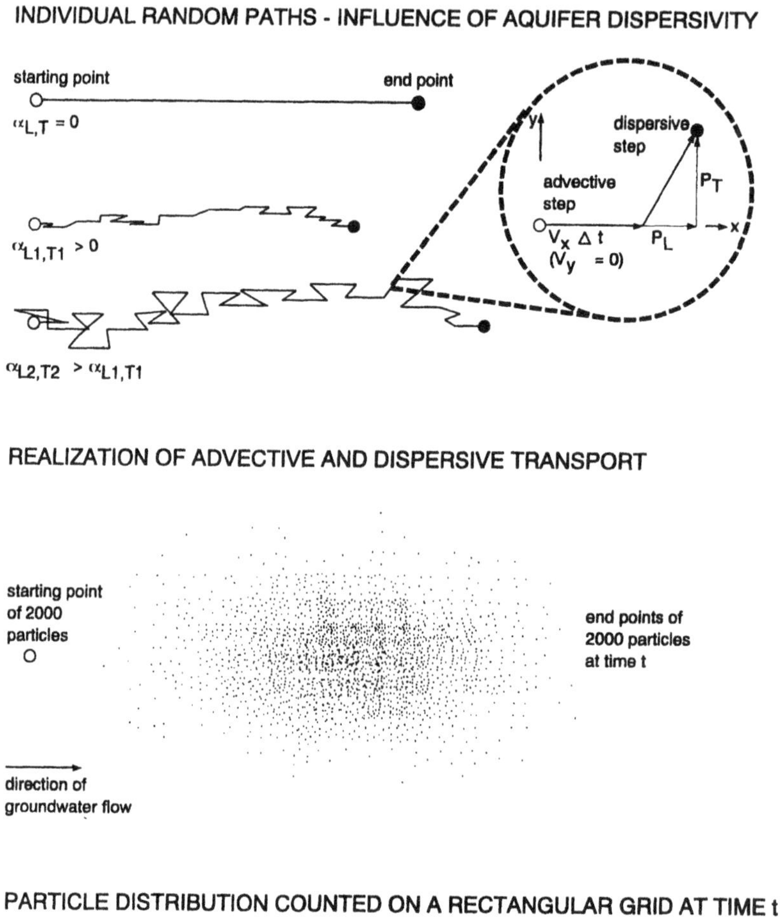

0	0	0	0	0	0	0	0	0	0	0	1	0	0	0	0	0	0	0	0	0
0	0	0	1	0	0	1	2	0	0	0	1	1	0	0	0	0	0	0	0	0
0	0	1	0	0	1	2	3	1	1	2	4	1	1	2	3	1	1	0	0	0
0	2	2	3	2	3	4	11	14	16	9	11	10	5	2	4	2	3	1	0	0
0	1	3	8	4	11	14	16	14	33	19	27	32	12	9	8	6	4	2	0	0
0	1	2	5	13	14	24	27	30	54	50	49	46	33	29	20	7	5	3	1	0
0	1	5	3	9	20	21	42	53	57	60	54	45	33	23	25	8	4	2	0	0
0	2	3	7	6	16	19	35	28	43	44	56	41	31	13	16	11	6	3	1	0
0	1	1	2	4	9	12	20	22	24	21	33	27	22	13	5	2	4	0	2	0
0	1	2	0	1	3	5	6	14	10	9	8	12	7	5	4	1	2	0	0	0
0	0	0	0	0	0	2	1	5	2	1	1	1	3	2	3	2	1	0	0	0
0	0	0	0	0	0	0	0	1	0	0	0	1	0	0	1	0	0	0	0	0
0	0	0	0	0	0	0	0	0	0	0	0	0	0	0	0	0	0	0	0	0

Figure 6.6 Realization of advective and dispersive transport using RW model assuming uniform groundwater flow.

port of 1 kg of contaminants with 1000 particles, each particle would represent 1 g. In other words, the summation of all particle masses gives the total contaminant mass in the aquifer.

RW models follow the advective and random movement of all particles representing the contamination plume shown in Figure 6.6, whereby the movements of individual particles are independent. Theory proves that the frequency distribution of particles in the limit of large particle number fulfill the advection-dispersion equation if the dispersion coefficients are independent of space (Ito 1951).

This condition is found only in uniform flow. In practice, the simple summation of advective and dispersive transport step is still a justified approximation when flow conditions change gradually. In highly variable flow fields such as flow in layered aquifers, a correction term has to be introduced in the RW model to compensate for the error (Uffink 1985; Kinzelbach and Ackerer 1986). This is due to the fact that in the simplified RW model, particles can move more easily into flow regions with low velocities than leave these regions as both the advective and the dispersive particle steps decrease.

The direct result of RW models is the particle (mass) distribution. The density of particles allows one to estimate concentrations. To convert particle clouds into concentration, one puts a rectangular grid over the plumes (in most cases identical to the one used in the flow simulation), and then the particles per cell are counted. The total mass of the particles within one cell, divided by the cell volume times the porosity, gives the average cell concentration. Concentration distributions are usually rough due to the stochastic nature of the transport simulation.

6.6 ILLUSTRATING STOCHASTIC TRANSPORT MODELING

In stochastic modeling the model user respects that natural transport systems are rarely described with certainty, and predictive prognosis becomes risky. The model user aims to compute expected values of concentrations and their respective variances rather than deterministic values.

In this book we are not concerned with a theoretical analysis of stochastic modeling but rather with its pragmatic side. We restrict stochastic modeling to "what if" evaluations involving, in particular, various possible interpretations of the subsurface hydrogeology. This is done by considering modeling in a Monte-Carlo framework (Schwartz 1977; Chu et al. 1987). The procedure consists of random generation of a number of equally realistic sets of model input data under consideration. Transport realization for each random data set is then computed by standard methods. In this regard the approach is a marriage of probabilistic theory of random model input and deterministic theory which studies transport. Note that a single random input parameter allows an indefinite number of realizations. In the practical application one restricts the number of computations to a realistic representative number. Statistical an-

MODEL INPUT DATA

Fixed values:
Porosity = 0.15
Time = 30 days
Aquifer thickness = 10 m

Random values:
15 kg < mass input < 30 kg
0.2 m < dispersivity < 0.6 m
1 m/day < velocity < 1.2 m/day

Case	M	v	αL
1	23.775	1.165	0.201
2	17.612	1.061	0.340
3	16.371	1.198	0.484
4	16.786	1.106	0.259
5	24.026	1.090	0.204
6	15.856	1.104	0.266
7	28.140	1.108	0.513
8	21.931	1.156	0.528
9	29.952	1.053	0.545
10	27.602	1.140	0.445
11	17.213	1.114	0.397
12	21.398	1.031	0.257
13	27.325	1.163	0.587
14	17.333	1.065	0.277
15	25.234	1.025	0.493
16	27.225	1.085	0.489
17	29.240	1.094	0.407
18	27.704	1.197	0.420
19	26.088	1.168	0.382
20	22.514	1.139	0.278

MODEL RESULTS

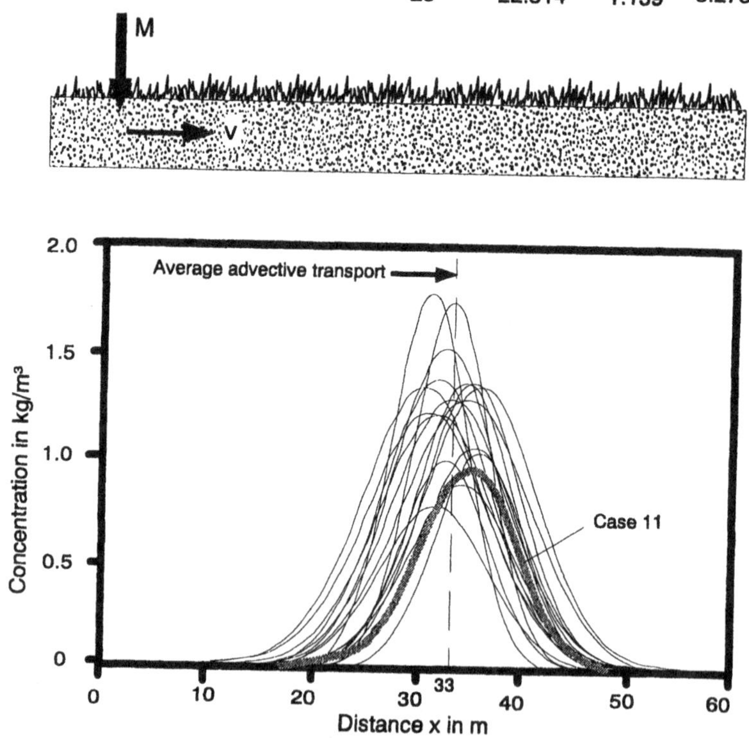

Figure 6.7 Example of stochastic modeling.

alysis of the results from all realizations considered leads to expected values of concentrations.

The approach is best illustrated using the analytical model for one-dimensional longitudinal transport in the case of an instantaneous point source (Equation 4.27). Assume that transport velocity v, longitudinal dispersivity α_L, and initial mass input M are known only as estimates within a given range. For each parameter we can generate random model input data that fall within the given range (Figure 6.7).

For 20 randomly determined combinations of v, α_L, and M, (cases 1 to 20 in Figure 6.7), realized concentration distributions are computed for a travel time of 30 days (Figure 6.7). The figure indicates that absolute concentrations are not predictable but a good approximation can be made of the range where the concentration is likely to be. Average expected concentrations, as well as information about uncertainty in concentration predictions, can be quantified by simple calculus.

The example in Figure 6.7 illustrates a simple stochastic approach. Computational effort increases tremendously in real field studies. Hence stochastic transport analysis in practical model application is most often restricted to few interpretations of nature. It is a powerful tool, even though predictions are not absolute. In addition to stochastic modeling deterministic simulations can be applied to an evaluation of worst-case scenarios.

7 Dimensionality of Flow and Transport Problems

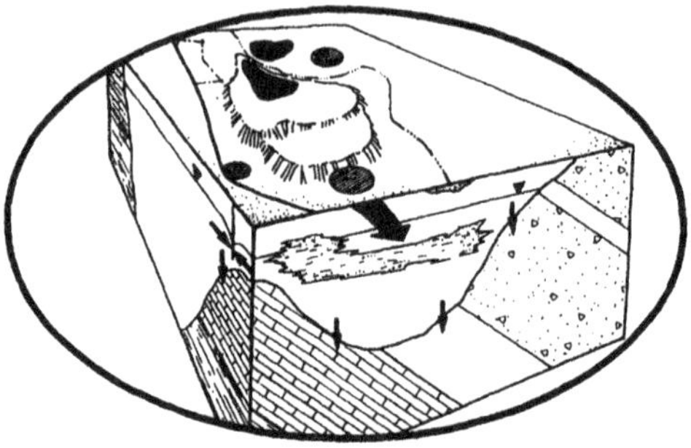

In specifying a groundwater flow and transport problem the model user usually focuses on the dominant flow directions and processes without considering conditions at a local scale. This approach is analogous to that taken in developing the theory of groundwater flow in which Darcy's law uses average flow instead of solving for the complex three-dimensional flow pattern in the pore network.

This chapter discusses selecting the appropriate number of dimensions for a groundwater model. Alternate approaches to simplifying a modeling problem, by reducing the number of dimensions, are presented with their advantages and disadvantages. Examples from field cases are also presented. This chapter:

- Discusses the need for simplifying the characteristics of the real aquifer system
- Presents seven approaches to reducing model dimensions
- Summarizes the impact of neglecting dimensions on transport predictions

7.1 NEED FOR SIMPLIFYING THE MODELING PROBLEM

Typical groundwater regimes are extensive, complex, and only partially understood. The general purpose of a groundwater model is to provide answers about the groundwater system in time to aid in groundwater management decisions. The real aquifer system must be simplified to make the problem tractable to efficient analysis. There are no particular rules on how to simplify a given set of site conditions for input to a model, although the model-selection procedure discussed in Section 8.3 may assist in the process. The relative importance of individual flow or transport directions to the investigated problem allows the model user to simplify by reducing the model dimensions.

Figure 7.1 illustrates the number of dimensions of various model types. Three-dimensional models closely approximate natural conditions, but modeling efforts may be extensive. The limiting factor in preparing a three-dimensional model is often the availability of detailed site-specific data. The multi-layer model represents stratified aquifers as a combination of two-dimensional layers linked by leakage. Two-dimensional models neglect flow and transport components in either the horizontal or vertical direction. Thus they yield predictions in two dimensions, averaged in the third dimension. One-dimensional models simplify further, often permitting analytical solution to the problem under investigation. Zero-dimensional (black box or compartmental) models focus solely on balancing inputs and outputs from the modeled area.

In each case the number of dimensions used in the model formulation simplifies the field situation. In many cases such simplifications are warranted and are within the range of other uncertainties in the model application, but they should be discussed in the model report. The most successful models that have been prepared to date are two-dimensional, saturated, groundwater flow, and conservative-solute transport models (National Research Council 1990). However, even less successful models assist in the decision-making process. The experienced model user chooses an acceptable trade-off between model simplicity and representation of the real world. Choice of a suitable approach is based on understanding of the dominant flow directions and transport mechanisms; less important aspects of the problem are neglected. The assumptions and judgments used in choosing simplifications affect the overall reliability and utility of the resulting model. These assumptions should be presented as an integral part of the documentation of model development.

Several basic differences exist between the simplification of flow problems and transport problems. The number of dimensions required to represent the transport problem often exceeds that needed to represent the flow problem. This situation is illustrated in Figure 7.2 which illustrates an ideal case. The figure shows a point contaminant source in uniform, horizontal groundwater flow. Close to the source the concentration gradients are such that the transport is three-dimensional. A two-dimensional transport model would, for the region close to the source, neither represent peak concentrations or actual concentration distributions. With increasing distance from the source, vertical concentration gradients decline due to dispersion, and the transport problem may be represented in two dimensions. Throughout this same domain, the overall flow system is one-dimensional in nature.

Another major difference between flow and transport simulations is that the time scale for transport is much greater than that for flow. Consequently

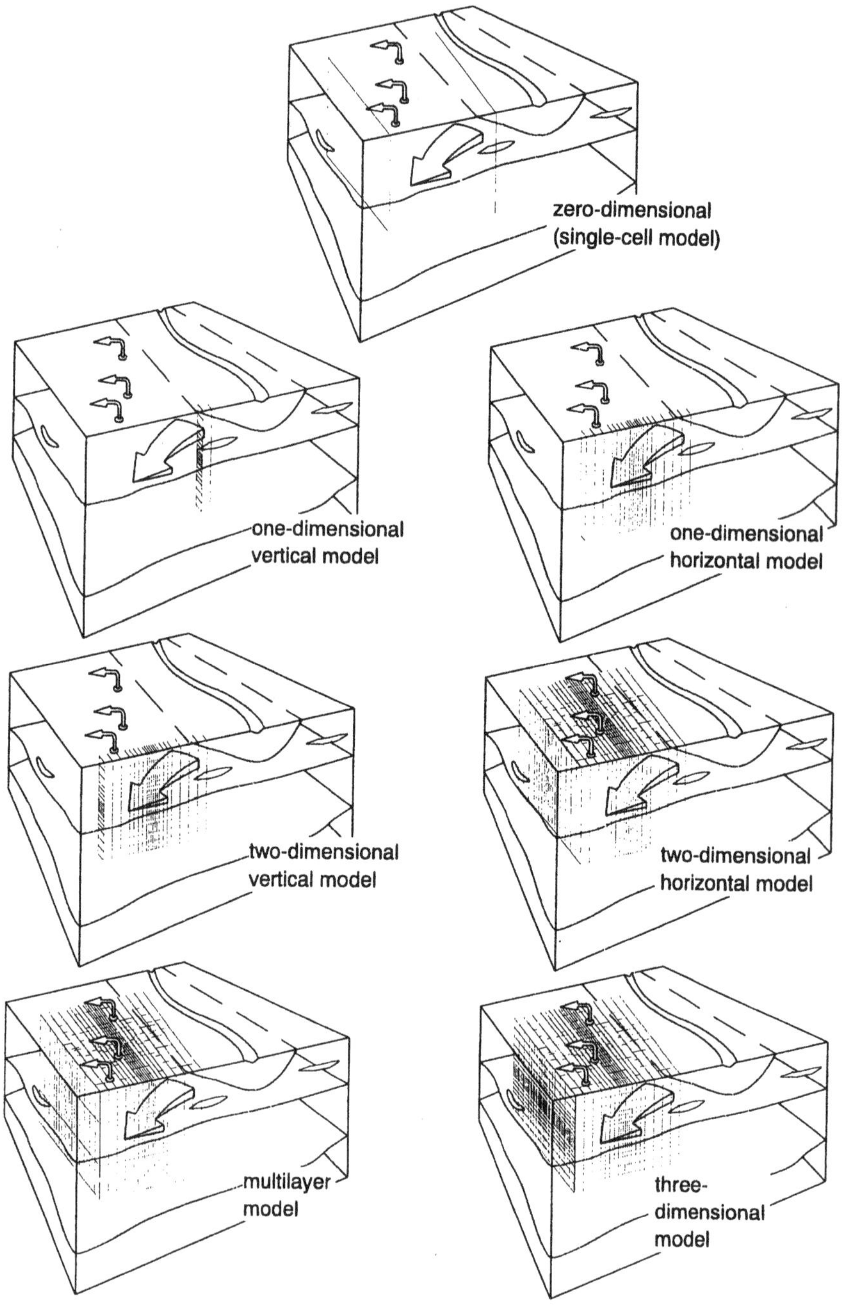

Figure 7.1 Common dimensional simplifications.

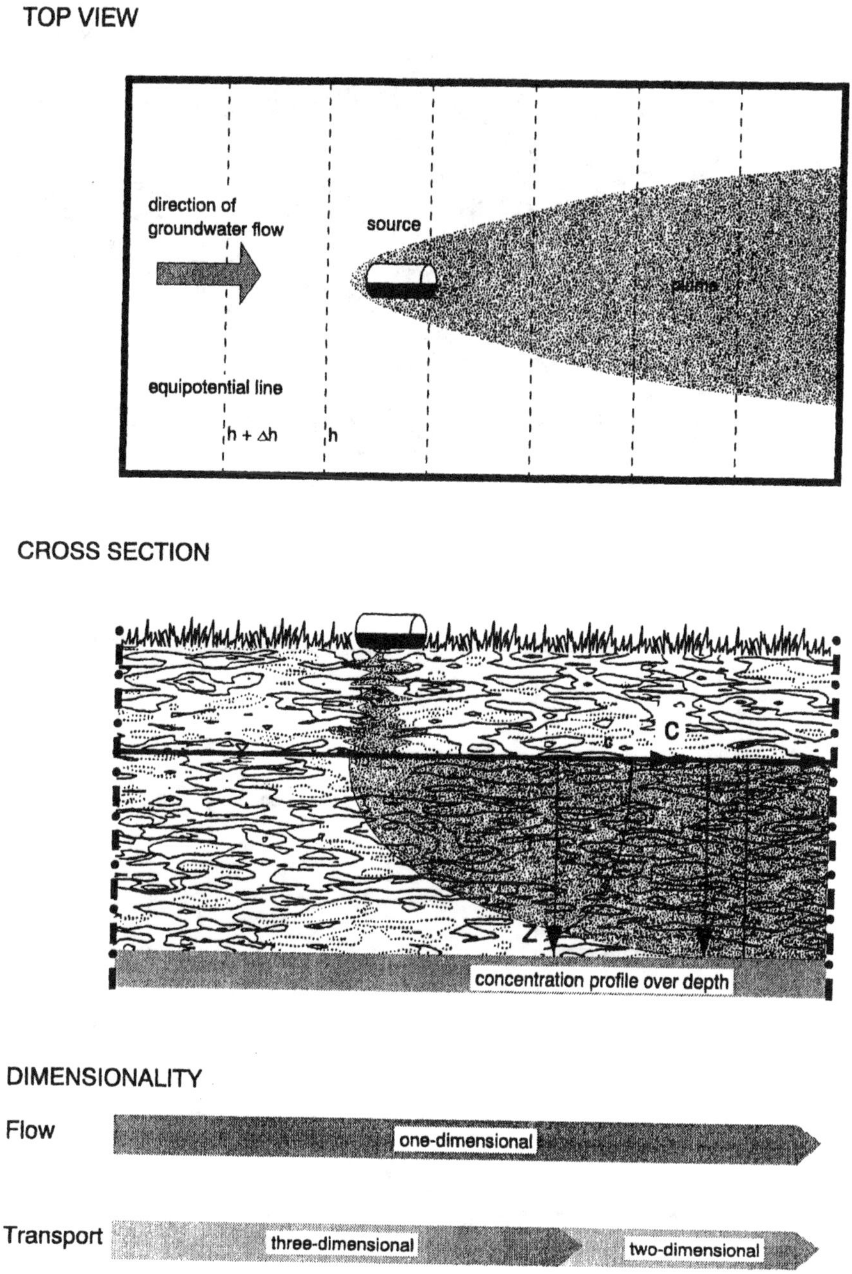

Figure 7.2 Example of different space dimensionality of flow and transport.

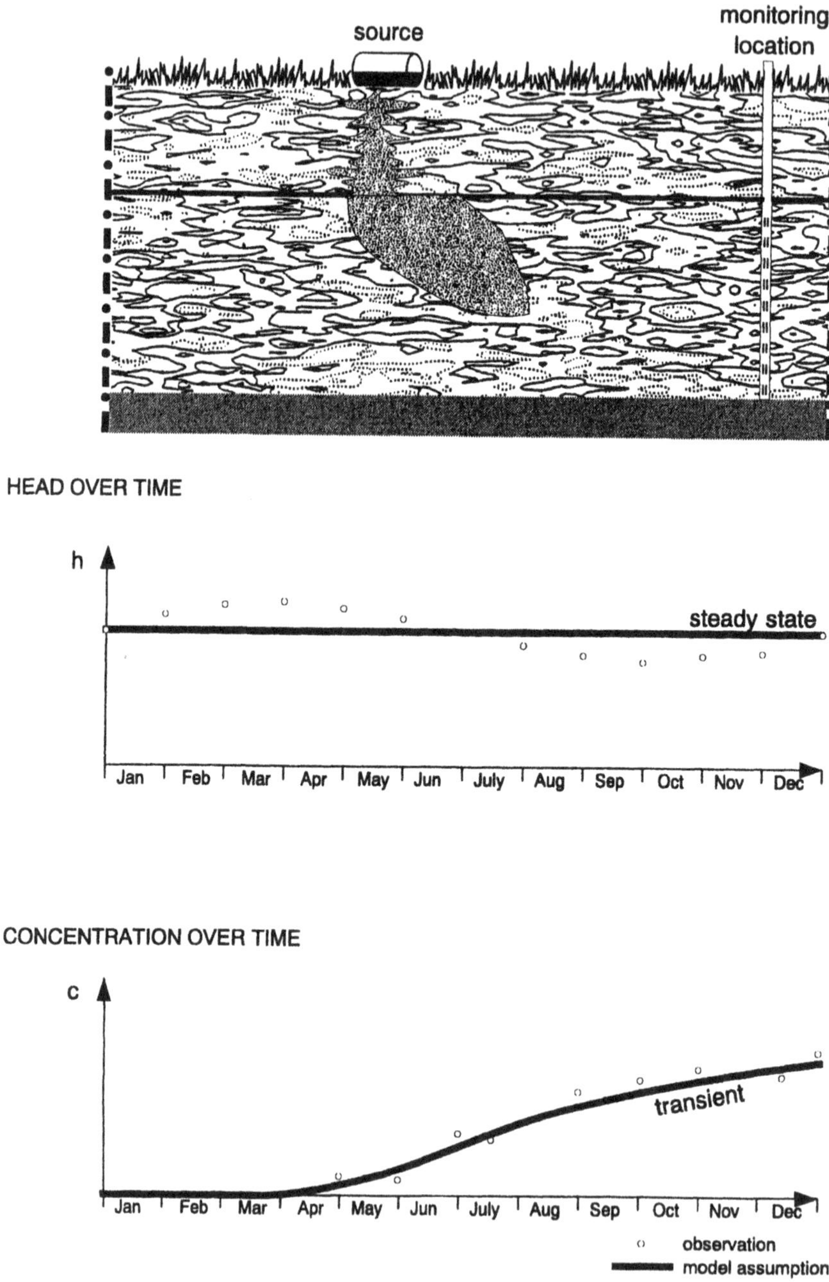

Figure 7.3 Example of different time dimensionality of flow and transport.

steady-state flow conditions can sometimes be assumed as the basis for transport simulations. An example of such a situation is shown in Figure 7.3.

Apart from the number of dimensions selected, simplification may be needed due to the scale of the groundwater problem. A large-scale problem may be represented by several smaller-scale models, or a small-scale model interfacing with a large-scale model. Ideally a model user should have a library of models of varying complexity, so that the model simplification selected is based solely on the characteristics of the problem to be solved.

7.2 APPROXIMATING SITE CONDITIONS

Seven approaches to simplifying actual conditions are presented in the following sections:

- Black box model
- Depth-integrated model
- Vertical cross-sectional model
- Multi-layer model
- Three-dimensional model
- Phased model
- Combination of models.

7.2.1 Black Box Model

Black box, zero-dimension, or compartmental models are discussed in Sections 5.1 and 6.1. In a black box approach the details of the groundwater regime are treated as a mystery, and attention is focused solely on balancing the inputs and outputs to the domain considered. This approach assumes that full-mixing and uniform properties can be assigned to the aquifer, which is represented as a single control volume. This approach is applicable when a global view of groundwater mass balances are required. Preparation of a black box model, which may be rapidly accomplished, is often an early step in a modeling study. The resulting mass balances may be used to assess the available data and the global adequacy of both the conceptual model to represent site conditions and the numerical model (if one is prepared). A field application of a black box model is presented in Appendix A.1.

7.2.2 Depth-Integrated Model

In a depth-integrated model the three-dimensional equations for flow and transport are averaged in the vertical dimension to give a horizontal, two-dimensional model of the aquifer system. This approach is sometimes known as the *hydraulic approach* or the *aquifer approach*. In this type of model it is assumed that the groundwater flow is essentially horizontal; that is, vertical flow components can be neglected. Hydraulic aquifer properties, such as transmissivity and storativity, are introduced as mean characteristics averaged over depth.

In practice, the depth-integrated model is often chosen for regional-scale flow models. Considering an aquifer system at regional scale, the system may cover an area of 10 by 10 km. The saturated thickness of the aquifer, however, may be as little as 10 m. Plotting the aquifer geometry on paper, without exaggeration, one would have to draw a line 1 mm thick and 1 m long. For a geometry such as this, neglecting the vertical flow components is often justified, thus reducing the problem to two dimensions.

The assumption of horizontal flow (i.e., hydrostatic conditions) holds true for a horizontal, homogeneous, or horizontally-stratified confined aquifer of uniform thickness, and in the absence of strong sinks and sources. The assumption is also appropriate for aquifers of variable thickness if the variations are small by comparison with the average thickness. Also leaky and unconfined aquifers may be well-represented by a two-dimensional model if the transmissivity of the aquifer greatly exceeds that of aquitards, and the slope of the potentiometric surface is small. This last condition is the basis for the Dupuit assumption for two-dimensional flow in unconfined aquifers. It can be shown that if

$$i^2 \ll 1 \tag{7.1}$$

where i represents the slope of the water table, then the error in making the two-dimensional assumption for flow will be small (Bear 1972).

Horizontal flow cannot be assumed in the neighborhood of features inducing vertical flow such as losing or gaining rivers, drains, partially penetrating wells and point sources such as leaky ponds. However, horizontal flow can be assumed, as a rule of thumb, at a distance of about 1.5 to 2 times the aquifer thickness from these sinks and sources (Bear and Verruijt 1987).

The depth-integrated model approach yields reasonable results for regional flow problems but may be inappropriate for transport problems. The primary reason for this is that depth-integrated concentrations provide little insight into peak concentrations and may not reflect actual mass flux at a given location. However, depending on the focus of a model, the depth-integrated approach may be applied to field cases in which mixing throughout the full saturated thickness of the aquifer may be assumed and water-quality stratification is unimportant. For example, if predicted concentrations are required for water supply wells with long open intervals or that stress a significant depth of the aquifer, then a depth-integrated concentration may well be an appropriate representation. The distance from a contaminant source at which full vertical mixing can be assumed is given (Sayre 1973) by

$$L \geq 0.5 \frac{m^2}{\alpha_z} \tag{7.2}$$

where

L = distance at which vertical mixing can be assumed in [L]
m = aquifer thickness in [L]
α_z = vertical dispersivity in [L] (some vertical dispersivity values are provided in Appendix D.3)

Examples of the number of space and time dimensions needed for flow and transport problems are compared in Figures 7.2 and 7.3. Research indicates that vertical mixing is smaller than horizontal mixing and in some cases is in the same range of magnitude as molecular diffusion (Sudicky 1986).

In transport modeling, the assumption of depth integration is appropriate for thin saturated thicknesses, fairly isotropic aquifers, and diffuse sources of ambient density. Also large-scale or long-term contaminant simulations may sometimes be adequately approximated by a depth-averaged model. Such an example is shown in Figure 7.4. This example focuses on water quantity management in an alluvial groundwater system. The Erolzheimer Feld refers to an alluvial aquifer along the river Iller, South Germany. Intensive investigations, including modeling studies, have been carried out to assess use of the groundwater reservoir as a drinking water supply (Mehlhorn and Flinsbach 1983).

Groundwater modeling was used to quantify the system's response to pumping and to identify hydrogeological controls. In view of the aquifer geometry with an aquifer thickness of less than 20 m, a length of about 23 km, and a maximum width of about 5 km, a two-dimensional depth-integrated modeling approach was selected. An FD model was applied, subdividing the model area into 120 by 22 cells. Figure 7.4 illustrates the FD grid in terms of the model input data for topography. The depth-integrated modeling approach was successful in simulating the key flow mechanism at the site, predicting observed aquifer responses, and evaluating future water-management practices. This two-dimensional approach may not be justified if:

- The aquifer is relatively thick or stratified.
- The source is relatively small (horizontally or vertically) or dense, that is, with greater than 1% density difference between source liquid and ambient groundwater (Mackay and Cherry 1985)

The following example illustrates the depth-averaged approach to a transport problem. Several spills of organic chemicals occurred at an electronics plant. The underlying aquifer is transmissive, but with a thin saturated thickness (0 to 7 m), and it overlies relatively impermeable bedrock, as shown in

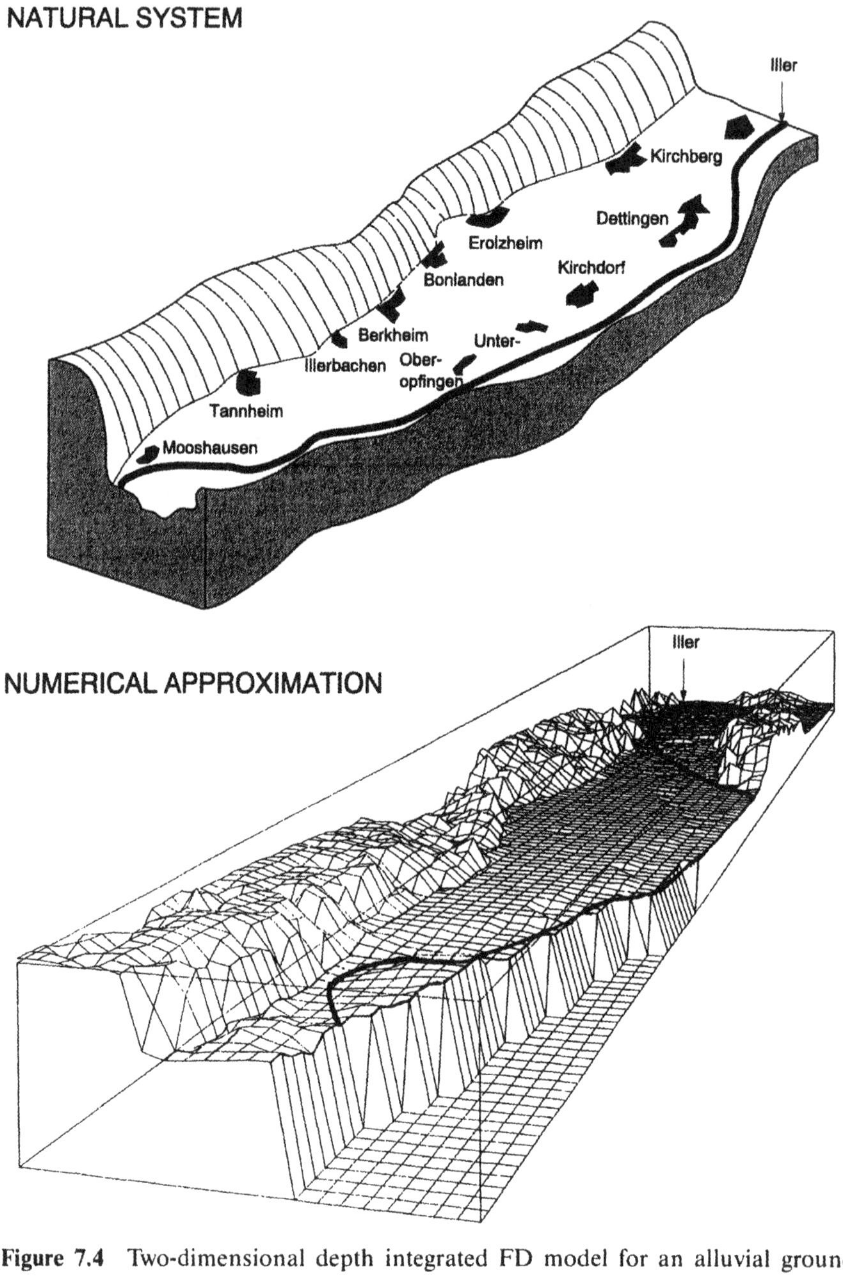

Figure 7.4 Two-dimensional depth integrated FD model for an alluvial groundwater system.

Figure 7.5. The depth to the water table varies seasonally between 1 and 3 m. An assortment of areal and point sources were identified as likely sources of contamination. The primary question to be resolved by modeling analyses is what distribution and pumping rates are most effective for a set of extraction wells.

A horizontal, two-dimensional groundwater flow and transport model was selected because the groundwater flow is essentially horizontal and vertical mixing of the contaminant was a good approximation in such a thin aquifer. This modeling approach also well represents the fully screened interceptor wells, the areal source, the variable surface recharge, and seasonally varying boundary conditions. This model did not address the following factors:

- Migration in the unsaturated zone, which was not considered important due to the shallow water table and semivolatile contaminants.
- Contaminant distribution near point sources, which were less problematic than the areal sources.
- Possible density-induced stratification of the contamination.

Since a small volume of chemicals was spilled over a long period of time, and low water concentrations were observed, density effects were ignored in designing the interceptor wells. A different model would have been appropriate if modeling analyses for the design of source remediation were to be analyzed.

Other typical uses of two-dimensional, horizontal models include the following:

- Use of a two-dimensional flow model for the *far field* (far from the source), linked with a three-dimensional flow and transport model for the *near field* (near to the source).
- Use of a flowline as the base of a two-dimensional model (rather than the base of the aquifer) so that contaminant concentrations are not artificially diluted.

For this last case a suitable flow line may be found using a vertical plane model (which is discussed in subsection 7.2.3). However, this flow line may not remain stationary if the flow pattern changes with time. Also some models allow vertical averaging over different depths for the flow and transport portions of the calculations.

7.2.3 Vertical Cross-sectional Model

Vertical cross-sectional models, or slice models, consist of two-dimensional models oriented vertically. The model is of unit thickness in the direction not simulated. The assumption used in this approach is that all conditions in this

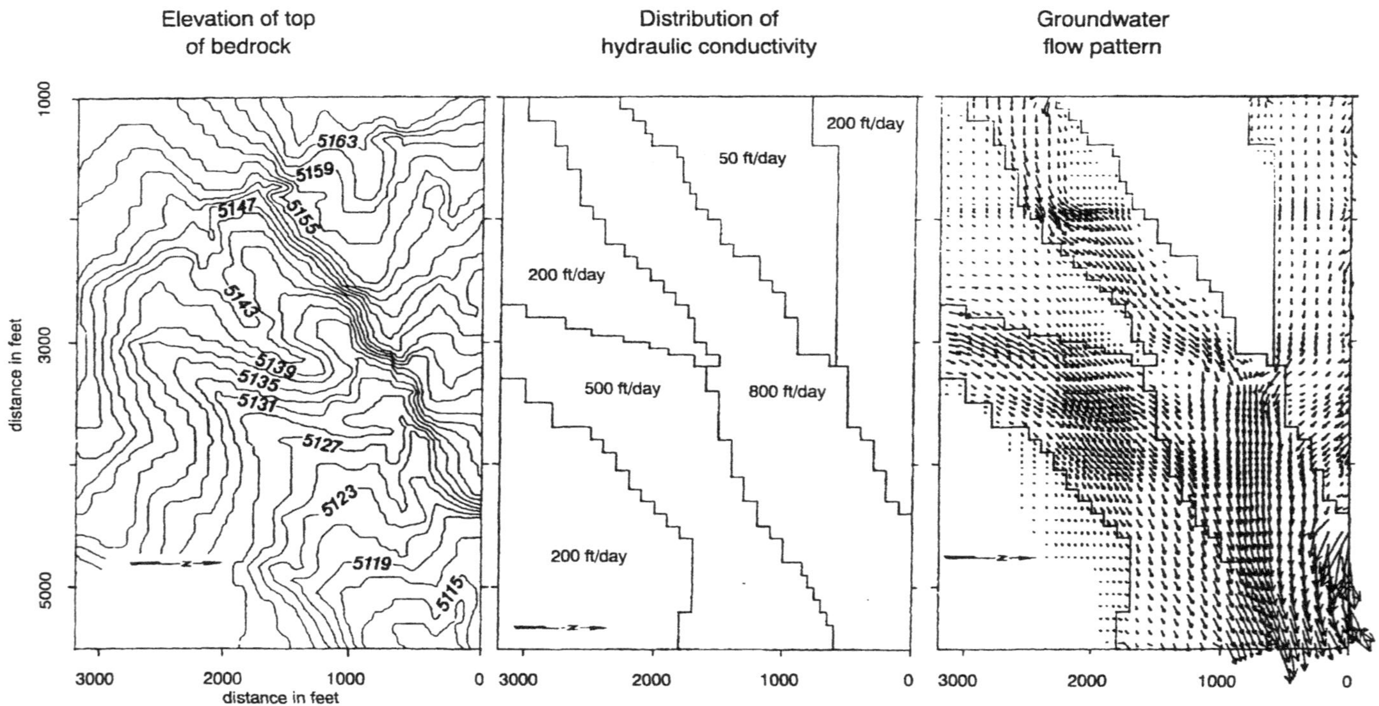

Figure 7.5 Two-dimensional depth integrated FD model for a shallow, alluvial groundwater system.

third dimension are the same as that simulated in the slice. Such a model must be oriented along a flow line (or its horizontal projection), to conserve mass. Point sinks and sources cannot be accurately simulated and would, in this approach, become line sinks or sources and thus be exaggerated in effect. An example of the effects of neglecting dispersion in the third dimension is shown in Figure 7.6. The source, a leaking pond, is predicted to result in a higher concentration contaminant plume, traveling deeper into the aquifer for the two-dimensional simulation than for the three-dimensional cases. Models with a vertical dimension (cross-sectional or three-dimensional models) are needed in cases where gravity is the dominant driving force for groundwater flow. Several such examples are shown in Figure 7.7. The predominant applications for vertical cross-sectional models are unsaturated-zone flow and transport problems and density-coupled or heat-coupled simulations. An illustration of model application along a flow line is given in Figures 8.10 and 8.11.

The simulation of the free water surface in an unconfined aquifer may be difficult to solve because it is a nonlinear problem with a boundary of unknown location. If a variably saturated code is used, then the location of the water table is automatically calculated. If a fully saturated model is being used, then one approach is to perform the following steps:

- Estimate the water-table elevation and supply a zero-flow boundary condition to the model at that location.
- Predict the resulting heads, and compare the predicted heads at the assumed water table with the elevation.
- Adjust the estimated water-table elevation to match the predicted heads.
- Repeat this procedure until the predicted heads at the water table equal the elevation, within a specified tolerance.

Some models have the added problem that a model cell that lies above the water table cannot rewet. In this case large cells accommodating all likely positions of the water table can be used to give at least a rough estimate of the water-table elevation.

Two of the advantages of the vertical cross-sectional model are that it:

- Permits consideration of vertical flow components, such as infiltration and percolation.
- Simulates the essential elements of flow and transport in a typical slice (e.g., in considering movement of a saltwater wedge).

The disadvantages of the vertical cross-sectional model are that boundary conditions, sinks, and sources are more difficult to simulate, and errors due to misconceptions may be less easily detected in the predicted results than in depth-averaged models. For example, a vertical cross-sectional model may appear to produce realistic predictions after calibration, but the calibration

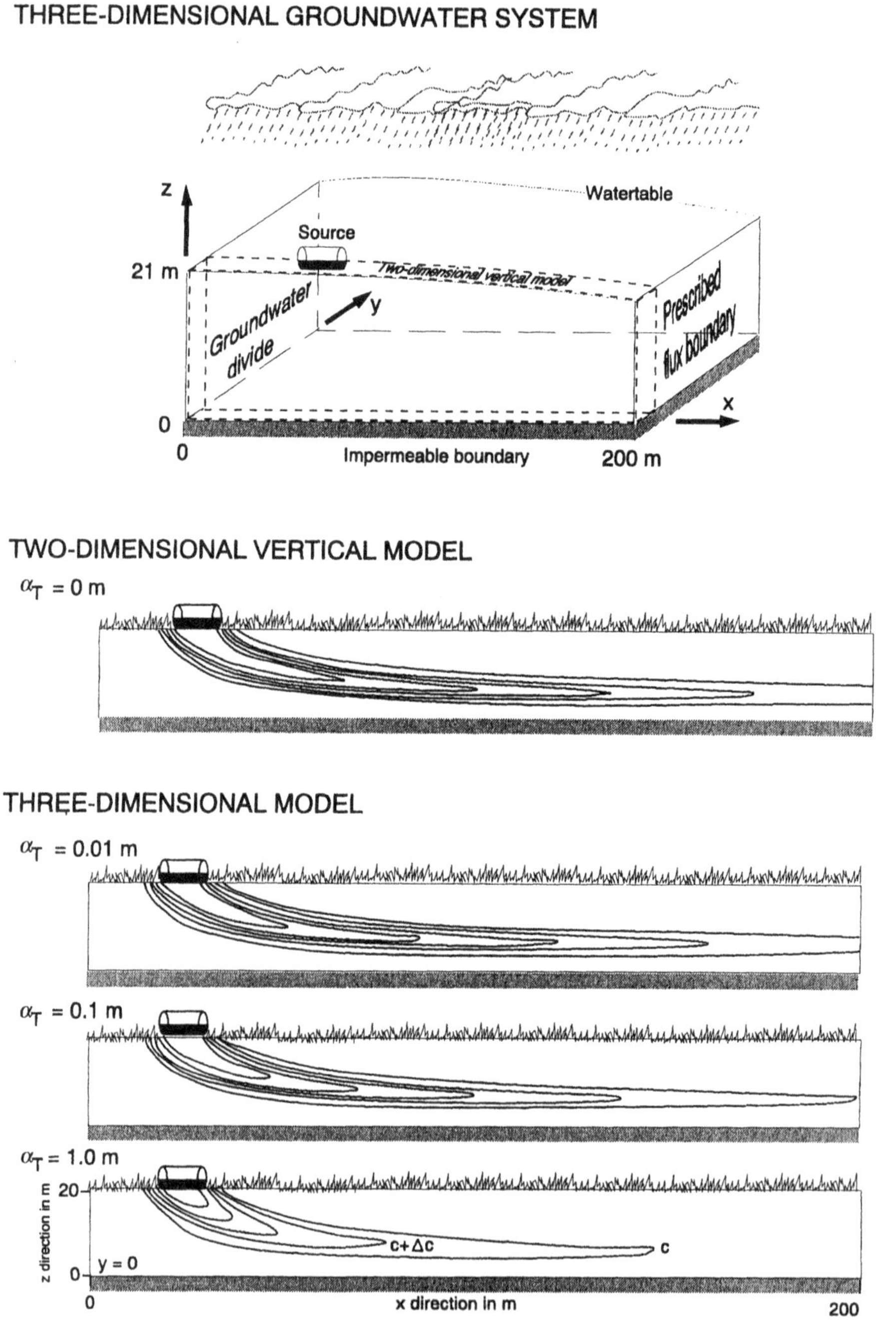

Figure 7.6 Comparison of two-dimensional and three-dimensional model results (after Frind 1988).

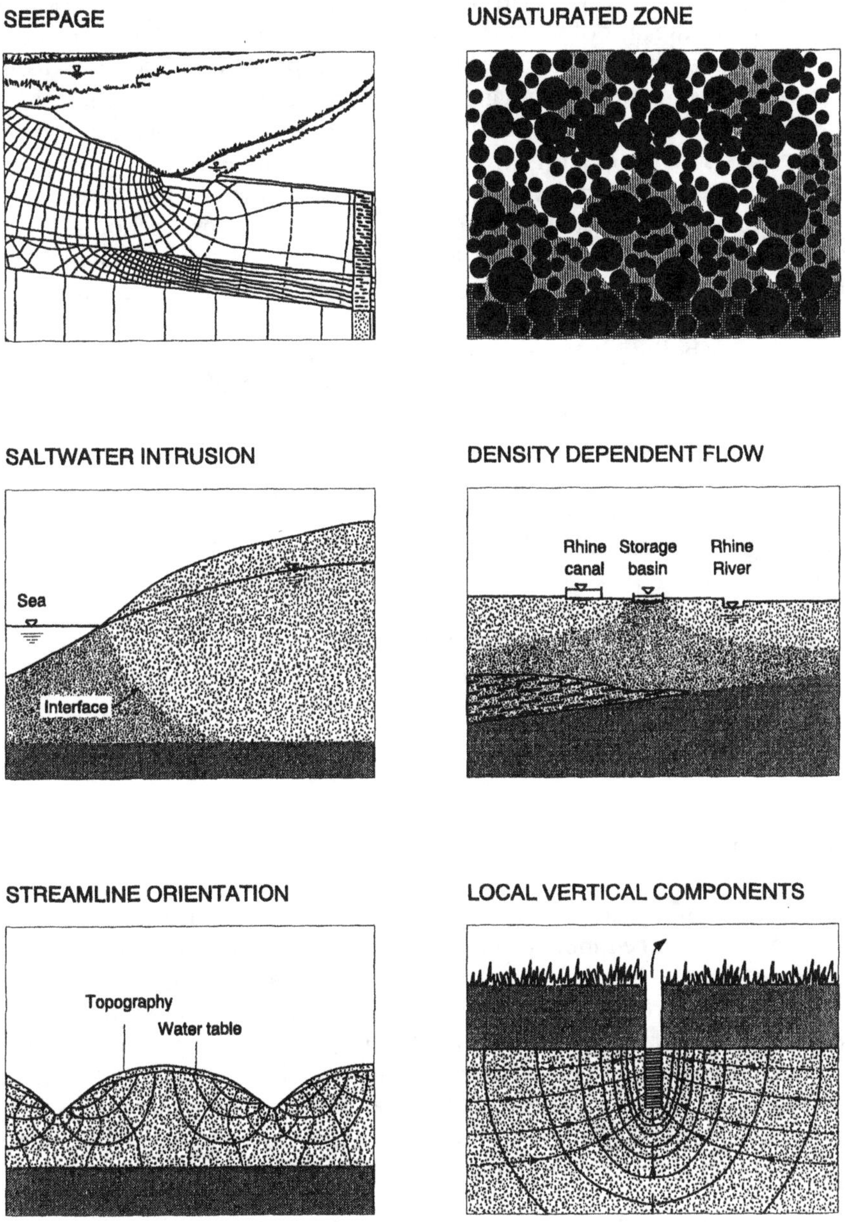

Figure 7.7 Typical applications of two-dimensional vertical models.

parameters (usually dispersivities or adsorption parameters) may be skewed to account for the lack of the third dimension. Section 7.3 includes more discussion of this point. It is possible to approximate point sinks, such as wells, by calculating the fraction of total groundwater withdrawal in the section simulated or as a fixed head rather than a flux boundary. Neither of these approximations is satisfactory for all situations. Alternatively, a wedge-shaped, vertical cross-sectional model may be used to represent a single source or sink. An example of the practical application of a vertical cross-sectional model follows: Suppose that the effect of site selection on the environmental risks of a coal-fired power plant has to be evaluated. One of the overall risks is chemical seepage into the groundwater due to leaching of solid wastes. The likely concentrations of pollutants at various sites shown in Figure 7.8 are to be modeled and analyzed. Transport and attenuation of the pollutants in both the saturated and unsaturated zones are of concern because both adsorption and residence time in the saturated zone, and advection and dispersion in the saturated zone, need to be considered. A vertical cross-sectional model is used to simulate the primarily vertical flow and transport in the unsaturated zone. The results of this model in the form of flow rates and concentrations at the water table are supplied as sources to a two-dimensional, horizontal-plane model of the saturated zone. Some of the model predictions are shown in Figure 7.8. Other vertical cross-sectional modeling applications are described in subsection 7.2.7 and in Appendixes A. 4 and A.5.

The vertical cross-sectional model is a powerful tool, but it requires more expertise and site-specific data (e.g., unsaturated-zone properties and vertical stratification data) than the two-dimensional horizontal-plane model.

7.2.4 Multi-layer Model

A multi-layer model consists of a vertically stacked sequence of two-dimensional depth-averaged models that are linked together by sources and sinks. This type of model is quasi-three-dimensional in that horizontal and vertical flow and transport components are simulated. There are different types of multi-layer models, but typically in flow-only models, flow in intervening aquitards is approximated by a leakage term. Also flow in the aquifer layers is approximated as being horizontal, whereas flow through aquitards is considered to be vertical. Multi-layer transport models usually consider the aquitards explicitly because adsorption and travel times through the aquitards are of concern.

The advantages of the multi-layer approach are that variations in transmissivity of layered aquifers may be readily simulated by varying thicknesses of model cells, and undulating aquifers may be accommodated even in a finite-difference scheme. In this way a complex layered sequence of aquifers or units may be represented by a reasonably small number of model cells. Also vertical stratification within aquifers can be simulated by multiple layers.

The disadvantages of the multi-layer approach are:

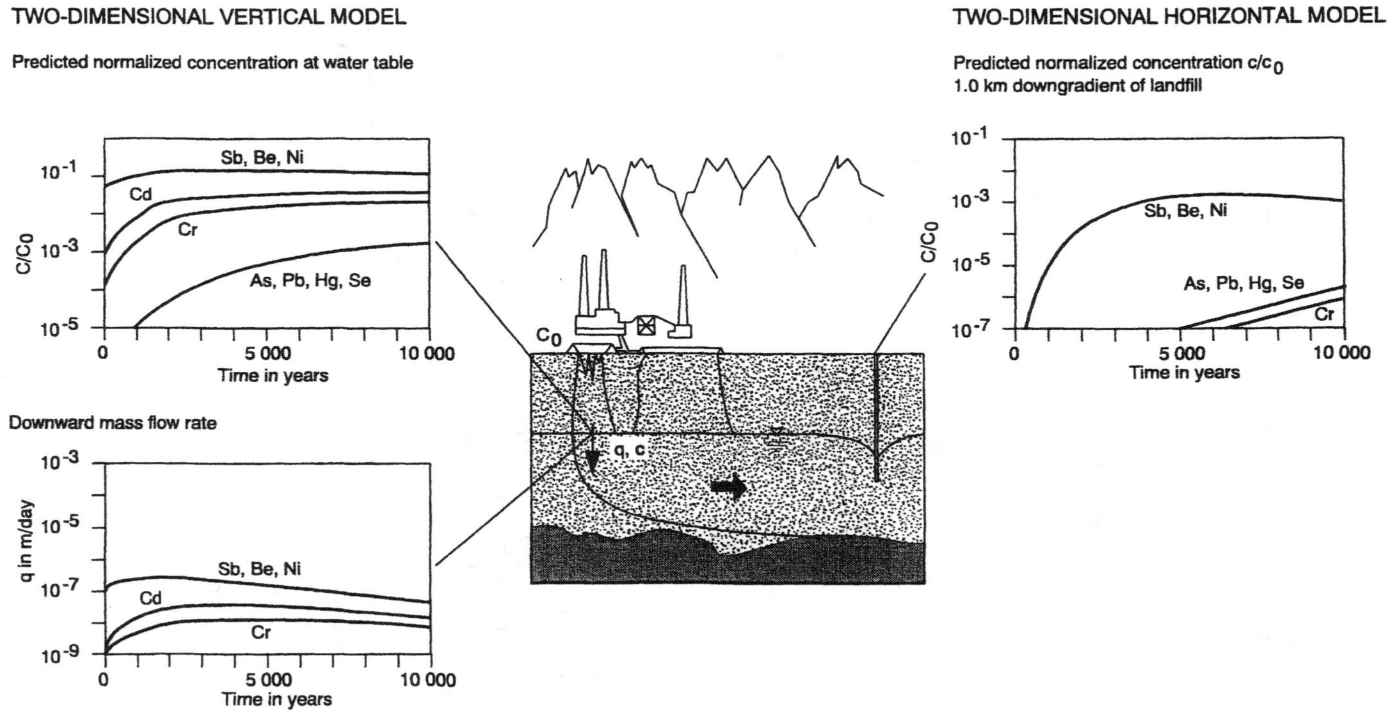

Figure 7.8 Combination of models to estimate transport of trace metals from ash ponds and landfills of a coal-fired power plant.

- Aquifer pinch-outs may not be well-simulated.
- Steeply sloping (i.e., greater than 10% grade) aquifers may not be accurately simulated; the assumption of flow perpendicular to the calculation cell faces is violated and mass imbalances result. For unconfined systems the Dupuit assumption of small slopes to the water table is violated.
- Calculation cells for nonhorizontal or nonuniform layers may not be approximated as rectangles (for area and flux calculations, etc.).

Another difficulty is sources or sinks of water or contaminant that exist in several of the model layers. For example, extraction wells or water injection wells may be open to several aquifers. Alternate approaches to approximating this situation are presented in subsection 8.4.3. Ideal applications for multilayer models include sites with many thin aquifers and aquitards that vary in thickness and are not horizontal. A typical application is shown in Figure 7.9 in which the prediction of water levels in multiple, interconnected aquifers was required. An entire groundwater basin was considered, with zero lateral recharge. The model was used to design a dewatering well field and a pumping scheme for the mine as it developed. Wells screened over multiple hydraulic units were simulated. In this particular project only groundwater flow was considered, and the aquitards were simulated using leakage rates.

7.2.5 Three-Dimensional Model

Fully three-dimensional models incorporate all three spatial components of flow in all model cells. These models may consider both saturated and unsaturated zones, or they may neglect changes of storage in the unsaturated zone and simulate recharge directly at the water table. Many model cells may be used to represent each hydraulic unit, thus enabling simulation of stratification of flow or transport. The data handling, grid generation, and analysis of predicted results for the volume of data generated for and by three-dimensional models are often handled by pre- and postprocessors. Applications that necessitate three-dimensional modeling analysis often involve one or more of the following:

- Thick aquifers
- Multiple aquifers
- Steeply sloping aquifers
- Dense contaminants
- Multiple or multilevel sinks or sources
- Significant anisotropy or fracturing

The main advantage of a three-dimensional model is that simplifying spatial assumptions and heterogeneity are not necessary, and multiple layers, ver-

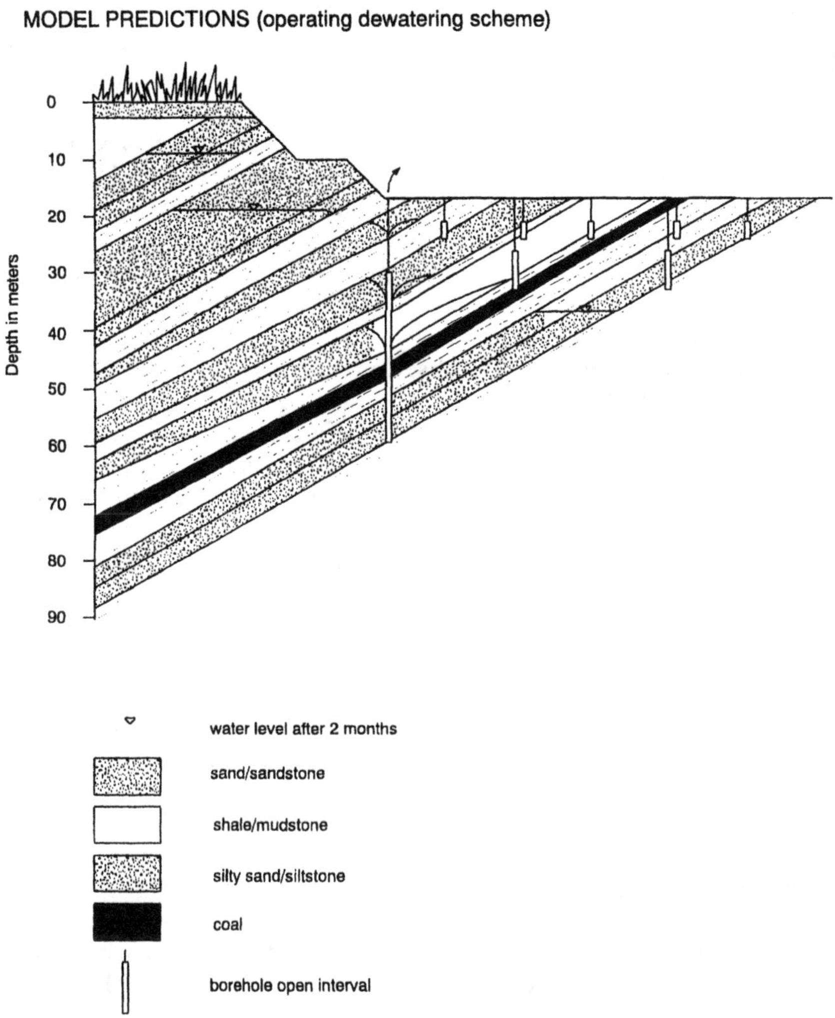

Figure 7.9 Evaluation of dewatering scheme for a coal mine using a multi-layer model.

tical variations, and point and areal sources can be accommodated with less simplification. The disadvantages of a three-dimensional modeling approach are the time, expense, and data needs for defining and calibrating a three-dimensional model. For this reason three-dimensional models are often reserved for only the most complex parts of a site, or for situations that cannot be successfully simulated by simpler models. A description of a typical three-dimensional model application follows:

The problem to be considered by groundwater modeling analysis is to identify the likely sources of groundwater contamination and to design groundwater remediation. The situation involves a dense contaminant that has contaminated two aquifers to a depth of more than 30 m. Groundwater in the upper aquifer flows perpendicular to the regional flow in the lower aquifer. The flow patterns are illustrated in Figure 7.10. Leakage between the two aquifers is driven by the rate of withdrawal from deep water supply wells; the direction of this leakage varies seasonally due to changing supply well pumpage. Contaminant sources at various depths potentially contribute to the observed plumes, shown in Figure 7.10. Remediation schemes involving wells screened at differing depths were tested via modeling analysis, as seen in Figure 7.10. Much extrapolation and interpolation of site-specific data was required to supply the model with a full data set. However, steady-state and transient water level data as well as extensive water-quality data were available to test the validity of the model. The observed concentrations were low enough that explicit density coupling, that is, coupling between rather than within time steps, between flow and transport calculations did not result in error. The assumption of explicit coupling reduced computational effort significantly.

Other three-dimensional applications are described in subsection 7.2.6 and in Appendixes A.4 and A.5.

7.2.6 Phased Models

A phased modeling approach is one in which a sequence of successively more complex models is used as the extent of site-specific data increases and the need for specific answers develops. The assumption underpinning this approach is that even the initial simple models based on limited data are useful, although not entirely definitive.

Assessments of contaminated sites frequently commence with a limited site-characterization program. A model user may be faced with selecting an appropriate model for use throughout the life of the project on the basis of sparse site-specific data. In such a case, a phased modeling approach may provide the best solution. The advantages of using phased models follow:

- Modeling analyses appropriate to the level of site-specific data available may be used throughout the project.
- An evolving conceptual model may be used, resulting in a model better integrated with site characteristics and remedial plans.
- Model analyses of parameter sensitivity, etc., may be used to identify data gaps throughout the planning and implementation of the field program.
- Other personnel involved in using the modeling results may gain insights on the site, and interact with modeling decisions during the course of the project.

The disadvantages of a phased modeling approach are that it requires the application of several models over a period of time, which is not a traditional

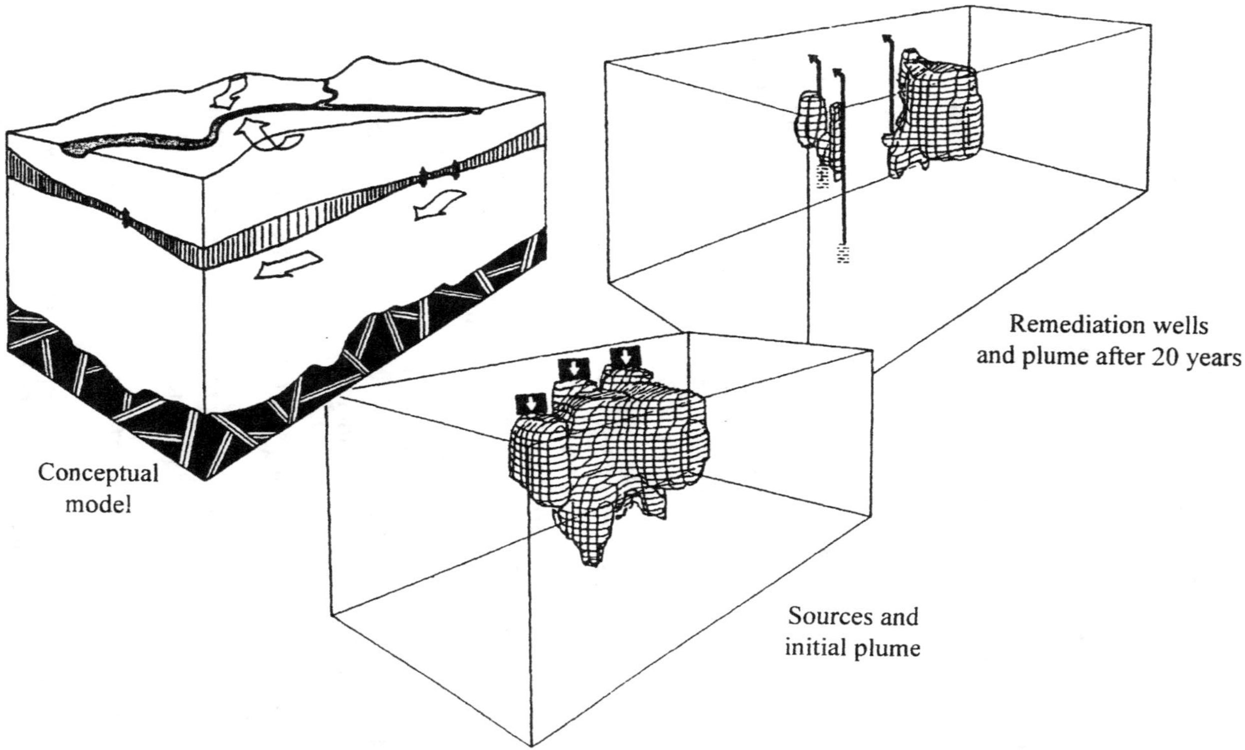

Figure 7.10 Contaminant migration and remediation in a multi-aquifer system using a three-dimensional model.

approach. An example of a phased model follows. An additional example is provided in Appendix A.5.

The example involves an inactive mine site at which a 50-acre tailings pile generated metallic leachates over more than 40 years. The purpose of the groundwater modeling analyses was to compare the effectiveness of remedial alternatives.

The tailings pile is situated adjacent to a river in a glaciated valley, as shown in Figure 7.11. The pile overlays unconsolidated glacio-fluvial materials and is bounded by terminal and recessional moraines. Groundwater flow in the vicinity of the pile is complex and three-dimensional. Flow occurs along the adjacent river valleys, upward to the rivers, and downward from the pile. Sampling indicates that a contaminated zone has developed beneath the pile and adjacent to the river. A sequence of models were used to predict the rate of contaminant migration for both existing conditions and in response to various remedial alternatives.

Initially, limited data were available. A simple two-dimensional, cross-sectional model, along a flow line, was used to estimate contaminant velocities and the potential extent of contamination. This is shown in Figure 7.12. As the field program progressed, the stratigraphy became better defined, and a more detailed multi-layered (quasi-three-dimensional) model was developed to evaluate parameter sensitivities and make a preliminary evaluation of alternatives (Figure 7.12). Eventually sufficient site-specific data were available to permit three-dimensional analyses with more complex heterogeneity and anisotropy, and variable infiltration rates. Also the comparative costs of the remedial alternatives warranted the additional effort of three-dimensional analyses to support the selected remedial approach. The three-dimensional model was used to provide a detailed evaluation of possible remedial measures (Figure 7.12). In each case the sophistication of the model framework is based on the amount of available data.

The results from a phased modeling analysis also allow general questions such as the following to be considered:

- How do modeling results improve with the addition of more field data?
- Is there a point of diminishing return in collecting additional data?
- Do more complex models provide more accurate results?

Results of the flow simulations for the three models are compared in Figure 7.13. This figure shows that the flow simulation accuracy improves with both the quantity of data and the area sampled. It is not clear whether additional data would have significantly improved the model predictions beyond that already achieved. Using the correlation coefficient as a measure, the simulation accuracy improves with model complexity (based on the number of model nodes), but the rate of improvement decreased beyond 10,000 nodes, as shown in Figure 7.13. In this case modeling flow with the three-dimensional

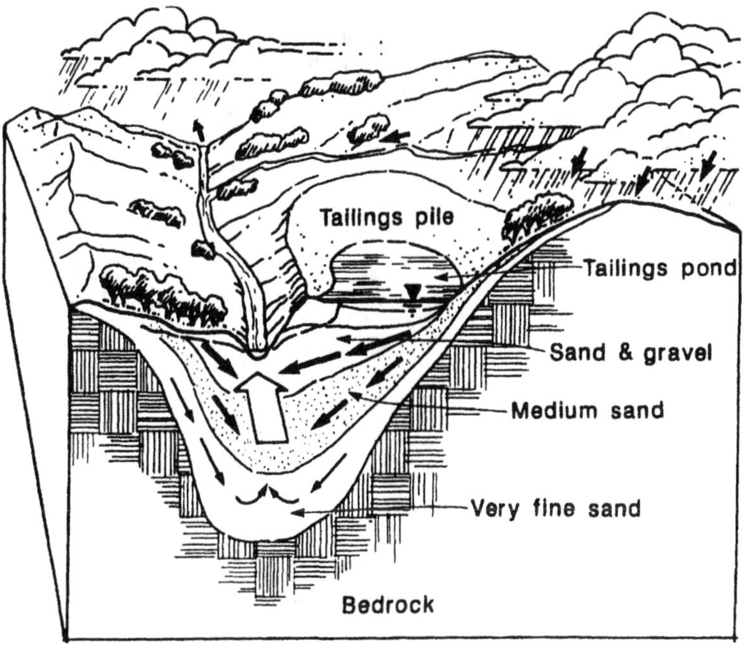

Figure 7.11 Investigation of remedial alternatives for a 50-acre tailings pile: Conceptual model.

model does not produce significantly more accurate results than those produced by the multi-layered model.

Transport predictions from the three models for two of these cases—no action or a capped tailings pile—are illustrated in Figure 7.14. The main concerns were whether contaminant concentrations might worsen in the future and whether capping the tailings pile would be an effective remedial action. All models showed that conditions could be expected to worsen unless remediation was undertaken. The multi-layer model appears to underestimate the effectiveness of capping the pile (measured by relative decrease in mass flow to the river) by comparison with the three-dimensional model predictions. In contrast to the conclusions for the flow modeling, the preparation of the more complex model was assessed to be worthwhile for the transport predictions for two reasons:

- Additional site-specific data were represented by the finer discretization.
- The predicted results differed from the two-dimensional results.

Unfortunately, a conclusive model audit comparing model results against field data was not possible in this case because of additional, poorly defined factors affecting river water quality.

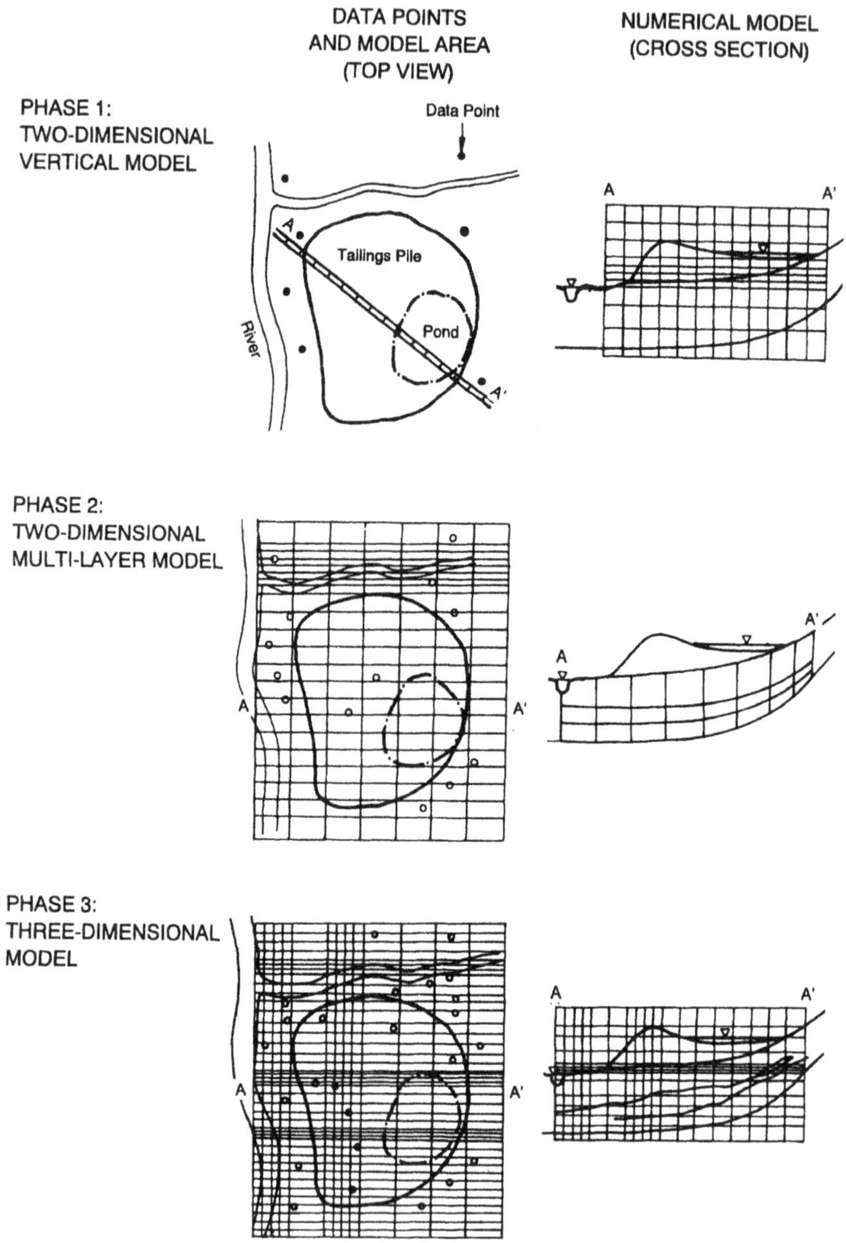

Figure 7.12 Investigation of remedial alternatives for a 50-acre tailings pile: Phased modeling approach.

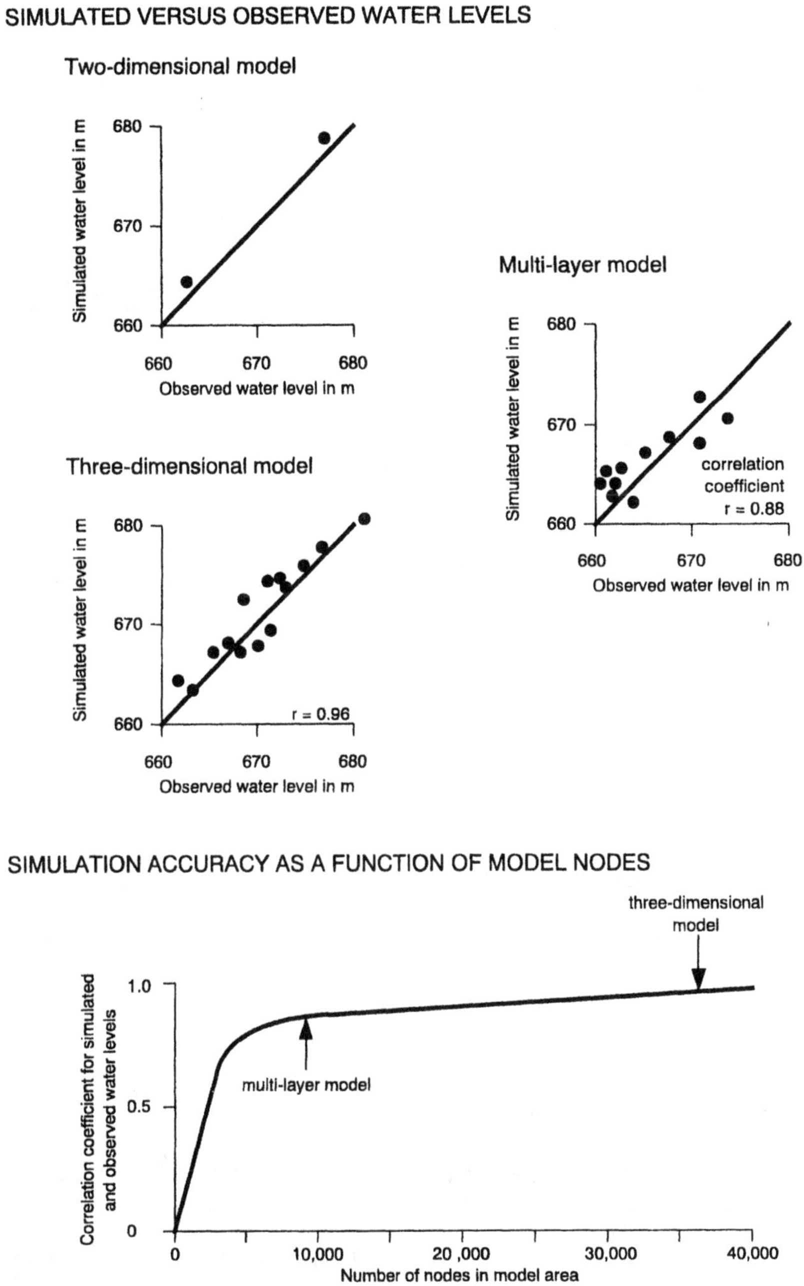

Figure 7.13 Investigation of remedial alternatives for a 50-acre tailings pile: Variation of flow simulation accuracy with model framework.

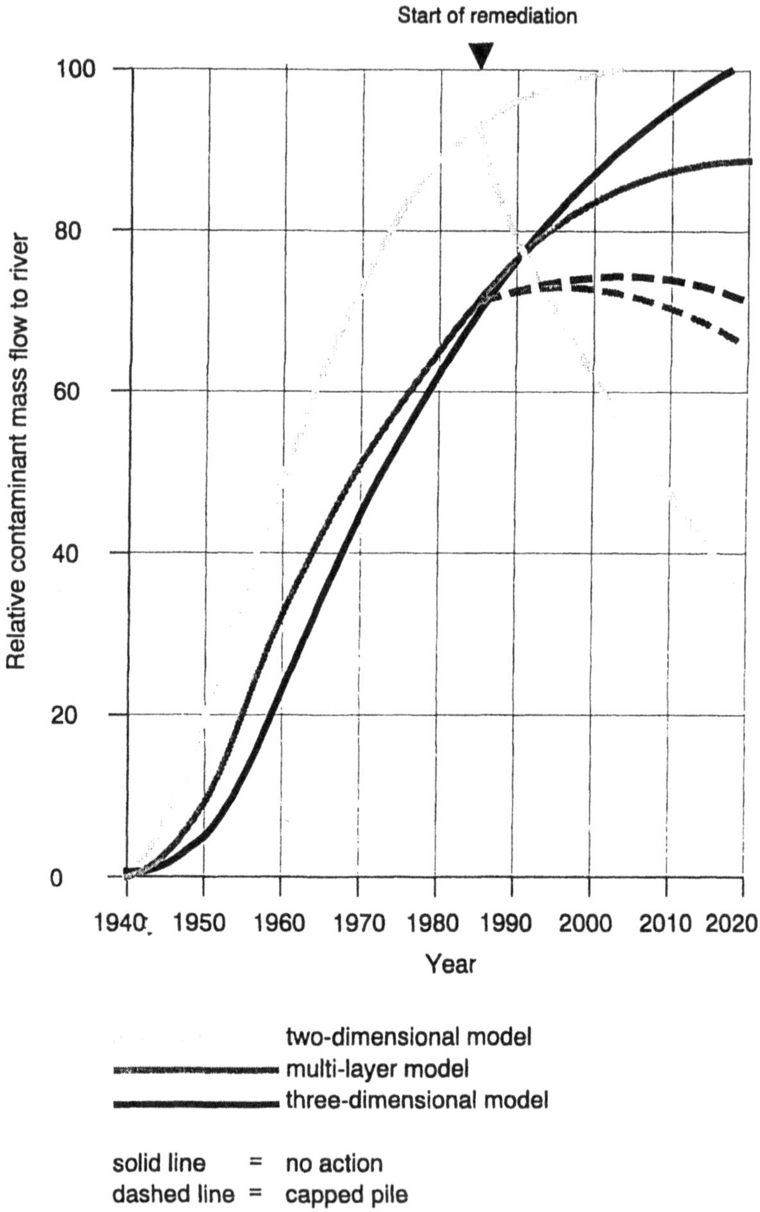

Figure 7.14 Variation of transport simulation results with model framework.

7.2.7 Combination of Models

It might seem as though three-dimensional modeling analyses, which employ fewer simplifying assumptions, are the ideal model choice. However, due to lack of data or due to the modeling effort involved, less sophisticated models should also be considered. An experienced modeler will choose the approach best suited to approximate the modeling problem while using the least effort. This philosophy can lead to an approach in which a combination of simplified models, rather than a full three-dimensional model, are used to represent the flow and transport system. For example, the flow system may be subdivided into well-defined units in which one- or two-dimensional modeling approaches can be justified. Model predictions from one model form the boundary conditions for the neighboring model. The following example illustrates such an approach, and an additional illustration is provided in Appendix A.5.

Modeling analyses were used to evaluate seepage from the proposed subgrade disposal of highly acidic uranium tailings in a pit, as shown in Figure 7.15. Perched and confined aquifers underlie the 30-m deep pit. Seepage into both aquifers and sand lenses intercepted by the pit walls are a potential concern. The model was also required to provide information on the effects of alternate liner and drain designs. This appears like a complex problem requiring a large three-dimensional model. However, a sequence of three models, culminating in a two-dimensional, horizontal-plane model, provided simulations that adequately represented the key mechanisms.

Initially, a one-dimensional, vertical, variably saturated groundwater flow and transport model was used to assess seepage from the backfill to the perched aquifer, corresponding to conditions in the center of the pit. Conditions around the pit walls were stimulated with a vertical cross-sectional, variably saturated flow and transport model (top illustration in Figure 7.15). These two models were used to select liner and drain designs to keep contamination of the perched aquifer to acceptable levels. Then attention was focused on the confined aquifer underlying the pit; contamination of this aquifer could be readily addressed with a two-dimensional, horizontal-plane model (bottom illustration in Figure 7.15). This project is reported in more detail in Sharma et al. (1983).

7.3 IMPACT OF SIMPLIFICATION ON TRANSPORT MODELING

The consequences of selecting a specific modeling approach were discussed in Section 7.2, but there are some points relating to simplification of transport simulations that are of general application. They are presented in this section, together with examples.

In many cases flow models can be more readily simplified than transport models for the following reasons:

- Groundwater recharge and discharge is often areal and therefore two-

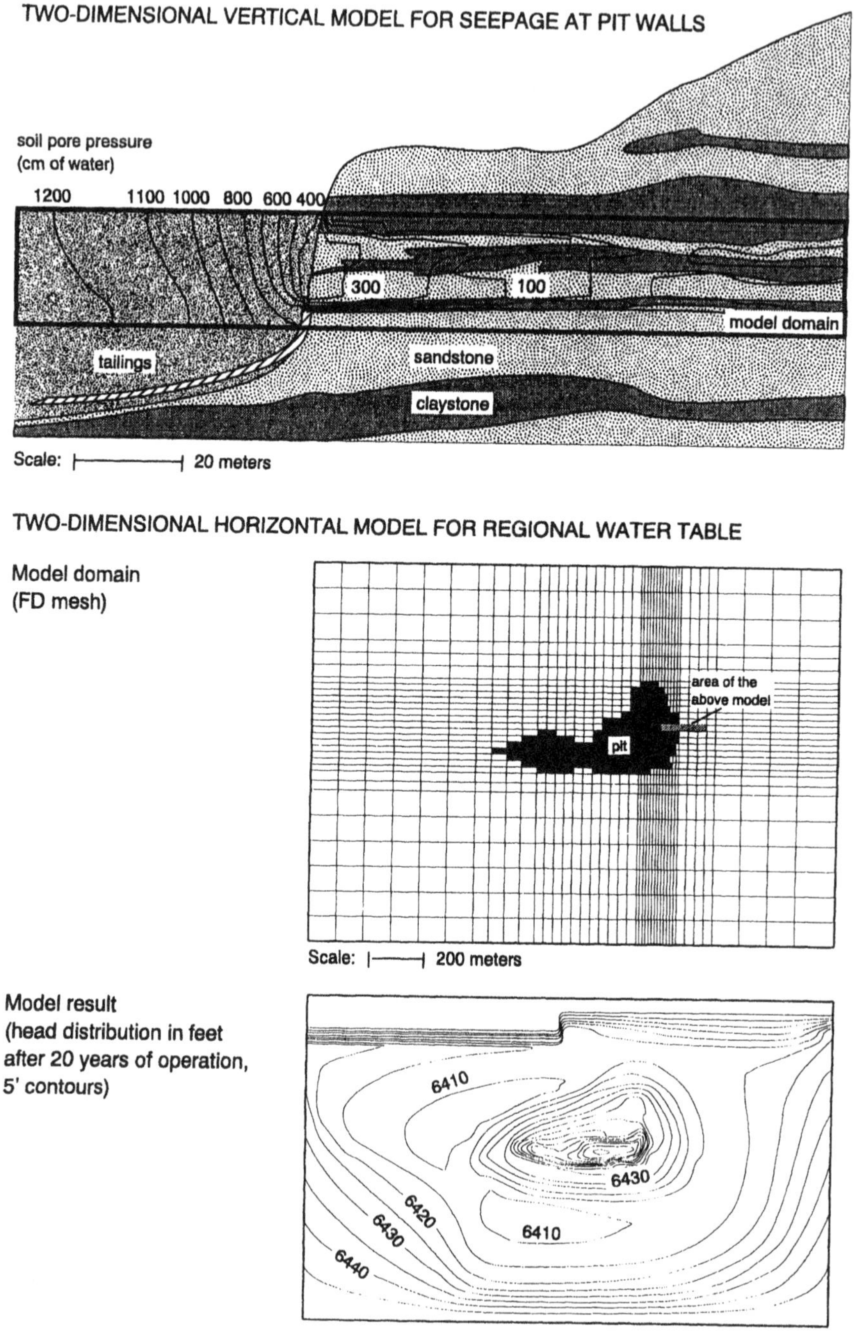

Figure 7.15 Combination of models for subgrade disposal of acid tailings.

dimensional in effect, whereas transport sinks and sources tend to be point sources with a three-dimensional effect.

- The overall approximate groundwater flow pattern is of interest, whereas stratification and other details of solute behavior are of concern.
- Small components of the flow mass balance, for example, local vertical hydraulic gradients, may be of little importance in a flow model but a key concern in a transport model.
- Multiple transport mechanisms on different spatial and temporal scales require detailed consideration.

These general statements are illustrated in the case described in subsection 7.2.6 in which a two-dimensional model adequately described site flow conditions, but a three-dimensional model was worthwhile for the transport calculations. Neglecting a dimension in a model can result in the prediction of a distorted plume. In general, by trying to simulate a three-dimensional system by a two-dimensional model (or a two-dimensional system by a one-dimensional model), the dispersion parameters will be artificially distorted. For example, Domenico and Robbins (1984) show that by equating the solutions for maximum plume concentrations in one-dimensional and two-dimensional analytical models, a scaled-up dispersion coefficient is calculated for the one-dimensional model. The scaled-up dispersion coefficient is proportional to the distance traveled from the source. The results for a specific case of two-dimensional contaminant transport, listed in Table 7.1, illustrate this point.

The problem for a model user then is that when using an $n - 1$ dimensional model to represent an n dimensional system, a scaled-up dispersion coefficient may be calculated, but this coefficient will vary in space and time. The large dispersion coefficients commonly calibrated in transport models may partly reflect this scaling-up effect. Also this implies that an $n - 1$ dimensional

TABLE 7.1 One-Dimensional Model Parameters for Two-Dimensional Problem

Distance Traveled	Scaled-up Dispersion Parameters	
(m)	D_x (cm^2/sec)	α_x (m)
1	2×10^{-4}	0.02
10	2×10^{-3}	0.2
100	2×10^{-2}	2
1000	2×10^{-1}	20
10,000	2	200

Source: Domenico and Robbins (1984).

Note: D_x = dispersion coefficient [L^2/T], α_x = dispersivity [L].

model calibrated to a contaminant plume of a certain size may not accurately simulate smaller or larger plumes, such as is required to show plume remediation. Since dispersion is more limited in the vertical than in the horizontal plane, it is better to approximate a transport problem by a two-dimensional horizontal-plane model than by a two-dimensional vertical-plane model, solely from the standpoint of dispersion simulation.

Errors originating in simplification of the natural system are often mistakenly balanced by modifying dispersion coefficients or other model parameters. For example, vertical contaminant migration due to density effects may be poorly represented by increasing vertical dispersivity. Clean surface recharge may also be mistakenly invoked to generate the same apparent predicted concentration distribution as for a density-driven plume. The distorted model will produce different conclusions than a realistic model, even though the distorted model may provide "realistic" predictions for one set of conditions and point in time. The errors introduced by preparing a simplified model should be assessed in comparison with errors or uncertainties built into the model.

8 Applying Numerical Models

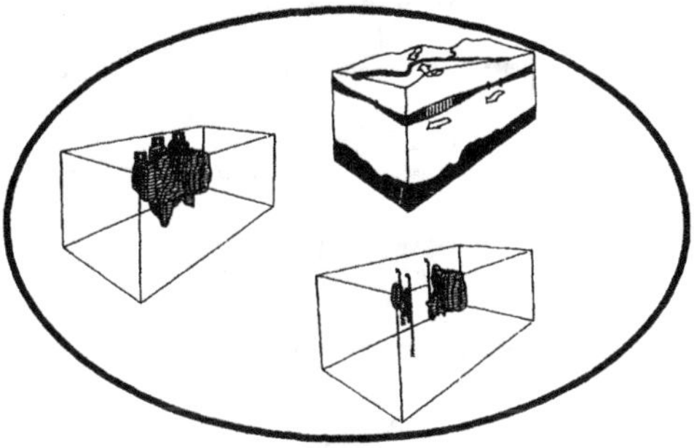

Numerical modeling is the most commonly used form of groundwater modeling analysis. The strength of numerical models is their ability to tie together data and physical principles into a coherent and useful picture of a site. Their weakness is their ability to hide gaps, errors, or misunderstandings inside a cloak of technical respectability. A model contains many levels of information at different scales, and different users will extract different results to fit their needs. Models are one of the best accompaniments to a characterization program in evaluating site conditions.

This chapter explains the steps usually required to prepare a realistic site-specific model (Figure 8.1). The steps of developing a numerical model, together with the related ramifications to the model predictions, are presented in the order they are normally undertaken. The emphasis of the following sections is on the practical everyday uses of models rather than on experimental, research-oriented, or untested approaches to modeling analyses. Many points that have been examined in previous chapters are reflected in this chapter to form one comprehensive review of model application. Additional guidance is provided in ASTM Standard D 5447-93 (ASTM 1993).

8.1 COMPILING DATA

The first step of a model study consists of collecting and evaluating relevant data on the flow system under investigation. Input data for the model are used for:

- Problem definition (material properties and geometry of hydraulic units)

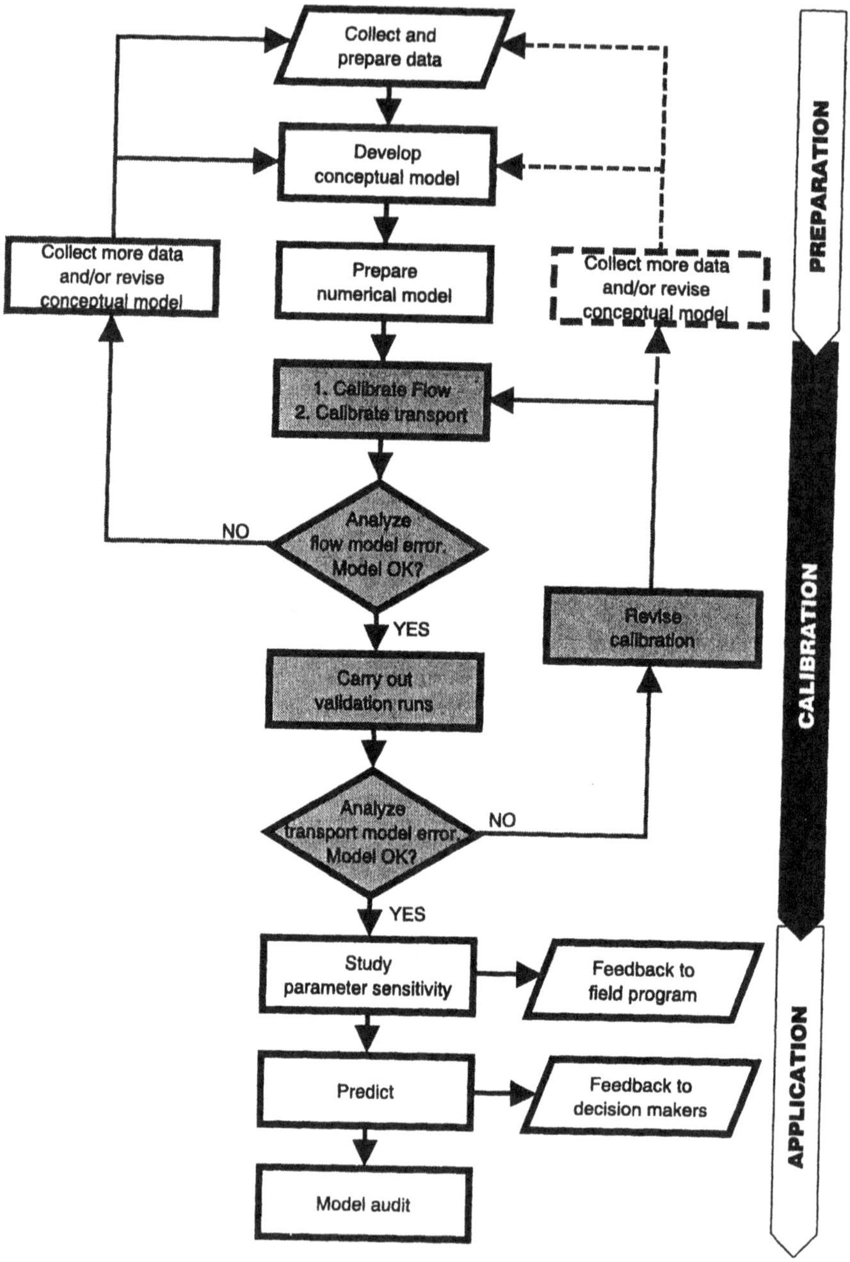

Figure 8.1 The groundwater modeling process.

- Numerical requirements (initial conditions, boundary conditions, and transient conditions)
- Modeling requirements (calibration, validation, and definition of alternate scenarios).

Input data for a variably saturated, variable-density, flow and transport model are summarized in Table 8.1. Data are separated into data defining the physical framework of the flow system and data defining groundwater recharge and discharge. An additional list of data needs and likely sources is given in subsection 8.4.5. Typical values for many of these input parameters are summarized in Appendixes B, C, and D.

Related groups of data should be summarized in the form of maps or plots, at a suitable scale and level of detail, overlaid on a suitable base map. A set of maps (see Figure 8.2) may comprise:

- Topographical map at a suitable scale depending on the groundwater system to be modeled. A simplified topographical map or map of significant site features can serve as the base map on which other data are overlaid.
- Hydrological map and/or vegetation map showing all surface water bodies such as lakes, streams, ponds, or channels. Springs, wetlands, and swamps can be included. The map should provide a clear picture of the surface drainage system. Morphological features can also be plotted on this map.
- Map indicating all wells and boreholes made for geological surveys, water supply, and site characterization. A suitable well or borehole identification will relate these elements to a separate database comprising related data such as well coordinates, screened interval, and other well design data.
- Geological map(s) together with cross-sectional plots or fence diagrams (Figure 8.3). Geological maps should show information such as faults, thickness and elevation of strata, depth of stream channels and changes in rock types. This information then can be related to groundwater occurrence and movement, possibly allowing identification of the hydrostratigraphy.
- Maps showing measured groundwater head distributions at different times. The interpreted potentiometric surface and monitor well locations are included on these maps.
- Land use map showing agricultural areas, recreational areas, industrial areas, irrigation canals and so forth. This map can be used in conjunction with the hydrological map to delineate recharge/discharge areas.
- Concentration distribution maps in horizontal and vertical planes, for several times, together with source locations. Historic land use maps or aerial photographs can also help with source identification.

TABLE 8.1 Typical Model Input Data

	Physical Framework
Aquifer type	Topography
	Geology
	Stratigraphy
	Aquifer geometry (Base, thickness, lateral extent)
	Lithological variation within the aquifers
Aquifer characteristics	Hydraulic conductivity/anisotropy
	Porosity
	Conductivity/Pore pressure relationship
	Saturation/Pore pressure relationship
	Specific yield
	Specific storage
	Dispersivity (for transport modeling)
	Soil bulk density (for transport modeling)
Aquifer boundaries	Location
	Prescribed head
	Prescribed flux
	Semipermeable boundary (Leakage factor, head in adjacent system)
	Hydrological Boundary Conditions
	INITIAL WATER LEVELS/GRADIENTS
Internal inflow/outflow	Recharge/discharge (Area, rate, duration)
	Extraction or injection wells (Location, rate, duration)
	Exchange with surface water (Surface water elevation, base of surface water, leakage factor)
	Groundwater Quality Boundary Conditions
Land use	Industrial, agricultural
Aquifer environment	Groundwater chemistry
	Background values of solutes/contaminants
	pH
Source	Location
	Area
	Source concentration
	Volume
	Duration
Contaminant	Solubility
	Density
	Viscosity
	Henry's law constant
	Octanol-water partition coefficient
	Diffusion coefficient
	Adsorption isotherm
	Decay rate
	Biodegradability

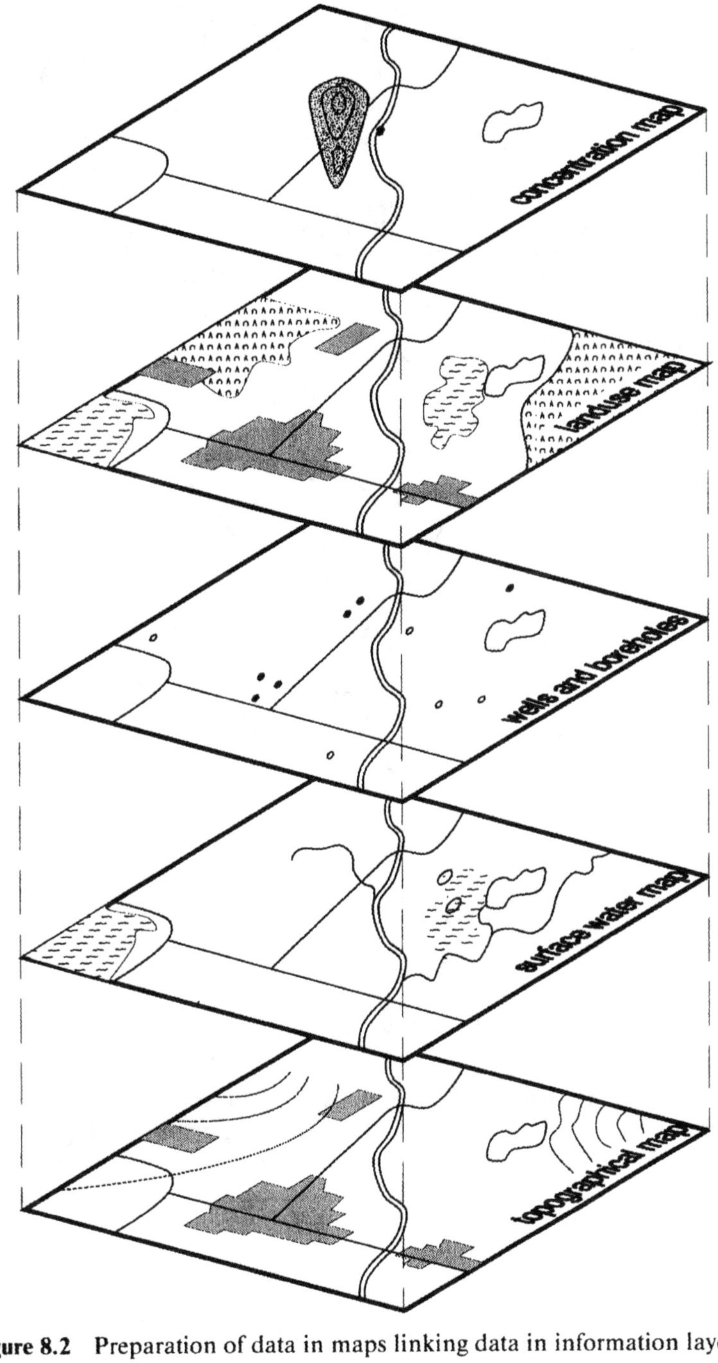

Figure 8.2 Preparation of data in maps linking data in information layers.

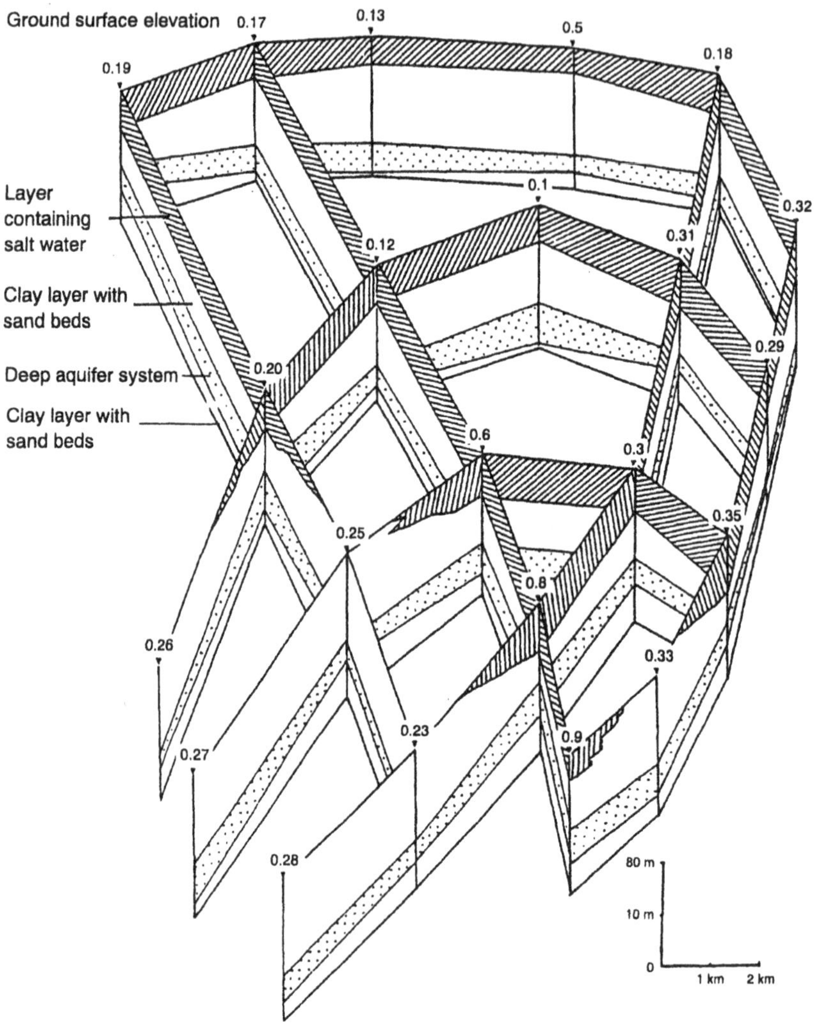

Figure 8.3 Fence diagram to illustrate stratigraphy.

- Additional maps might include structure-contour maps of aquifer base elevations or aquifer and aquitard isopachs, elevations of top and base of hydrogeologic units, thickness of units, hydraulic conductivity or transmissivity distributions, evaporation and rainfall contour maps, vertical hydraulic gradients and flow-direction maps.

Figure 8.2 shows a set of such maps overlaid as layers.

Model design, data collection and review often occur at three scales (Figure 8.4). At a regional scale, information is gathered to identify and define suitable model boundaries. Data at a local scale are reviewed to estimate the area within which accurate predictions are required. The actual size of the model

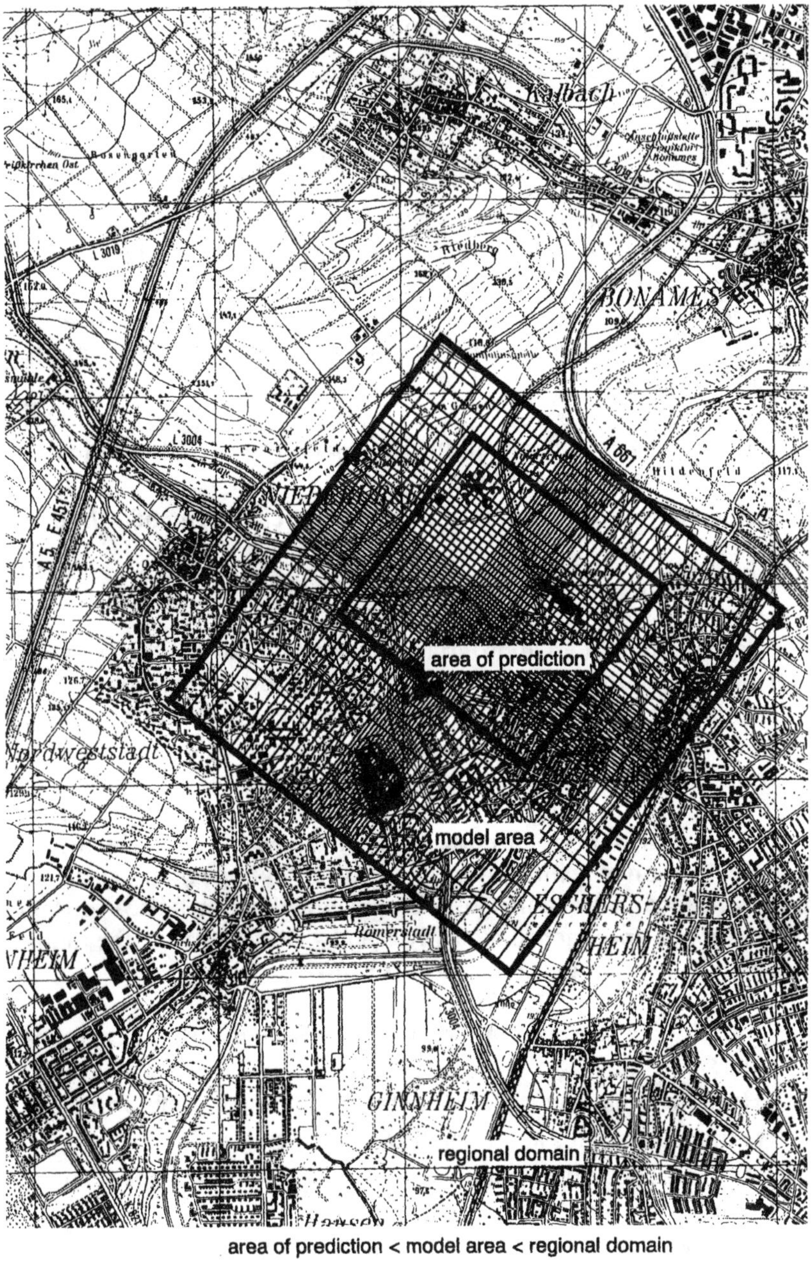

Figure 8.4 Collecting input data at different scale.

domain then ranges somewhere between the area of prediction and the regional boundary. Once the model domain is designed, future data evaluation concentrates on this area.

8.2 DEVELOPING A CONCEPTUAL MODEL

The second step in a modeling study is developing a conceptual model. A conceptual model is an idealization of the real world that summarizes the current understanding of site conditions and how the groundwater flow system works. It embodies all of the important features of the flow system, while incorporating simplifying assumptions. The three purposes of developing a conceptual model are:

- Develop a better understanding of site conditions, and be able to communicate this understanding.
- Define the groundwater problem for development of a numerical model.
- Aid in selecting a suitable numerical model.

The development of an appropriate conceptual model is the key to a successful modeling study. The conceptual model will evolve with the addition of new data; the first conceptual model may not be the last. The term "conceptual model" is sometimes used to refer to the entire conceptual model, model selection, and model setup process. These other aspects of developing a site-specific model are discussed in the following sections of this chapter.

The conceptual model is the point where all relevant ideas are incorporated into an idealization of reality. Concepts that initially appear farfetched can be saved for future consideration if the first conceptualization fails. A conceptual model cannot be approached logically, step by step. However, the following questions may be used to get the process started:

- Is the physical system understood well enough to be cast into to a model?

  Have field data been interpreted and cast into a useful format?

  Is there consensus about the flow system and geologic setting?
- What are the technical (and also perhaps political, economic, or legal) issues to be addressed?

  Can they be answered by modeling analyses?

  Do the technical issues require a high degree of accuracy in predictions?
- What are the physical and chemical processes that should be incorporated?

  Confined and/or unconfined conditions?

  Saturated and/or unsaturated flow?
  Miscible or immiscible transport?
  Geochemical reactions or decay?
  Density-dependent flow and transport?

- How many dimensions are needed in the model, and which dimensions are the most appropriate?
  Heterogeneity and/or anisotropy or multiple aquifers?
  Point or areal sources or sinks?
  Density-dependent or bouyant flow and transport?
  Remediation to be simulated?
- How will the model be calibrated?
- Is a modeling analysis acceptable to all parties involved?
  Modelers and technical team?
  Clients or reviewers?
  Regulatory personnel?
- Will modeling be cost-effective at this site?
  Could analytical or mass-balance models suffice?
  Will additional field data be required anyway and make a modeling analysis superfluous?

Once you consider questions like these you can develop the conceptual model. The three steps in developing a conceptual model are (1) explore and summarize the key mechanisms governing groundwater flow and chemical-species transport at the site, (2) develop the assumptions and simplifications required to make the real situation tractable to analysis, and (3) establish the framework of the model (number of dimensions, type of model).

The conceptual model itself may take various forms. Different approaches are suited to different flow systems and model objectives. Figure 8.5 illustrates different types of conceptual models:

- Flow diagram indicating the interactions between different elements of the problem (Figure 8.5*a*).
- Mass-balance summary (Figure 8.5*b*).
- Geological cross sections labeled with key processes (Figure 8.5*c*).
- Three-dimensional diagram summarizing site conditions (Figure 8.5*d*).

This stage in a modeling analysis is difficult because it involves collating and interpreting many disparate pieces of information, together with various opinions and experience from other sites, into a cohesive picture capable of being analyzed numerically. Inexperienced modelers may need to seek experience with comparable sites from other modelers, hydrogeologists, hy-

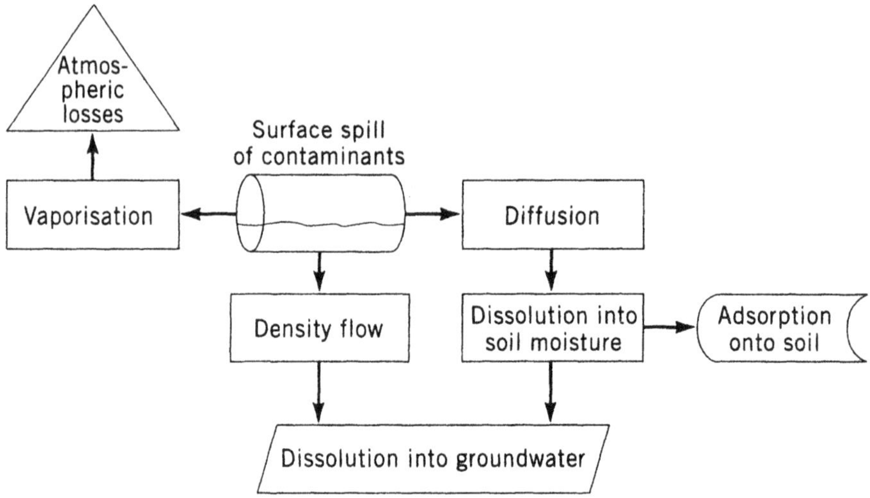

Mass balance for flow

Water in storage	=	7.4×10^8	m^3
Change of water in storage	=	-3.4×10^2	m^3/d
Inflow from rivers	=	1.5×10^8	m^3/d
Inflow at wells	=	2.2×10^2	m^3/d
Inflow by infiltration	=	2.2×10^7	m^3/d
Inflow by boundaries	=	2.8×10^7	m^3/d
Outflow at rivers	=	0.0	m^3/d
Outflow at wells	=	0.0	m^3/d
Outflow by evaporation	=	0.0	m^3/d
Outflow at boundaries	=	2.0×10^8	m^3/d
Error	=	-4.4×10^5	m^3/d

Figure 8.5 Conceptual model examples.

drologists, or project managers. Factors that modelers often overlook during this process of model development include:

- Effect on transport calculations of selecting too few dimensions (see Chapter 7).
- Neglect of significant transport in bedrock, or other low permeability units, despite negligible flow (see example case in Appendix A.5).
- Underestimation of pollutant source quantities based on observed dissolved concentrations (see subsection 8.4.5).
- Neglect of density effects (see Section 3.3).

PROCESSES IN CROSS SECTION

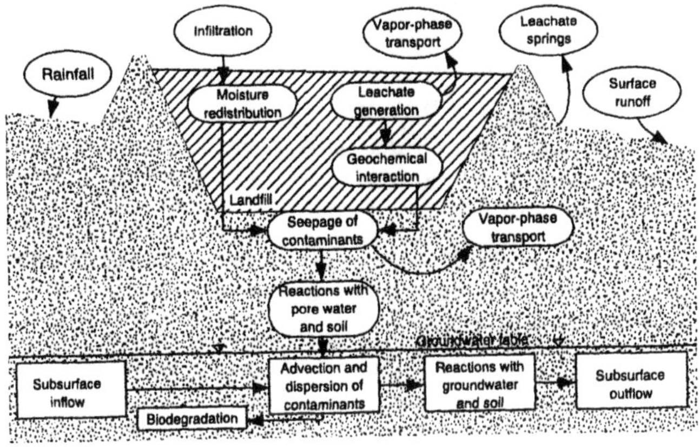

THREE-DIMENSIONAL SCHEMATIC DIAGRAM

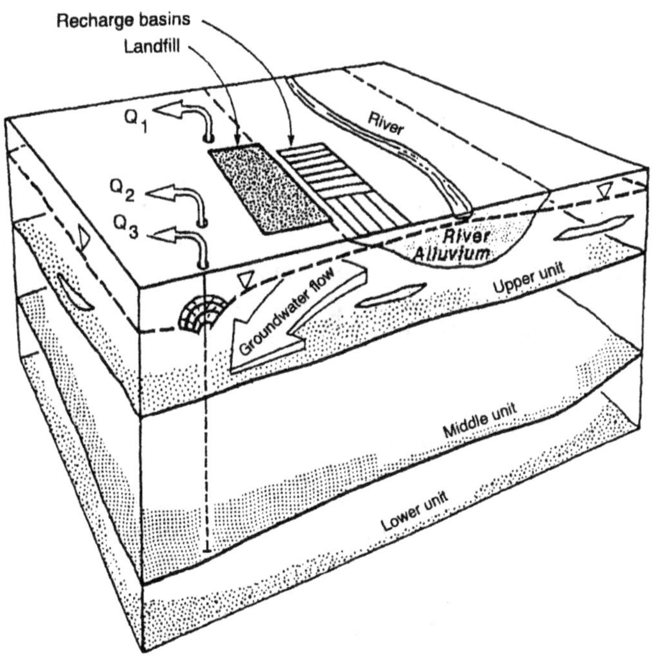

Figure 8.5 (*Continued*)

The observed behavior of a contaminant plume may demonstrate the key factors to consider when developing a conceptual model. For example:

- A plume with multiple high-concentration zones may indicate a pulsing source, multiple sources, aquifer heterogeneity, and so on.
- A plume with high concentrations at depth may indicate density effects, clean surface recharge, or fracture flow.
- A plume that has not migrated substantially may contain chemical species strongly retarded due to sorption, chemical reactions, decay, degradation, or volatilization.
- A plume in which bacteria density correlates with chemical concentration may indicate a biodegradable contaminant.
- A plume that migrates at the base of an aquifer or along the slope of the bedrock, rather than in the direction of groundwater flow may indicate density effects.

Review of the modeling approach that you developed, with all interested parties, on completion of the conceptual model helps to ensure overall model acceptance.

8.3 SELECTING A MODEL CODE

The third step in the modeling process is to select a code. The process of selecting an appropriate numerical model is discussed in various papers and reports, including van der Heijde and Park (1986) and Bond and Hwang (1988). The selection process cannot be entirely rationalized, and often involves nonscientific issues, but general considerations are listed in the following subsections. The selection approach discussed in these sections has been developed with and for the U.S. Environmental Protection Agency (EPA), and has been successful in gaining model acceptance in many projects in the United States.

8.3.1 Defining Model Objectives

Objectives criteria are used to distinguish models designed for general studies from those designed for detailed analysis. In groundwater transport the objectives of general studies are to:

- Develop understanding of the main factors governing contaminant transport at a particular site.
- Rank sites by means of quick, simplified analyses.

Such studies may be used in the initial stages of a modeling study, or when site-specific data are sparse, or when general, worst-case or relative conclusions only are required. Models selected on the basis of these objectives include analytical models, analog models, empirical models, mass-balance models, and simple numerical models.

The objectives of detailed modeling studies are to:

- Identify data gaps and guide the field program (including placement of monitoring wells).
- Predict the present and future concentrations in groundwater.
- Compare the effectiveness of remedial alternatives.
- Characterize source areas and release history.
- Optimize remediation, engineering design, or monitoring network.
- Characterize uncertainty and/or perform cost/benefit analyses to support management decisions.
- Assist with negotiations or litigation by providing a focal point for discussions.

Numerical models are often selected for application in a detailed modeling study because they can simulate a more realistic and detailed picture of the site. The predicted results are correspondingly more specific and reliable. Detailed studies may be regional or local in scale. A regional study may address questions such as basinwide water management, travel time to and concentrations at distant points, and the effects of external factors (e.g., other well fields, other pollutant sources, neighboring flow systems, and water bodies) on flow and transport. A local study might address questions such as planning a monitoring network, flow and transport adjacent to pollution sources, and designing remedial alternatives. Local models generally produce high-resolution results in time and space, whereas regional models may define general trends and may incorporate local models into a wider framework.

8.3.2 Selecting a Model on Technical Criteria

Technical criteria are used to match site characteristics with a model of appropriate capabilities. The chosen model should be capable of adequately representing the governing flow and hydrogeology processes and hydrostratigraphy of the site, as far as is known. These governing processes may include:

- Flow conditions (confined/unconfined, horizontal/vertical, saturated/unsaturated).
- Stratigraphy.
- Variations in time and space of boundary conditions (e.g., recharge and discharge) and contaminant sources.
- Porous, fractured, or karstic media.

- Spatial variations of material properties (both natural and engineered properties).
- Single or multiphase flow.
- Flow or transport affected by fluid density or temperature.

8.3.3 Implementing the Model

If several selected models satisfy the objectives and technical selection criteria, then one can use implementation criteria to narrow the choices. Implementation criteria involve the following considerations:

- Has the model been peer-reviewed and is it available for further review, or is it a research model or closely held proprietary model unavailable to public agencies or third parties?
- Is the model well-documented and are users supported?
- Has the model been verified against results from analytical solutions, laboratory results, other verified models, and field data?
- Has the model been applied successfully at similar sites?
- Is the model easy to implement on available computer systems and relatively easy to use by the modeler(s)?
- Does the model have a good track record of acceptability to reviewing/regulatory agencies, and do peer reviewers and experienced users regard the model favorably?

The objectives, technical and implementation criteria can be summarized as four key questions:

- Can the model adequately simulate site conditions?
- Can it satisfy the objectives of the study?
- Is the model verified and reasonably well field tested?
- Is the model well-documented, peer reviewed, and available?

Sometimes application of all these criteria will disqualify all known models. Usually the choice of the "best" model consists in a compromise among the criteria listed above. Today a considerable number of practical models are available. Hence model selection should be based on the selection criteria rather than on subjective arguments such as familiarity with, or availability of, a specific model.

8.4 SETTING UP A MODEL

Model setup is the fourth step in the modeling process. Model setup and calibration often constitute 50% to 70% of the total modeling effort. Model setup entails selecting the model domain, discretizing data in space and time, defining boundary and initial conditions, and assembling and preparing model input data. Choice of the domain and discretization affect the physical and numerical resolution and level of effort (cost) of the modeling study.

8.4.1 Selecting a Model Domain

Selection of the optimum model domain involves balancing the following factors:

- The domain should cover the entire area of interest, including areas that may be affected by future chemical-species transport, and should encompass the effects of internal disturbances (e.g., aquifer pumping or injection, or seepage from impoundments). Future transport can be roughly estimated by calculating transport velocities and retardation factors or by analytical solution. If the entire chemical-species plume is not included in the model domain, overall mass balances will not be available.
- The boundaries of the domain should take advantage of natural groundwater boundaries such as rivers, lakes, drains, groundwater divides, edge of aquifer, boundary between adjacent pumping centers, coastline, groundwater recharge/discharge area or boundary location distant (in hydrologic terms) from the area of interest. Note that rivers, lakes, and drains are not always groundwater boundaries and that groundwater divides may move over time or with depth.
- The model domain should be oriented parallel to the primary groundwater flow direction (at least in the area of primary interest, e.g., highest concentration plume) to reduce numerical dispersion (see also Section 6.3).
- Available data should adequately define conditions throughout the domain selected.
- Domain size should be minimized to reduce computational effort.

Frequently the ideal domain size for a groundwater flow problem is much larger than the domain required for a groundwater transport model. In this case a sequence of two models (see also subsection 7.2.6), a problem-oriented finite-element grid (see also Section 6.3), or a compromise between the two domain sizes may be selected.

8.4.2 Discretizing a Model in Space and Time

This section discusses the factors that commonly affect the selection of model discretization. Table 8.2 lists these factors and how they generally affect selection of the model grid.

TABLE 8.2 Common Factors Affecting Model Discretization

Factor	Aspect of Model Discretization Affected
Modeling objectives	Domain size and areas of finer resolution
Area and duration of interest	Domain size
Location of sources and sinks	Finer discretization zone
Heterogeneity and anistropy	Orientation and refinement of grid
Particle velocity and retardation	Cell size and domain size
Natural boundaries	Limit on cell size to simulate boundary
Numerical stability	Limit on ratio of cell sizes
Numerical accuracy	Limit on cell sizes
Computational effort	Limit on total number of cells
Resolution of flow field	Finer discretization where high gradient
Resolution of concentration distribution	Finer discretization where high gradient

We can select the time and space discretization, that is, the time steps and cell (element) sizes, for numerical models by optimizing the following aims (listed in order of importance):

- Enhance model solution stability and convergence.
- Increase model resolution.
- Minimize numerical dispersion.
- Minimize computational requirements for memory, storage, and run-time.

How to meet these aims by the choice of appropriate model discretization is discussed in this section.

Model solution stability and convergence can be improved by the selection of time steps and calculation-mesh cell sizes that are consistent with one another (see Chapters 5 and 6 for related stability considerations). For example, it can be shown that for a two-dimensional, unsteady flow problem, stability will be ensured by the following condition (Bear and Verruijt 1987):

$$0 < dt < \frac{S}{2T} \frac{dx^2\, dy^2}{(dx^2 + dy^2)} \tag{8.1}$$

where

S = storativity in [1/L]
T = transmissivity (of aquifer or cell, depending on dimensions of model) in [L^2/T]

dx = cell width in the x-direction in [L]
dy = cell width in the y-direction in [L]

Other criteria are presented below.

Accurate predictions require selecting a cell size sufficiently fine to represent local variations in hydraulic head or concentrations, and defining time steps small enough to represent temporal variation of conditions. While variable cell sizes allow for greater flexibility, highly variable cell sizes can introduce a loss in accuracy (Bear and Verruijt 1987) and stability.

Numerical dispersion, or unnatural spreading of a chemical-species plume, occurs due to the neglect of higher-order terms in the Taylor series expansion of the finite-difference or finite-element formulation of the governing equations. It can also occur due to inappropriate space or time discretization. The potential for numerical dispersion varies with alternate schemes for approximating time and space derivatives. We can minimize unwanted dispersion by selecting appropriate calculation-mesh cell size, mesh orientation, and size of time steps. The following subsections discuss how to make appropriate selections. We test for the existence of numerical dispersion by applying a finer mesh and time step and comparing the coarse and fine-scale predicted results. Computational efforts are minimized by reducing, as far as possible, the total number of calculation cells and time steps in a given calculation. Minimizing computational effort is counter to accuracy requirements, and a compromise must be made, with accuracy in the area of interest being the primary goal.

In discretizing a model, the orientation of the model, space discretization and time discretization must be considered. These selections are discussed in the following subsections.

Orienting the Model Grid

The orientation of the model grid, which most often naturally accommodates all of the factors influencing choice of orientation, is a mesh-oriented parallel with the large-scale geologic features. In nearly all cases the optimum model orientation does not coincide with primary compass directions, roads, or plant boundaries. The following factors affect mesh orientation (in order of importance):

- Hydrologic, hydrogeologic, and geologic features at the site. Representation of key features such as rivers, streams, impoundments, faults, and other natural boundaries can be simplified through appropriate orientation of the mesh. For example, a fault zone that affects groundwater flow is best represented by cells oriented parallel and perpendicular to the fault.
- Predominant groundwater flow direction. Numerical dispersion due to the groundwater velocity being split into components parallel to the calculation-mesh axes is minimized if the mesh is oriented along the

direction of predominant groundwater flow. If the flow direction varies within the model domain, alignment of the calculation mesh with the flow direction in the primary area of interest is optimum.

- Anisotropy of hydraulic properties. Since the hydraulic conductivity is expressed in the model as components aligned with the axes of the calculation mesh, inaccuracies are reduced by choosing a mesh oriented coincident with the conductivity tensor.

Choosing the Model-Grid Cell Size

Selecting the model grid is comparable to selecting a net for fishing; the openings in the net must match the size of the "fish" (heterogeneities and predictive details) to be captured. This is illustrated in Figure 8.6, which shows an example borrowed from atmospheric modeling. Figure 8.6 shows a sequence of three different discretizations of world topography. Using a grid of 500 by 500 km, mountain ranges are barely identified. However, a grid of 100 by 100 km allows land elevations to be distinguished in far greater detail.

Consider the following factors when choosing a model grid:

- Degree of heterogeneity in hydraulic or transport parameters, and in boundary conditions.
- Model domain size.
- Predicted resolution required to meet modeling objectives.
- Restrictions imposed by computational resources.

Similar constraints apply to vertical discretization, with the added consideration of stratification due to density effects, recharge, and shallow or deep sources or sinks of water or contaminant. In general, the accuracy of the predicted results improves with finer calculation meshes, but computational time and space requirements increase correspondingly.

For transport problems, as demonstrated in Section 6.2, the cell size that minimizes numerical dispersion can be calculated using the cell Peclet number (Pe) which is the ratio of the advective to diffusive terms in the transport equation. The cell Peclet number is defined as the dimensionless ratio (Equation 6.11)

$$\mathrm{Pe} = \frac{v_x \, dx}{D_x} \tag{8.2}$$

where

v_x = particle velocity in the x-direction in [L/T]
dx = cell size in the x direction in [L]
D_x = dispersion coefficient in [L^2/T].

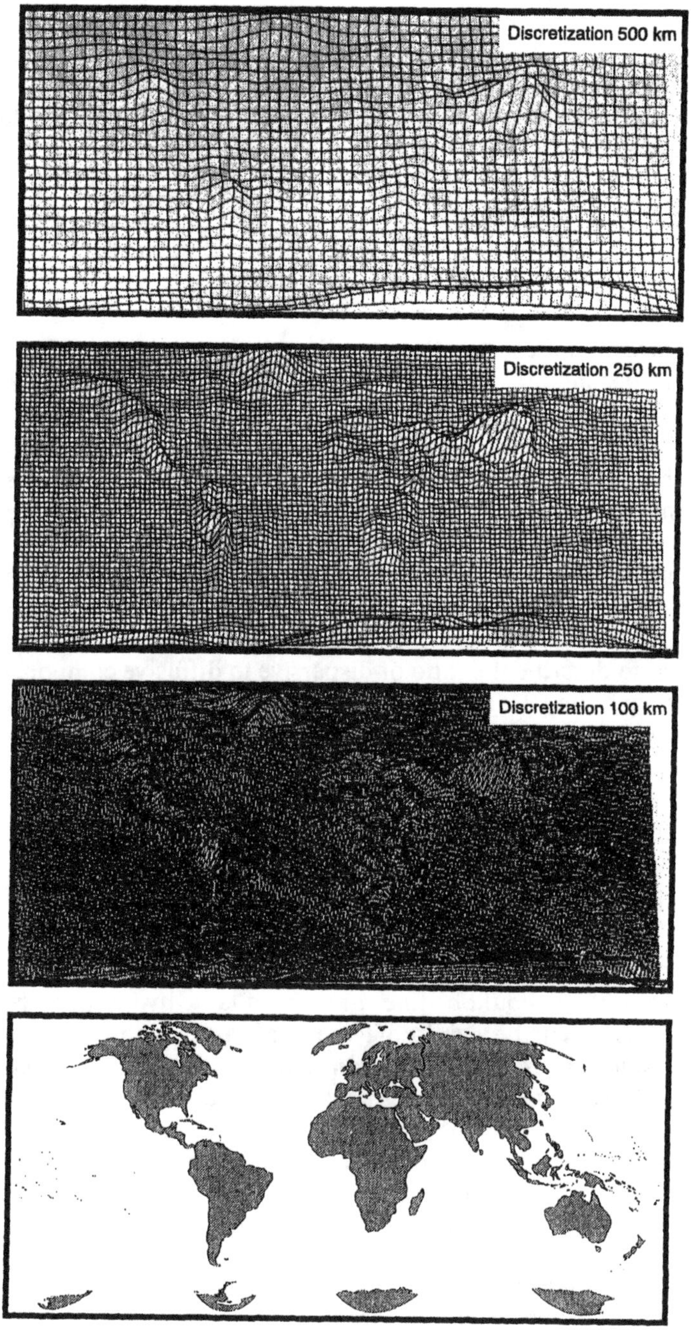

Figure 8.6 Discretization of the world topography using decreasing grid spacing (after FAZ, June 1, 1994).

Since the dispersion coefficient can be defined empirically as

$$D_x = v_x \alpha_x \tag{8.3}$$

where α_x is longitudinal dispersivity in the x-direction [L], the Peclet number can be rewritten and the Peclet criterion can be defined as

$$\text{Pe} = \frac{dx}{\alpha_x} \leq 2 \tag{8.4}$$

To ensure numerical stability and minimize numerical dispersion, the cell Peclet number should ideally be no greater than two (Pinder and Gray 1977); that is, the cell size should be no larger than twice the dispersivity. Since the dispersivity is a measure of the characteristic length of heterogeneities of the system, this criterion also makes physical sense. In practice, the Peclet number constraint is often relaxed outside the area of interest, where lower predictive accuracy is acceptable.

For unsaturated flow the Peclet number criterion has been shown to be nearer 0.5 (El-Kadi and Ling 1993), implying the need for a finer mesh in unsaturated analyses. Confusingly, the term "Peclet number" is also sometimes used to describe the ratio of dispersive to diffusive components of the transport equation (Bear 1979).

The use of geostatistical methods, such as kriging, to estimate data between data points is discussed in subsection 8.4.5. If model input data are being kriged to provide interpolations, then the associated semivariogram may be used to select an appropriate cell size. The cell size can be related to the range of the semivariogram:

$$dx < 0.3 \text{ range of semivariogram} \tag{8.5}$$

Mesh resolution should be finer in areas of large hydraulic or concentration gradients, and in the vicinity of features of particular interest (pumping or injection wells, rivers, sources of contamination, etc.). The mesh may be relatively coarse in areas of small gradients, in areas of less interest, or in areas where data are sparse.

The appropriate ratio of the lengths of the cell sides (cell aspect ratio) is calculated by comparing the travel time across the cell in each direction. Ideally the ratio of travel times should be unity, though ratios of up to 10:1 may be used without introducing significant error. In general, the greater the variability in cell size, the greater is the computational effort required in generating a convergent solution. Gradual variation in cell sizes, for example, cell size increases between adjacent cells no greater than a factor of 1.5, will facilitate model convergence, increase stability, and reduce inaccuracy. A sample calculation mesh and the criteria used in generating it are shown in Figure 8.7 and Table 8.3.

Space discretization can be more flexible in FE models than FD models because FE models are not constrained to a rectilinear grid and can conform more closely with the geometries and flow directions in the model. A carefully designed FE model can satisfy most space discretization requirements using fewer elements than an FD model. In FE models each node and element must be numbered. The FE numbering scheme has no effect on model accuracy, but it does strongly affect the storage requirements and execution time. The bandwidth of the coefficient matrix is linked to the node numbering scheme, and model execution time is proportional to the square of the bandwidth:

$$\text{Bandwidth} = 2 \times (\text{largest difference between node numbers for any element} + 1) \tag{8.6}$$

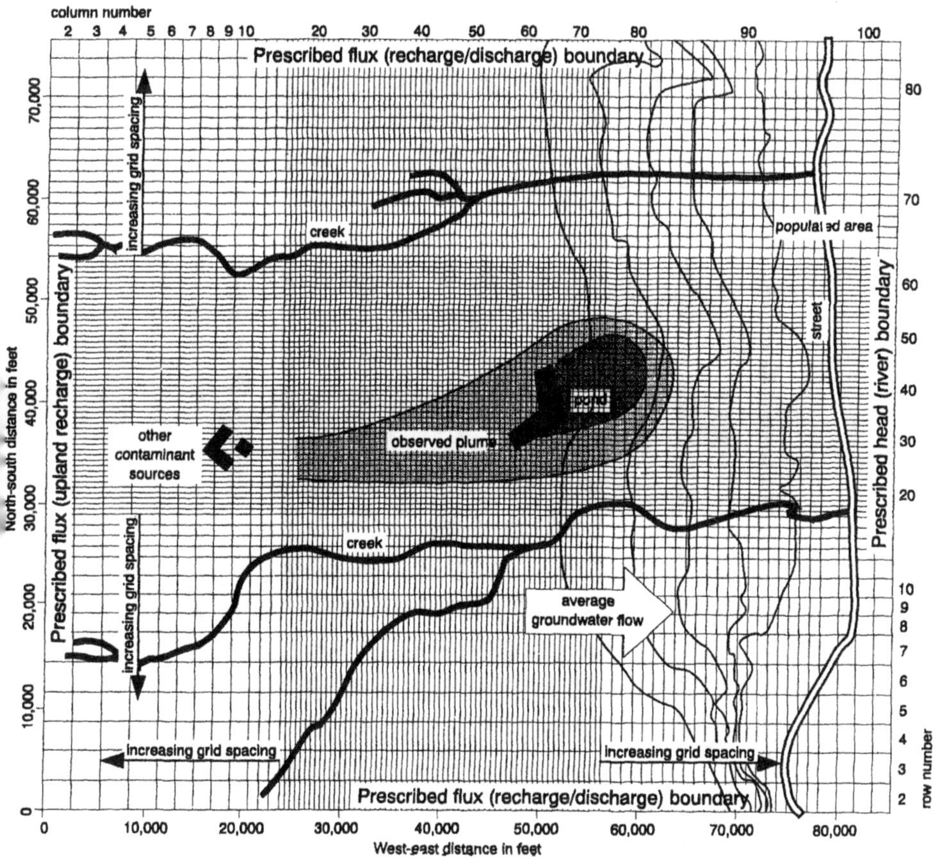

Figure 8.7 Example calculation grid–grid selection criteria.

TABLE 8.3 Example Calculation Grid–Grid Selection Criteria for Figure 8.7

Category of Criterion	Criterion	Purpose of Criterion	Model Grid
Orientation of calculation grid	Along primary flow direction	Minimize numerical dispersions due to grid orientation	Yes
Location of boundaries	Using suitable boundary conditions Far from area of prediction Encompasses plume	Prevent boundary effects from influencing solution	Yes
Cell aspect ratios	$1{:}10 \leq \Delta x : \Delta y \leq 10{:}1$	Minimize convergence problems	$0.125 \leq \Delta x : \Delta y \leq 8{:}0$
Cell Peclet number	$Pe \leq 10$	Minimize numerical dispersion due to space discretization	$Pe < 5$ in area of plume (Pe up to 37 at boundaries)
Cell Courant number	$\Delta t \leq$ travel time across cell	Minimize numerical dispersion due to time discretization Maximize numerical stability	Not relevant in steady-state calculation
Concentration gradients and hydraulic gradients	Fine cells near ponds and sources	Resolve relatively steep gradients	Yes
Smoothness of discretization	$0.5 < (\Delta x_n : \Delta x_{n+1}) < 2.0$	Minimize convergence problems	Yes
Total number of calculation cells	Minimize number of calculation cells	Minimize calculation time	8282 cells

The bandwidth can be minimized by numbering nodes sequentially along the shortest grid distance, as shown in Figure 8.8. In this example, numbering nodes sequentially along the horizontal yields a bandwidth of 9, whereas numbering vertically yields a bandwidth of 5. Numbering is more complex in real field studies. An example of grid numbering for a field application is also presented in Figure 8.8. Element numbering may also be done so as to group consecutive elements in the data input file. For example, in layered models, element numbering may follow material layers so that consecutive elements have the same hydraulic properties.

Selecting Time Step Size

Two kinds of time interval are used in models: stress periods (during which boundary conditions are constant and between which boundary conditions vary) and time steps (during which model calculations are made). This section discusses time step selection. Time steps are required for transient calculations. Factors affecting choice of time step include stability considerations, numerical dispersion in transport calculations, time variation of boundary conditions and time-related modeling objectives. In general, the smaller the time step, the more accurate are the predicted results. Too small a time step results in excessive computation time, while too large a time step results in an excessive number of iterations required to reach a mass-balanced solution, and possibly numerical dispersion or instability.

Criteria for selecting time steps for both flow and transport calculations are presented below. A typical approach to selection of spatial and temporal discretization is to use the dispersivity to select mesh size, and to use this mesh size to calculate an appropriate time step.

The time step (dt) that minimizes numerical dispersion can be calculated using the cell Courant number (Co); the ratio of the advective to time-dependent terms in the transport equation. The cell Courant number, and Courant criterion, are defined as the dimensionless ratio (see also Equation 6.8):

$$\mathrm{Co} = \left| v \frac{dt}{dx} \right| < 1 \tag{8.7}$$

To minimize numerical dispersion and maximize numerical stability, the cell Courant number should be no greater than unity for the smallest cell (Pinder and Gray 1977). This criterion can be interpreted physically as a restriction that transport of a particle across a cell should occur in one or more time steps (Bear and Verruijt 1987). For nonlinear problems the choice of time step is critical because the stability of the solution is even more sensitive to time step size than is the case for linear solutions. Alternate forms of the Courant number have been proposed for unsaturated flow problems (El-Kadi and Ling 1993).

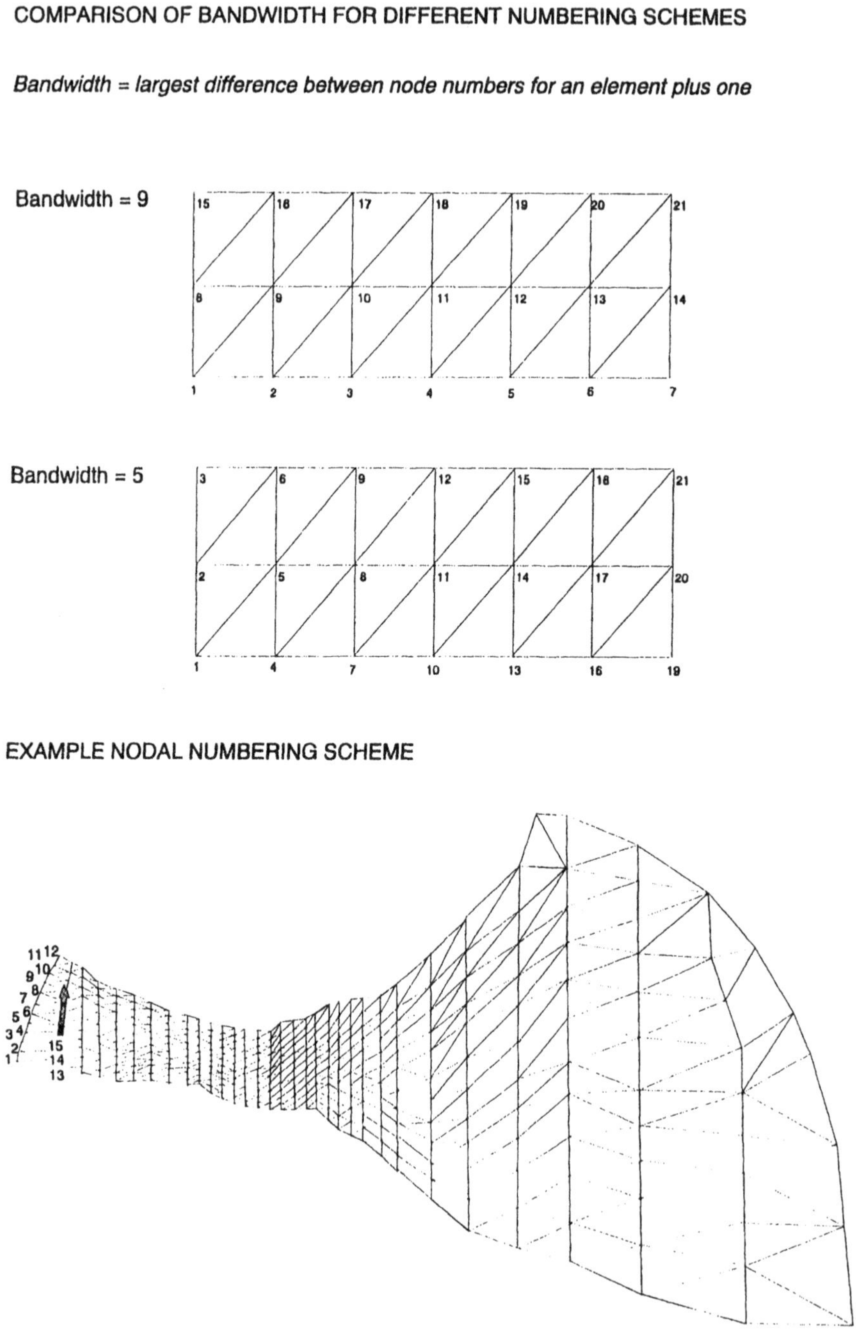

Figure 8.8 Dependency of bandwidth on nodal numbering scheme.

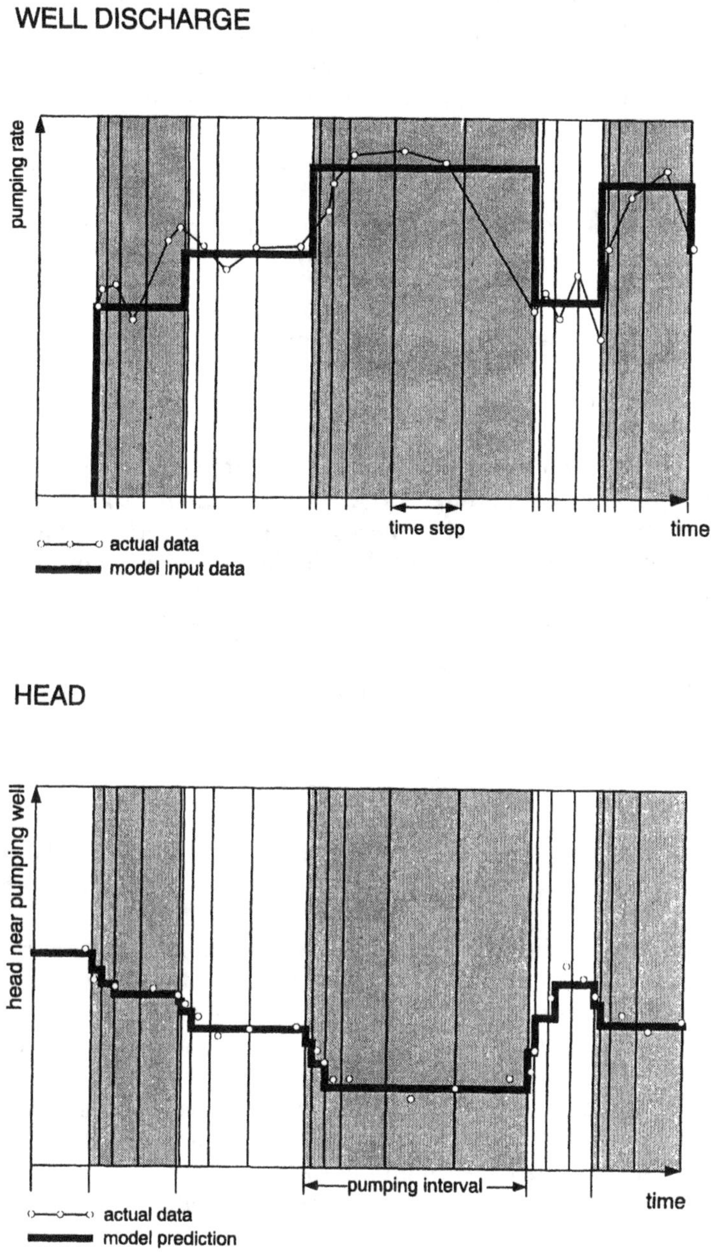

Figure 8.9 Example of time discretization.

Errors arise in the model predictions due to truncation, as discussed in subsection 8.4.1. These errors must decrease over the course of the calculations or the predictions will become unstable. For an unsteady flow problem, solved by an explicit finite-difference approximation, Equation 8.1 presents the condition for stability linking time step and cell size. In addition, time steps should be small during times of changing hydraulic or concentration boundary conditions, such as initiation of water withdrawal (Figure 8.9) or initiation of contaminant sources. Gradual variation of time steps throughout the calculations will facilitate model convergence and improve model stability.

8.4.3 Defining Model Boundary Conditions

A model boundary is the interface between the model calculation domain and the surrounding environment. Boundaries occur at the edges of the model domain and at other points where external influences are represented, such as rivers, wells, leaky impoundments, or chemical spills and so forth. Boundary conditions are expressions of the effect of the external world on the model domain, and they are required to complete the description of a flow or transport problem. The mathematical expression of the boundary condition is required for a well-posed problem. There are three major types of boundary conditions, all of which may vary with time (Table 8.4). These boundary conditions each have differing degrees of constraint upon the model solution, and implications to the ease of developing a balanced model solution.

Understanding Hydrologic Boundary Conditions

Commonly encountered hydrologic boundaries, include surface water bodies, seepage faces, water table, impermeable boundaries, recharge boundaries, and local sources or sinks. Each of these boundary conditions is discussed below.

Surface water bodies, such as rivers, lakes, canals, sea coast, impoundments, and drains, can be represented by prescribed pressure or prescribed flux boundaries. A prescribed pressure (or head) boundary is appropriate if the model calculation cells are small enough to define the volume of the surface water body accurately, the head in the surface water is known, and any intervening hydraulic resistance can be defined. This boundary condition is useful when the groundwater or surface water levels are subject to fluctuation, since the resulting flux between groundwater and surface water will be calculated rather than prescribed. A prescribed flux boundary is appropriate if the flux between surface and groundwater is known.

Seepage faces adjacent to surface water bodies or along drains, tunnels, and the like, may be represented by specification of the head or pressure in the surface water in combination with model cells of high hydraulic conductivity (relative to the aquifer) in the air adjacent to the potential seepage face. Seepage flow may be controlled by the presence of a semipermeable layer, in which case a third-kind boundary condition is applicable.

TABLE 8.4 Model Boundary Conditions

Boundary Type	Boundary Name	Common Applications	Constraints on Solution	Effects of Boundary Condition on Solution
First kind, or Dirichlet boundary	Prescribed pressure, hydraulic head or concentration	Lakes, rivers, springs, constant-head wells, seepage faces	Most constrained	Easiest to solve
Second kind, or Neumann boundary	Prescribed flux of head or concentration	Impermeable boundary, water divide, streamline, infiltration, evaporation, sinks, and sources	Moderately constrained	Moderately difficult to solve
Third kind, or Cauchy boundary	Semipermeable or head-dependent flux	Leaky rivers, drains, seepage faces	Least constrained	Most difficult to solve

The water-table location may be a prescribed pressure boundary in models restricted to vadose-zone or confined-aquifer simulations. The water table is the surface at which pore-water pressure is equal to atmospheric pressure (usually assumed to be zero pressure) or the hydraulic head is equal to the elevation head.

Impermeable boundaries, such as groundwater divides, boundaries parallel to flowlines, or impermeable fault zones, can be represented by a special case of the prescribed flux condition—a zero-flux boundary. Low-permeability zones, such as aquitards, slurry walls, and liners, can be represented by low-permeability cells, with the provision that at least two model calculation cells are used to represent such a zone. This is because of the permeability averaging that occurs in calculating flow between neighboring cells.

Recharge boundaries, such as precipitation, infiltration, evaporation, transpiration, and other areal sources, can be represented by a prescribed flux boundary. This type of boundary may be subject to other constraints, such as excessive infiltration leading to ponding and/or runoff, evaporation being a function of the depth to the water table, or transpiration related to the depth of the root zone.

Local sources and sinks, such as injection or extraction wells, and point sources can be represented by a prescribed flux boundary. However, to approximately represent a point source or sink in a two-dimensional, vertical-plane model, a flux cannot be prescribed without distorting the predictions. An alternative is to use a prescribed head boundary in which a translation from the head in the well to the equivalent head in the model cell (as discussed by Beljian 1988) is performed. In the case of an extraction or injection well open to several aquifers, there are three possible approaches:

- Represent the cased volume of the well by small, highly permeable cells and specify the source/sink cell at the pump elevation.
- Calculate the contribution from each aquifer proportional to the transmissivity of each zone and specify several prescribed flux cells.
- Use an empirical analytical solution to divide pumpage between layers.

In the case of a deep groundwater basin, perhaps up to 3000 m deep, the base boundary condition selected should reflect known conditions and the purpose of the model. For example, a layer of lower permeability, suggested by field data or lithologic or structural characteristics, could be used to define a shallower effective depth of the model (location of zero-flow boundary). In the absence of such a layer, a fictitious depth of the aquifer base would be specified. Boonstra and de Ridder (1981) suggest using, as a first approximation, an effective depth of one-fourth to one-eighth of the average distance between major streams draining the basin.

Figure 8.10 shows an example of the effects of a prescribed head versus a prescribed flux boundary. The model focused on the unsaturated zone. In this case two alternate model domains were used to illustrate that a local model

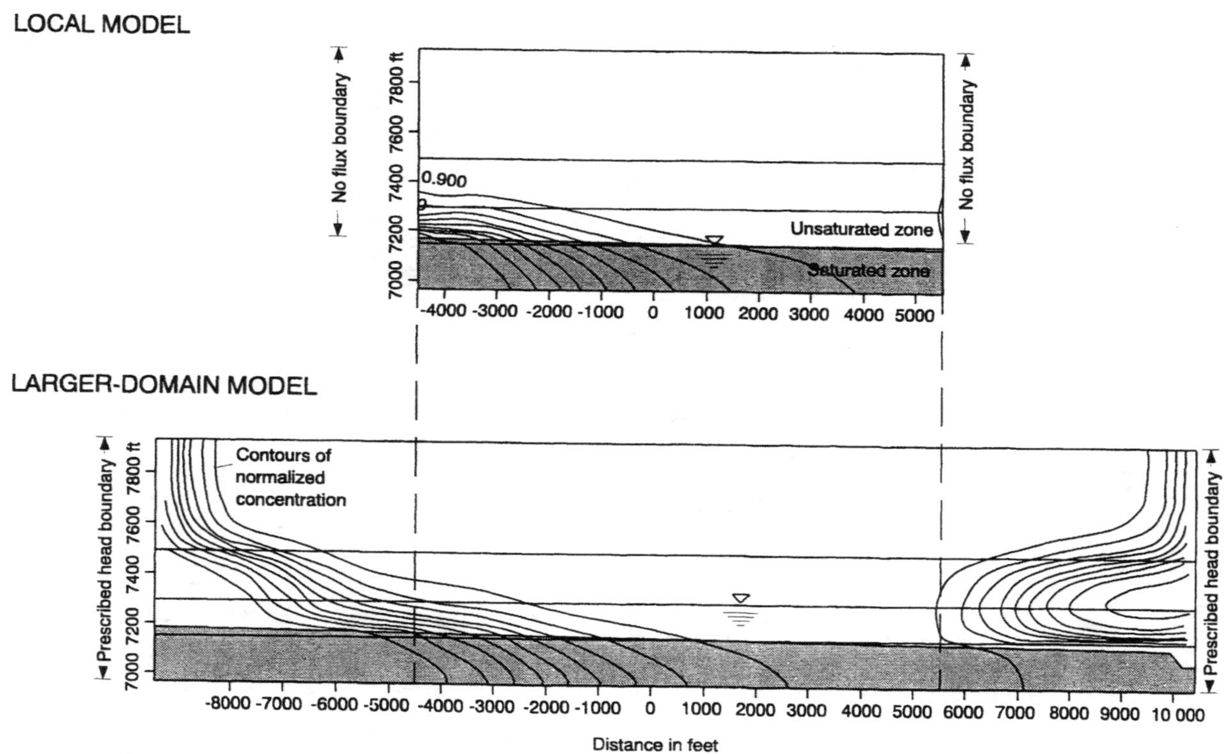

Figure 8.10 Investigating the effect of alternate boundary conditions in flow modeling.

with zero lateral flow boundary conditions in the unsaturated zone produced the same flow predictions as a larger-domain model with prescribed head boundary conditions.

Understanding Transport Boundary Conditions

Commonly encountered concentration boundaries include leaky structures, leachate generation, immiscible contaminant sources (NAPLs), groundwater recharge/discharge, evaporative discharge, injection wells, and surface water bodies. Each of these boundary conditions is discussed in Section 3.4 as well as below.

Leaky structures, such as leaking landfills, ponds, drains, or infiltration beds, can be represented by a prescribed flux boundary with associated concentrations. Similarly injection wells can be simulated by a prescribed flux boundary with a constant concentration in the injected water.

Some materials (residual NAPL, mine waste, etc.) generate contaminated leachate, acting as passive polluters of through-flowing groundwater. Such sources may be represented as prescribed concentration boundaries, with a possible upper limit on the chemical mass that can be mobilized.

Immiscible contaminant sources, such as solvents, coal tar, or paints, can be represented in several ways, two of which are (1) in a multiphase model the contaminant source can be represented explicitly; or (2) in a solute model the source can either be represented by a prescribed concentration boundary or as a prescribed flux boundary, with concentration equal to the pure-phase solubility of the chemical species (or use Raoult's law for solubility in a mixture). If the prescribed concentration boundary is chosen, there is no check on the total mass of solute entering the model domain. The source then is assumed to be infinite during the span of the model simulation. This is often a reasonable assumption because the mass available is usually much greater than the mass dissolving. If the prescribed flux boundary is chosen, then the rate of dissolution of the chemical species must be estimated. The dissolution rate is often a model calibration parameter because chemical to groundwater mass transfer coefficients, immiscible source contact area, and the effect of the groundwater velocity on dissolution rates are poorly understood (Cherry 1990; Mercer and Cohen 1990).

In most situations the effects of a source rather than the source itself may be known or measured. In this case the source may be represented as a black box (effective source) by a prescribed mass flux or concentrations at some down-gradient point. The danger of this approach is that the effects of the source may not be fully known and the source thereby misrepresented.

Groundwater recharge/discharge, such as metal-laden recharge from mining areas, geothermal upwelling, recharge from agricultural irrigation or feed lots, or recharge from domestic or industrial areas, can be represented by a prescribed flux boundary with known concentrations.

Surface water bodies can result in clean or contaminated recharge or act as sinks for polluted water. Usually the surface water body can be well represented by a prescribed head and concentration boundary (see the discussion at the beginning of this subsection). However, in the case of a perched surface water body a prescribed-flux or third-type boundary is appropriate.

Specification of inappropriate boundary conditions will constrain the model solution and result in incorrect predictions (Franke et al. 1987; Frind and Hokkanen 1987; see Figure 8.11). Due to the lack of complete field data about boundary conditions, assumptions will be necessary. These assumptions must be supported by site data and the results of the model. Not all combinations of boundary conditions are acceptable; some will lead to unstable or nonunique solutions. For example, for a steady-state problem, specified-flux conditions on all boundaries will not lead to a unique solution.

Common misuses of boundary conditions include the assumption of a zero-flux boundary at an aquifer/bedrock boundary when in reality significant transport may occur in the bedrock, or specification of prescribed head (or pressure) conditions at many model boundaries resulting in a highly constrained model solution.

8.4.4 Defining Initial Conditions

The initial conditions describe the distribution of heads (or pressures) and concentrations throughout the model domain at the start of the simulation. Errors in initial conditions will propagate through a transient solution, causing unrealistic predictions. The initial conditions for a steady-state simulation are important mainly to save computational effort in reaching a solution. However, the initial conditions for a transient problem strongly influence the predicted results. The initial conditions supplied to a transient run should be the result of a steady-state flow or transient flow and transport simulation of background flow and transport conditions, which will give a mass-balanced starting point. Background conditions will include many of the features of the domain such as long-term pumpage, agricultural recharge, metal-laden upland recharge, or salt-laden geothermal upwelling.

Initial conditions for the unsaturated zone might be the pore pressures calculated based on a steady infiltration or evaporation rate. The assumption of hydrostatic pressures in the unsaturated zone often leads to unrealistic suction pressure and relative hydraulic conductivity, which translates into incorrect water fluxes in the vadose zone.

Specification of an observed concentration distribution (e.g., an interpreted plume based on observed concentrations) as initial conditions for a transient transport simulation often leads to erroneous predictions because a field program rarely measures the highest-occurring concentrations and differing interpolations would lead to widely varying predictions. Therefore it is better to use estimated source terms as a starting point for the transport model, even if the source is not well defined. These sources can be used to calibrate the transport model, and to predict the starting conditions for subsequent model

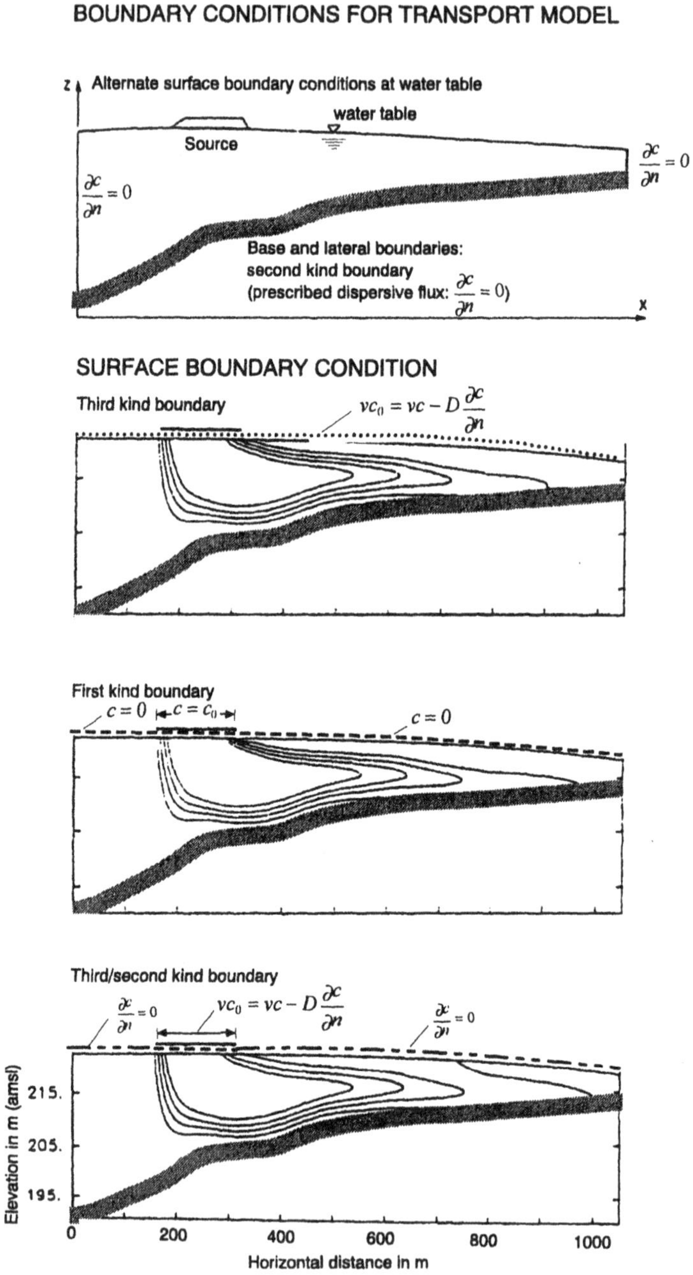

Figure 8.11 Investigating the effects of alternate boundary conditions in transport modeling (after Frind and Hokkanen 1987).

simulations. If the source term is unknown, one approach to source estimation would be to use an analytical inverse model (e.g., Domenico and Robbins 1985; Dale and Domenico 1990) to estimate sources based on the observed chemical plume. Other options are discussed in the next section.

8.4.5 Preparing Model Input Data

The selection and preparation of model input data are discussed in the following sections. The choice of individual model input parameters, based on literature data, is discussed in Appendixes B and C. The following three subsections discuss the types of input data required, how to estimate unknown or uncertain data; and a checklist for model calculations.

Listing Required Input Data

Input data for the model are used for:

- Problem definition (material properties and geometry).
- Numerical requirements (initial conditions, boundary conditions, time-stepping constraints, and spatial discretization).
- Modeling requirements (calibration targets, validation targets, and definition of alternate hypotheses and scenarios).

Input data for a typical, variably saturated variable-density, flow and transport model are summarized in Table 8.1. Typical values for many of these input parameters are presented in Appendixes B, C, and D. Data sources for a site model are listed in Table 8.5.

Estimating Uncertain Data

Field data provide local estimates of conditions, whereas a model requires input of data distributed over the entire model domain. Either model input data are zoned (with homogeneous values within each zone) or they have a continuum of input values. Overall model realism reflects the methods used to estimate model input data from field, laboratory, and literature data. Estimation methods range from use of parameter estimation or inverse models to trial and error methods. Inverse models range from analytical models for contaminant problems (e.g., Domenico and Robbins 1985; Dale and Domenico 1990) to three-dimensional, numerical parameter estimation models (e.g., Hill 1993). The advantages of inverse methods are:

- Results include the mean value and the parameter variance.
- Subjectivity is removed from the calibration process.
- All of the field data are honored.

TABLE 8.5 Common Data Sources for a Site Model

Model Input Parameters	Potential Source of Data[a]
Hydraulic conductivity	Slug, pump and packer tests, and published data
Distribution of hydrogeologic units	Boring logs, geophysics, and geologic maps
Specific storage	Slug and pump tests
Specific yield	Pump tests and porosity data
Recharge/discharge	Precipitation, soil properties, streamflow, pumpage records, elevation, vegetation maps, land use
Unsaturated soil properties	Permeameter tests
Initial water levels, gradients	Field water levels
Background concentrations	Field concentrations
Porosity	Soil analysis
Molecular diffusion coefficient	Published data
Dispersivity	Tracer tests, other field-tested models, or published data
Adsorption distribution coefficient	Batch and column tests, empirical equations for organics, and published data
Soil bulk density	Soil analysis
Density and viscosity	Published data
Contaminant sources	Material inventory, storage, and use, leachate tests, aerial photographs, etc.

[a] Other possible sources include published data on neighboring areas or similar sites.

The disadvantages of inverse methods are:

- Extensive time and computational effort are involved.
- Intuition based on training, experience, and knowledge of the site and other soft data are neglected.
- The model output may be erroneous, unstable or nonunique if the input data are sparse or of differing quality.

The trial-and-error method involves variation of uncertain input data, taking into account observed data, intuition, and analogies with other sites. The trial-and-error approach is the most widely used approach in practical applications. The advantages of this method are:

- Data of varying accuracy may be used appropriately.
- Soft data, such as inferences based on site-specific data, may be reflected in the input data.
- Less effort is required to understand results.

The disadvantages of the trial and error method are:

- The resulting model reflects, in part, the experience (or inexperience) of the modeler.
- There is no simple criterion for assessing satisfactory completion of this task (see Section 8.5).
- The uncertainty in the input data, and its reflection in accuracy of predictions, is not quantitatively defined.

There are a range of methods that fall between the inverse and trial-and-error methods and can be used to augment trial-and-error and parameter estimation. These techniques include geostatistics and kriging, indicator kriging, semiprobabilistic methods, search theory, equivalent media approximations, and observed plume evaluation. An explanation of each of these methods follows.

Geostatistics provides a method for characterizing a parameter distribution. Interpolated data, and the probable accuracy of estimation, can be calculated by kriging the observed data. This method, in its original form, uses the assumption that all data are spatially correlated, with the closest pairs of data being the most strongly correlated. Kriging can be readily undertaken if the data set is relatively homogeneous and smoothly varying, but kriging carries with it advantages and disadvantages similar to inverse modeling. In addition the variogram analysis accompanying kriging will provide information on the correlation length of the data, which is helpful in selecting model cell sizes (see subsection 8.4.2). Reducing interpolation uncertainty through collecting additional data can be assessed in advance, giving a rationale for extending or terminating a field program. Kriging should not be used to interpolate groundwater concentration data because the contaminant distribution typically violates two key assumptions of the kriging method: isotropic and stationary random fields.

Semiprobabilistic methods may involve use of pieces of other methods to suit the problem at hand. The following description provides an example of the application of a semiprobabilistic method. The problem involved model investigation of the effect of sand lenses as preferential flow paths for contaminant migration at a new facility. To identify all the sand lenses by means of a field-drilling program is very costly, but a deterministic modeling approach is not appropriate, since no contamination field data existed to corroborate a trial-and-error method to parameter estimation. Although other methods can be devised to suit differing needs, the method used to estimate model input data for this site was as follows:

- Borings across the site intercepted multiple sand lenses of varying thicknesses. These lenses did not correlate between borings.

- Sand lens geometry was observed, in road cuts, to follow a pattern of similar length-to-thickness ratios. This ratio was used to select the model cell aspect ratios.
- The typical thicknesses of the sand lenses were known from the existing borehole data. These data were used to generate a histogram of lens thickness versus number of occurrences of that thickness (probability distribution).
- A simple computer program was used to randomly generate material property distributions based on the sand lens thickness histogram, with a cutoff on the number of lenses based on the observed ratio of sand to other material in the site borings.
- A deterministic model for groundwater flow and contaminant transport was run for each random distribution (sometimes called a *realization*).
- Based on the predictions of travel distance from the deterministic model, a histogram of travel distance versus number of occurrences of that travel distance was prepared. Hundreds of realizations would be required to generate a set of results representative of the possibilities at this site. However, random- and stratified-sampling methods can reduce the number of runs needed by about an order of magnitude (McKay et al. 1979). This reduces the number of runs to a manageable number. It was shown that the possibility of a sequence of interconnected sand lenses providing a long travel path was remote.

Search theory or indicator kriging can be used to reduce uncertainty in parameters such as location of a geologic boundary, location of a plume boundary (helpful for capture-zone analyses), and location of a source boundary. That is, parameters, or targets, in which a present/absent or yes/no answer is required. The result of such an analysis is the probability of targets of different shapes and sizes existing, dependent on the distribution of drill-hole data (Savinskii 1965). This method could also be used to evaluate the worth of collecting additional field data (answering in advance the questions of how much the uncertainty will be reduced, and that density of drill holes are appropriate) as is described by Freeze et al. (1990).

Equivalent homogeneous media assumptions can sometimes be used for all or part of the model domain. The problem is to calculate the effective hydraulic conductivity that would result in the same predicted groundwater flow as the actual heterogeneous medium. A variety of methods for different, steady-flow situations are presented in Figure 8.12. These results use the assumption that all data are equally valid. There are no equivalent rules for transient flow, except that the effective hydraulic conductivity has a value that lies between the arithmetic mean and the geometric mean. For example, in an aquifer pumping test the effective hydraulic conductivity at early times approaches the arithmetic mean, whereas at late times the effective hydraulic conductivity approaches the harmonic mean (Gorelick and Hernandez 1990).

LAYERED SYSTEM

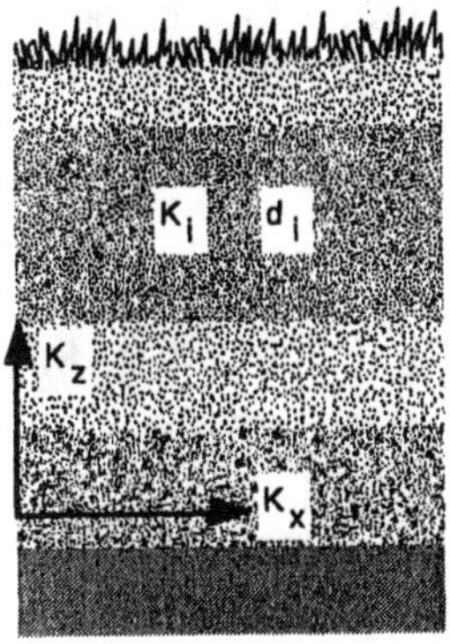

Arithmetic mean $$K_A = \frac{\sum_{i=1}^{n} K_i d_i}{\sum_{i=1}^{n} d_i} = k_x$$

Harmonic mean $$K_H = \frac{\sum_{i=1}^{n} d_i}{\sum_{i=1}^{n} \frac{d_i}{k_i}} = k_z$$

Geometric mean $$K_G = (K_1 K_2 \ldots K_n)^{1/n}$$

HOMOGENEOUS AQUIFER WITH ISOTROPIC CORRELATION LENGTHS

Two-dimensional uni-directional flow $$K_x = K_y = K_G$$

Three-dimensional uni-directional flow $$K_x = K_y = K_z = K_G\left(1 + \frac{\sigma_y^2}{6}\right)$$

HOMOGENEOUS AQUIFER WITH ANISOTROPIC CORRELATION LENGTHS

Two-dimensional uni-directional flow

$$K_x = K_G \exp\left[\sigma_y^2\left(\frac{1}{2} - \frac{1}{1 + \lambda_x / \lambda_y}\right)\right]$$

$$K_y = K_G \exp\left[\sigma_y^2\left(\frac{1}{2} - \frac{\lambda_x / \lambda_y}{1 + \lambda_x / \lambda_y}\right)\right]$$

Infinite domain uni-directional flow $$K_H < K < K_A$$

where

λ_x, λ_y = correlation length in x and y

σ_y^2 = variance of K

Figure 8.12 Equivalent homogenous medium approximations of hydraulic conductivity.

The source term supplied to a model is often controversial and uncertain. As a first approximation, the mass in an observed solute plume may be calculated by estimating the aquifer volume corresponding to a certain groundwater concentration, and then calculating the mass of contaminant in water and soil in that volume, using the following formulas:

$$M_w = C_w n V \tag{8.7}$$

$$M_s = C_s(1 - n)\rho_b V \tag{8.8}$$

where

M_w = mass of contaminant in groundwater in [M]
M_s = mass of contaminant on soil in [M]
C_w = concentration in groundwater in [M/L^3]
C_s = concentration on soil in [M/M]
n = porosity in [1]
V = volume of ground being considered in [L^3]
ρ_b = bulk density in [M/L^3]

This mass, however, is unlikely to adequately represent the source volume because the monitoring program will probably not measure the highest concentrations in the plume, a NAPL plume may exist, and the mass on the soil will be uncertain. In the case of one extensively monitored, contaminated site, only a few percent of the total volume of the contaminant source was calculated from the observed concentrations.

Despite the great interest in immiscible-phase or nonaqueous phase liquid (NAPL) modeling, practical models are still not available (see Table 9.3). In groundwater modeling classes at the Waterloo Centre for Ground Water Research, Cherry (1990) presented the following conclusions about immiscible contaminants:

- Dissolution of immiscible phases is not well understood, particularly in the vicinity of groundwater pumps (i.e., at high groundwater velocities), and as the residual immiscible mass is depleted.
- NAPL may be present in the groundwater despite saturated, dissolved concentrations not being observed because (1) contaminated and clean waters have mixed in the vicinity of wells, (2) the source is heterogenously distributed, (3) the source is becoming depleted, or (4) dispersion has occurred in the aquifer.
- Detailed vertical monitoring may be the key to delineating residual NAPL zones.
- The ratios and relative concentrations of chemicals in the NAPL and groundwater can provide insight into the source characteristics.

Checking Model Calculations

The following general list for checking model calculations is intended to maintain consistency and credibility in model predictions:

- Plot all input data provided to model and check for accuracy and consistency.
- Check the Courant number for appropriate time step size.
- Check the Peclet number for appropriate cell sizes.
- Check model stability and convergence behavior; corrections to previous solutions should ideally monotonically decrease with time after each change in stress.
- Check model flow and solute mass balances.

When checking the model solution's behavior, note that a well-converged flow solution is prerequisite to a well-behaved transport solution. If the model solution oscillates, diverges, or gives unrealistic results for reasons unrelated to factors identified in the list above, then underrelaxing the solution may damp oscillations (at the cost of greater computational effort). If the relaxed model solution still gives inaccurate results, then a sequence of runs in which the model input data are successively simplified will usually lead to a satisfactory solution. For example, a sequence of runs may involve:

- Remove density and viscosity effects.
- Remove unsaturated zones (or change unsaturated characteristics to resemble saturated coefficients).
- Simplify or remove some boundary fluxes.
- Simplify model layering.
- Remove anisotropy.
- Remove heterogeneity.
- Decrease number of dimensions.

At some point in this sequence the model solution will usually improve. The missing complexities can then be added back individually, leading to the source of the problem. Alternative approaches to model problem solving are suggested in Section 8.5.

A poor flow mass balance will usually result in a correspondingly poor solute mass balance. The size of an acceptable mass-balance error depends on the situation being simulated and on the objectives of the modeling analysis. However, flow mass errors of less than 1% and solute mass errors of less than 5% are commonly accepted.

A model embodies many disparate pieces of data and assumptions. While individual assumptions may appear reasonable, the net effect of the assumptions may be unrealistic.

8.5 CALIBRATING THE MODEL

Model calibration, step 5 of preparing the model, is the process of varying uncertain model input data over likely ranges of values until a satisfactory match between simulated and observed data is obtained. Calibration is needed to account for unmeasured, unknown, or unrepresented conditions or processes and uncertainty in measured input data. If an inverse model is used for parameter estimation, then this step of the modeling process is partially automated. However, if data need to be extrapolated, or the field data are unsuited to automatic calibration, then the more traditional approach of trial-and-error calibration is needed. Model testing prior to application in predictive mode is often split into two processes: calibration and validation. The field data are split into two data sets, often for different time periods at one site; one data set is used for calibration and the other for validation. Model validation is described in Section 8.6.

The general approach to the numerical modeling process is illustrated in Figure 8.1. The model calibration step requires the greatest effort. The calibrated model may evolve in a variety of ways. For example, the modeling analysis might begin with incomplete field data, and the model may be refined as additional data become available.

Alternatively, the model input data could contain some well-defined parameters and some highly uncertain data. Varying these uncertain data is the starting point for the calibration process. Typically parameters should only be varied within measured or likely ranges, parameter distributions should be limited to geologically feasible hypotheses, and parameter values in areas beyond the extent of field data can be varied the most.

It is quite possible that alternate, equally statistically valid, model calibrations can be developed from a single dataset (Brooks et al. 1994). This is most likely to happen in cases where minimal data are available and/or the model extends significantly beyond the area adequately characterized by field data. An example of the model calibration process is described in Appendix A.5. Model calibration usually involves most of the following steps, in the order listed:

- Specify calibration criteria and calibration and/or validation protocol. This step is often required in controversial or sensitive modeling applications. General expectations for model predictions should be developed before calibrating the model. Calibration criteria compare model-prediction errors with key components of the model mass balance. That is, a discrepancy between predicted and observed heads is compared to a key hydraulic gradient, as described below. Or, a discrepancy between predicted and observed concentrations is compared to the variability in observed concentrations in neighboring monitor wells. Model performance criteria might be:

  Paired-data testing (i.e., comparing predicted and observed values for corresponding locations in time and space). Common examples of such testing are *variance* between predicted and observed data should be less than about 10% of the range of observations, *standard deviation* between predicted and observed data should lie in the range 0.7 to 1.0 depending on the number and quality of data points, and *bias* between predictions and observations should be random rather than systematic.

Averaged paired-data testing (i.e., the predicted and observed values are averaged over space and/or time and then compared).

Frequency distribution testing (i.e., predicted and observed cumulative frequency distributions are compared).

- Calibrate the flow model before calibrating the flow and transport model (except in cases of density-coupled transport, in which the flow and transport calculations are inextricably bound together). Accept only convergent, stable, and well-balanced model calculations.
- Simulate natural background conditions (regional recharge, geothermal upwelling, natural mineralogic dissolution, etc.), predict the water levels and concentrations corresponding to preexisting conditions, and compare predictions with any available observations.
- Modify model assumptions and/or uncertain input data, within reasonable bounds, to obtain a realistic simulation. Specify model input data in ranges of values. Note the accuracy of these data so that changes made during the calibration procedure will concentrate on the most uncertain data while remaining within realistic bounds.
- Predict transient flow and transport conditions for the period of development up to the present (this process is also known as *history matching*). Ideally the transient calibration period should be as long as, or longer than, the period of future predictions to which the calibrated model will be applied.
- Evaluate the model predictions versus historical observations (Figure 8.13) The model evaluation should use as many pieces of information as possible (i.e., not just water levels and concentrations but also spring levels, river in/outflows, vertical hydraulic gradients, river concentrations, and any other relevant descriptive data).
- Decide whether additional model refinement is needed based on the calibration criteria you specified.
- Examine "calibrated" model input and output and evaluate whether:

  Input data individually and jointly make sense.

  Site-specific data cover the area predicted to be of concern.

  Model output indicate initial conceptualization was appropriate.

During model calibration model input data or assumption modifications should start by varying the most uncertain parameters or those that are the most critical to the predictions (e.g., hydraulic conductivity for flow or adsorption distribution coefficient for transport). It is a waste of effort to sys-

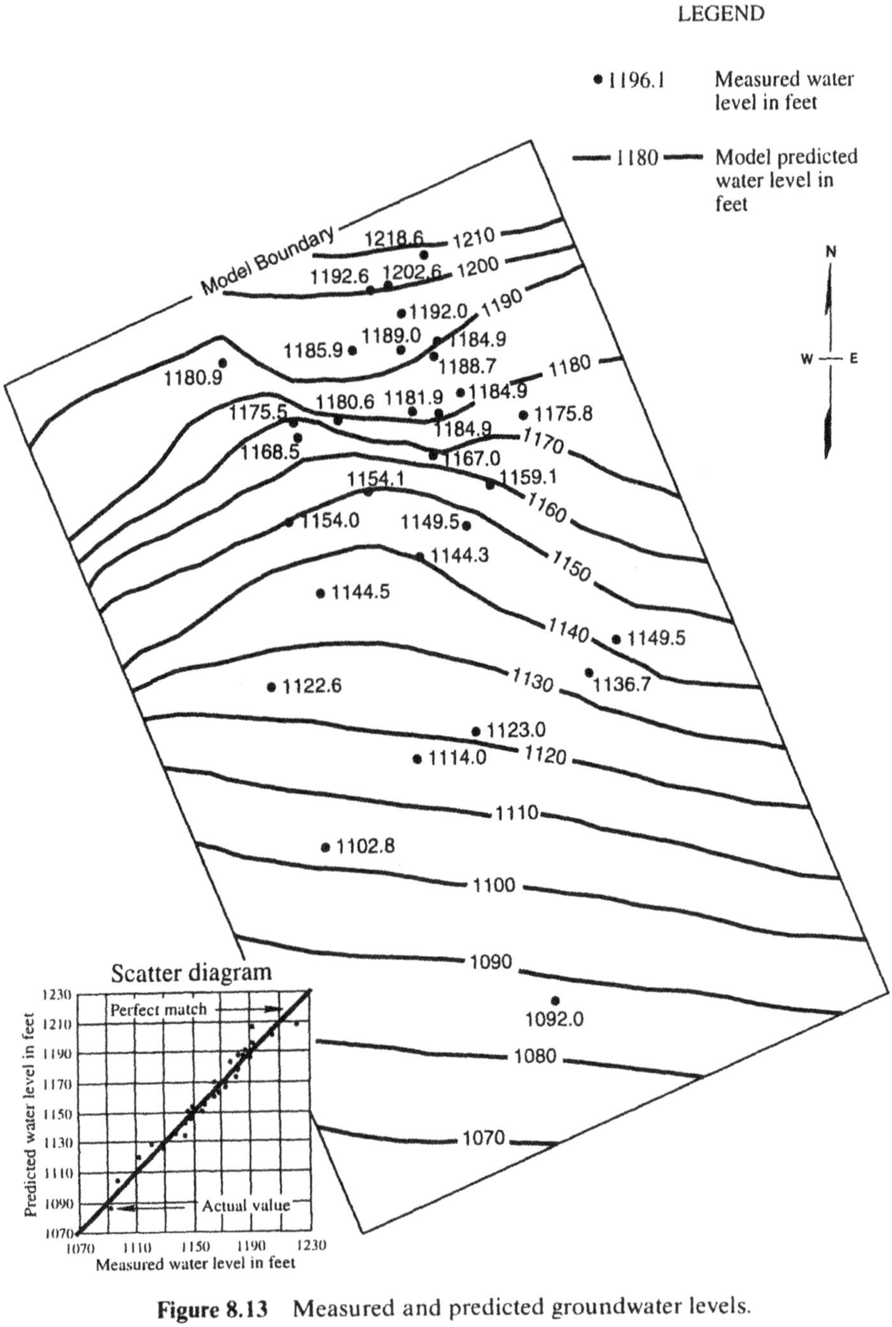

Figure 8.13 Measured and predicted groundwater levels.

tematically vary all parameters during calibration; instead, the most promising combinations of parameters should be tested and refined.

If the model predictions fail to meet expectations, then a model error analysis may be used to assess overall predictive bias (see Section 8.6), or, for more extreme cases the entire underpinnings of the model may have to be re-evaluated. This requires an assessment of modeling assumptions and errors.

There are five general sources of model errors:

- Mathematical model errors. These errors involve the physical and mathematical basis of the computer code, and its numerical framework. This framework should be appropriate to the situation to be simulated, with inherent assumptions honored.
- Conceptual errors. Among these are misconceptions about the governing mechanisms, boundary conditions, sources, and dimensionality of the problem (see Section 8.2 and Chapter 7).
- Input data errors. Input errors include mistakes in data entry (which should be checked by plotting all inputs with a nonsmoothing contour-plotting package), sets of assumptions about data that in combination do not make sense, measurement error, and levels of heterogeneity that either have not been identified and characterized in the field or cannot be represented in the model. The last two sources of error are unavoidable and should be taken into account when interpreting the model predictions.
- Numerical errors. Examples include truncation errors due to truncation of the Taylor series expansion of the finite-difference or finite-element formulation of the governing equations, roundoff errors due to the precision of numbers stored by the computer, and numerical dispersion due to discretization (see Section 6.2).
- Interpretation errors. These can take the form of misunderstanding of the predicted results particularly in the case of models without post-processors capable of providing meaningful summaries, analysis, and graphic output), misconception of the expected results (e.g., prediction of contaminant migration at an angle to groundwater flow), and comparison of spatially and temporally averaged model predictions with point observations.

If consideration of all of these sources of error does not reveal the source of the problem, then the process of model simplification described in Subsection 8.4.5 on checking model calculations can be used to assess the model and probably uncover problem areas. The discrepancies between model predictions and observed values can uncover factors missing from the model; see Section 8.7 for an expansion of this idea.

As is the case with weather forecasting, a model that does not meet calibration criteria is still useful. It can be used to assess the interactions of the mechanisms governing groundwater flow and transport at a site, to help to develop a better overall understanding of site conditions, and to assess the need for additional data. It cannot, however, be used to predict remedial design or to show regulatory compliance.

8.6 ANALYZING MODEL ERROR

The objectives of model error analysis are to quantity how well the model simulates the physical system and to identify problem areas in the model. The most basic measure of model error is the difference between predicted and observed values. It has been stated in U.S. Congressional Hearings that groundwater transport models cannot be expected to give better than order of magnitude accuracy. In specific model applications the predictive capability of the model depends on:

- The nature, extent, and reliability of data available with which to define the model.
- The intrinsic capabilities of the mathematical model to represent site conditions.
- The realism of the conceptual model used to prepare the site-specific model.
- The degree of bias introduced by assumptions made during the modeling process.
- The time and volume over which model predictions are used, as compared to the calibration and validation volumes and time periods.

The degree of uncertainty in data considered useful by geologists and hydrogeologists is often orders of magnitude greater than the data accuracy imposed by design constraints. Therefore there is often a discrepancy between the accuracy of data supplied to a model and the accuracy expected from model predictions. Once again, the model purpose will partly determine the level of acceptable error in a model.

The method typically used to quantity model error is to compute the difference between predicted and observed values (residual) at a point (e.g., monitoring location) and evaluate these differences. Other methods of comparing predicted and observed data have been described in Section 8.5 and in ASTM Standard D 5490-93 (ASTM 1993). The residuals may be illustrated as scatter diagrams of predicted values versus observed values (Figure 8.14) or histograms of residuals (Figure 8.14). The scatter diagrams, together with the computed coefficient of determination, indicate where the greatest discrepancies occur and whether there are a few major discrepancies or general disagreement between predictions and observations. Ideally a histogram of residuals should be normally distributed around zero, with an average residual of zero. If the histogram is skewed, the implication is that the model consistently over- or underpredicts the variable of concern. The histogram

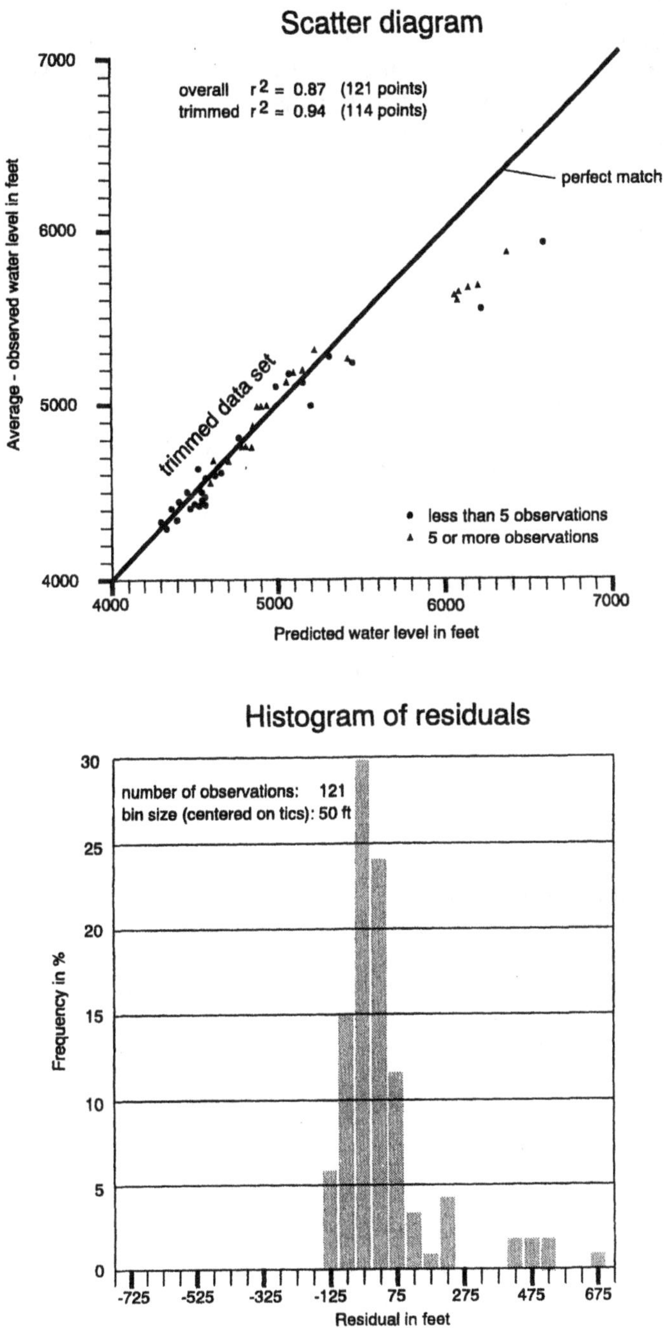

Figure 8.14 Model error analysis.

also indicates the percentage of residuals that are unacceptably large. A map of the residuals (which may be contoured) can be used to relate errors in prediction to the hydrogeolgic setting and indicate problem areas. Ideally residuals should be randomly distributed.

Other measures of the goodness of fit between predictions and observations (Loague and Green 1991) include

Maximum error (ME),

$$\mathrm{ME} = \max\left\{\left|P_i - O_i\right|\right\}_{i=1}^{n} \tag{8.9}$$

Root mean square error (RMSE),

$$\mathrm{RMSE} = \left[\sum_{i=1}^{n} \frac{(P_i - O_i)^2}{n}\right]^{\frac{1}{2}} \left[\frac{100}{\overline{o}}\right] \tag{8.10}$$

Coefficient of determination (CD),

$$\mathrm{CD} = \frac{\sum_{i=1}^{n} (O_i - \overline{o})^2}{\sum_{i=1}^{n} (P_i - \overline{o})^2} \tag{8.11}$$

Modeling efficiency (EF),

$$\mathrm{EF} = \frac{\sum_{i=1}^{n} (O_i - \overline{o})^2 - \sum_{i=1}^{n} (P_i - O_i)^2}{\sum_{i=1}^{n} (O_i - \overline{o})^2} \tag{8.12}$$

Coefficient of residual mass (CRM),

$$\mathrm{CRM} = \frac{\sum_{i=1}^{n} O_i - \sum_{i=1}^{n} P_i}{\sum_{i=1}^{n} O_i} \tag{8.13}$$

where

O = observed value
$\bar{o}$ = mean observed value
P = predicted value
n = number of values

ME, RMSE, and CRM tend to zero and CD and EF tend to one for perfect predictions. RMSE, EF, and CRM are relative measures useful to compare the calibration of alternative model runs. ME and CD are directly useful and more commonly used.

The degree of model error that is acceptable depends on several factors:

- Degree of natural heterogeneity or complexity of boundary conditions. A highly heterogeneous or complex domain may not be modeled precisely.
- Location, number, and accuracy of measurements. The "perceived model accuracy is inversely proportional to available data, actual model accuracy is directly proportional to available data" (Pinder 1990).
- Purpose for which the model has been developed.

Model sensitivity analysis (see Section 8.8) may also provide feedback on model error/predictive uncertainty due to uncertainty in model inputs.

8.7 VALIDATING THE MODEL

Model validation (sometimes called verification), the seventh step, is the process of demonstrating that the calibrated model is an adequate representation of the physical system. Model validation is a shortcut to gaining greater confidence in model predictions in the absence of uncertainty analyses. Validation is more common in hydrologic modeling for which time-history data are often available. ASTM (1993) discusses validation at length, including concepts of global versus local validation. However, ASTM (1993), while recommending validation to reduce the problem of nonunique solutions, also states that a calibrated but unvalidated model may be used for predictive analyses in conjunction with a careful sensitivity analyses. This step is rarely addressed in limited modeling studies because of the extra data and modeling effort required. However, in the absence of validation a model is untested beyond the exact conditions used in the calibration, and use of the model to make other than general predictions is questionable. A successfully calibrated model cannot be relied upon to provide accurate predictions in all cases (Freyberg 1988). Four possible approaches to model validation, the fourth method requiring the least effort on the part of the modeler, are listed below.

- Successfully predict alternate conditions. An example of this method is to calibrate using steady-state data and validate using transient data (Figure 8.15). This method has the advantage that the comparison data sets are independent, but additional parameters (such as storage parameters and fluxes) come into play in the transient calculations that may invalidate an

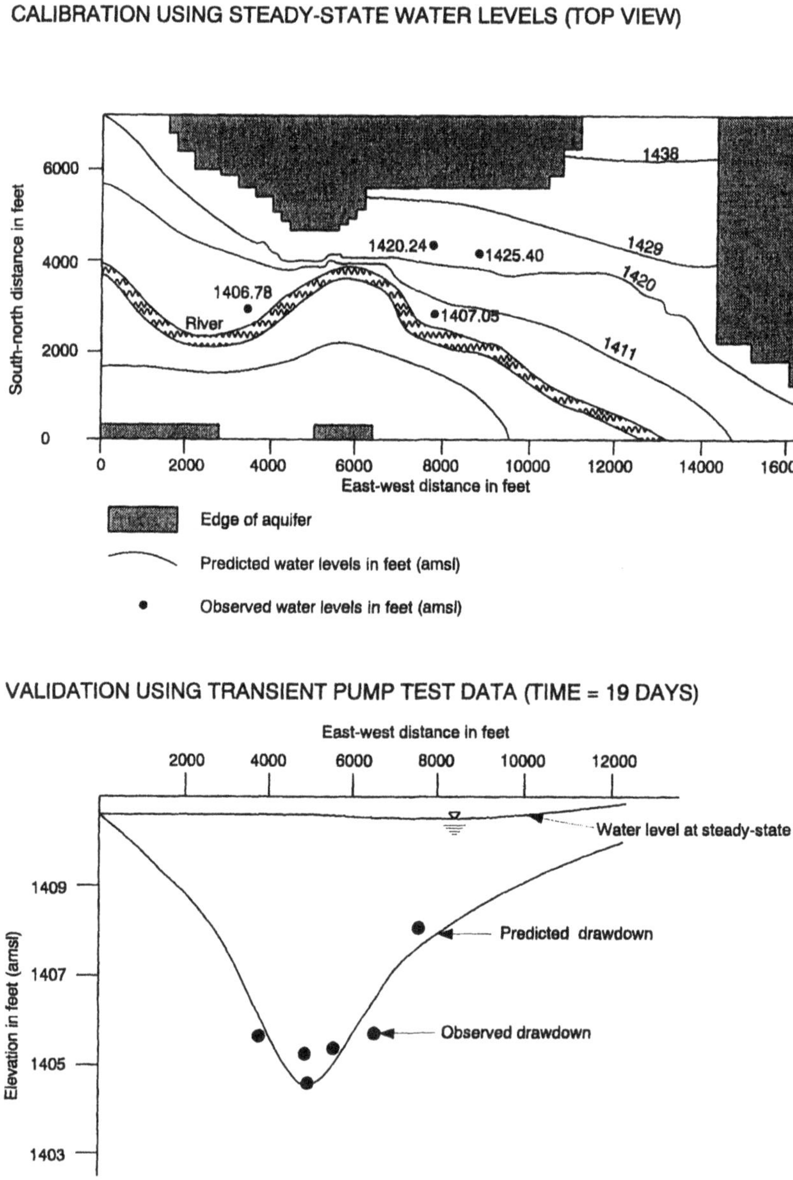

Figure 8.15 Validation using transient data.

apparently realistic steady-state model. An alternate approach is to use a part of a transient data set in calibrating the model, and the remainder of the data set, possibly subject to differing boundary conditions, in validating the model. This is the most commonly applied approach to model validation.

- Successfully predict existing conditions. In this method model validation uses comparison data not employed in the calibration process (Figure

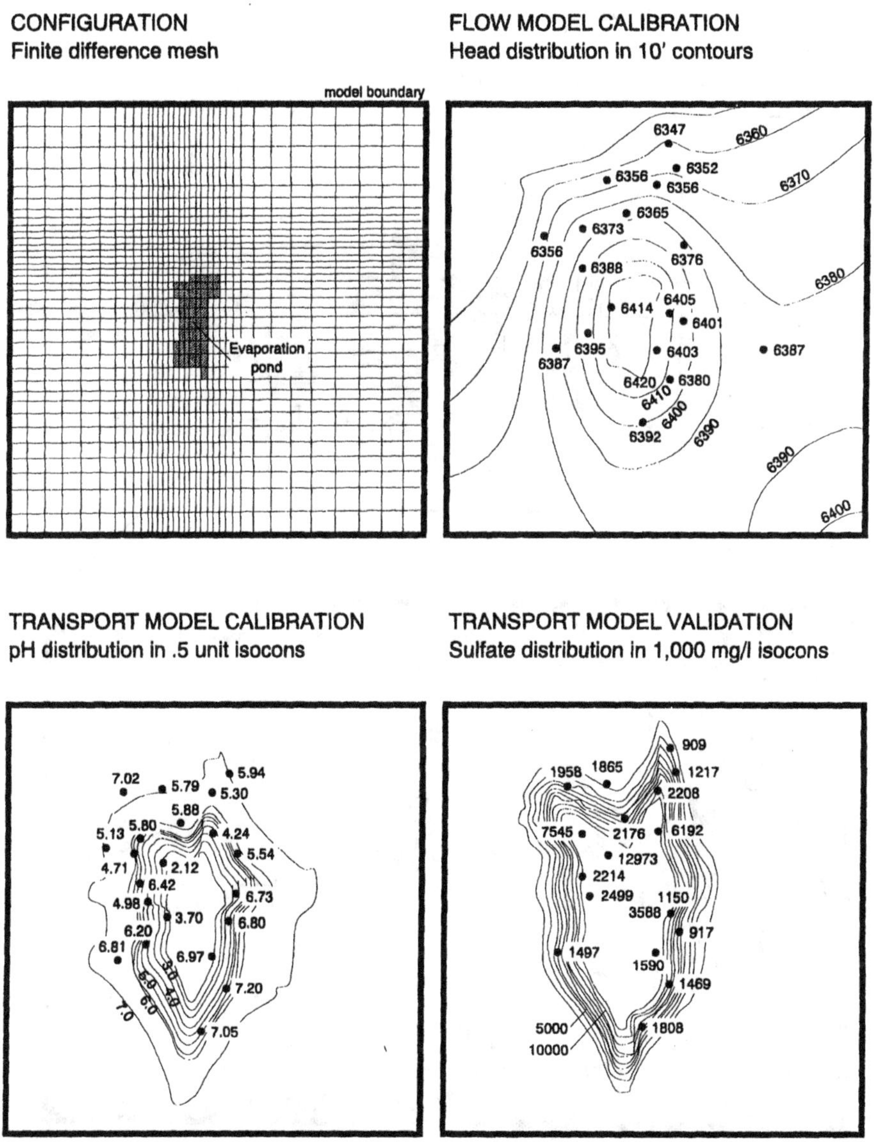

Figure 8.16 Validation using comparison data not employed in model calibration.

8.16). This method is the most useful but can only be used if there are sufficient data to describe the entire domain satisfactorily using half the available data; this is rarely the case. Even when there are sufficient data, this method is not ideal, since the calibration and validation data sets may not be entirely independent.

- Compare the model predictions with the results of other models of the identical situation (Figure 8.17). This method validates the numerical

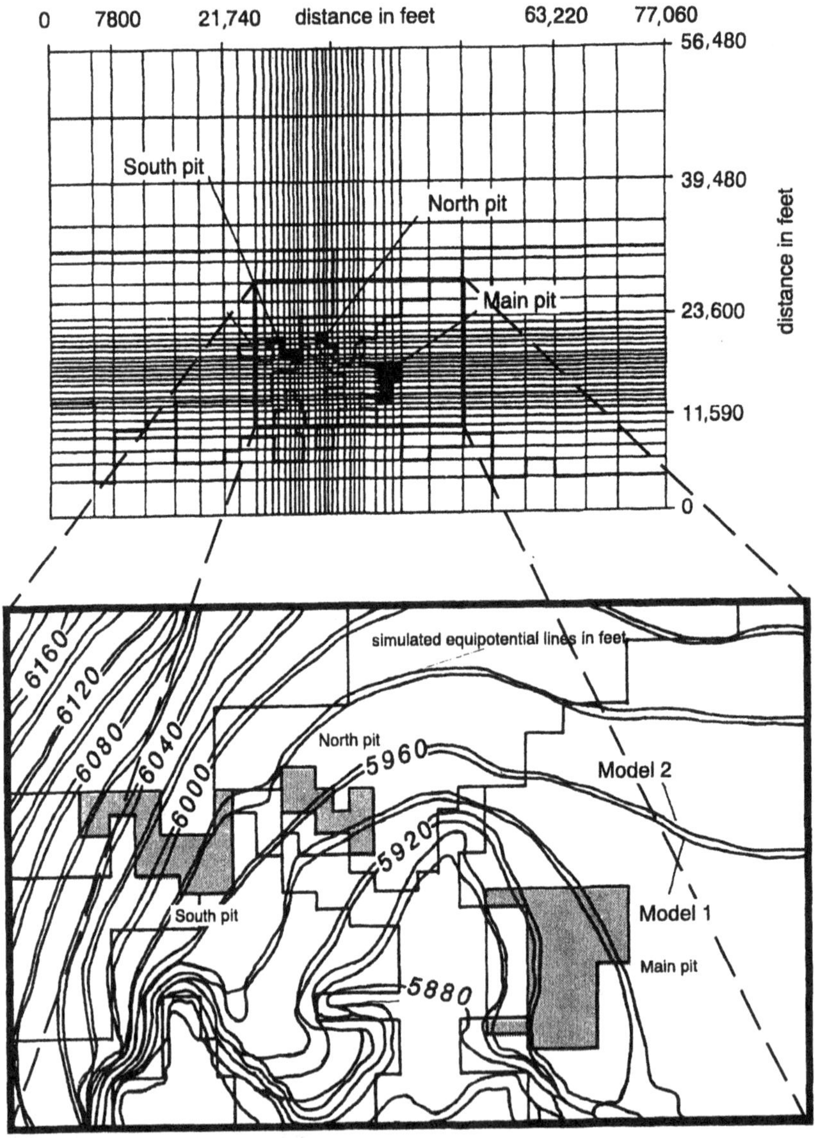

Figure 8.17 Validation by comparing different models applied to mine dewatering and reclamation.

model but not the conceptual model, since the same assumptions, boundary conditions, and input data are employed in the two sets of model simulations.

- Predict conditions for locations beyond the existing monitoring network or at future times, to validate the model when additional fieldwork is undertaken (Figure 8.18). This is the most persuasive argument in favor of a realistic model, and the most practical method for validating a transport model. This method requires that modeling analyses and fieldwork be conducted in parallel over a period of time.

In each of the above methods, you must consider measurement errors, precision, and completeness of the comparison data set when evaluating the model.

8.8 DEMONSTRATING MODEL SENSITIVITY

Model sensitivity analysis is the eighth step in the modeling process. The purpose of sensitivity analyses is to demonstrate the model responses to variations in uncertain input parameters. The model response to these variations is of interest because the range in the resulting predictions illustrates the level of model prediction uncertainty and, given a sensitivity case with results as statistically valid as the calibrated model, the nonuniqueness of the calibrated input data set. A systematic sensitivity analysis provides sufficient data to rank the input parameters in terms of their influence on the predicted results. The results of a sensitivity analysis can be used to:

- Identify sensitive input parameters for the purpose of guiding additional field data collection and, perhaps, focussing calibration efforts
- Define parameters to be used in uncertainty analysis.

A typical sensitivity analysis involves the following steps:

1. Assemble the input data together with their ranges of uncertainty. These ranges of values should have been assessed during the model calibration stage. The ranges may be based on the greatest extremes estimated from the:
   - Variation of observed values
   - Uncertainty in measurement
   - Range in literature values for similar conditions
   - Range in historical conditions
2. Rerun the calibrated model with each of the input parameters (ideally including grid size see Figure 8.19, and time step) individually varied to their maximum and minimum values. An ad hoc sensitivity analysis

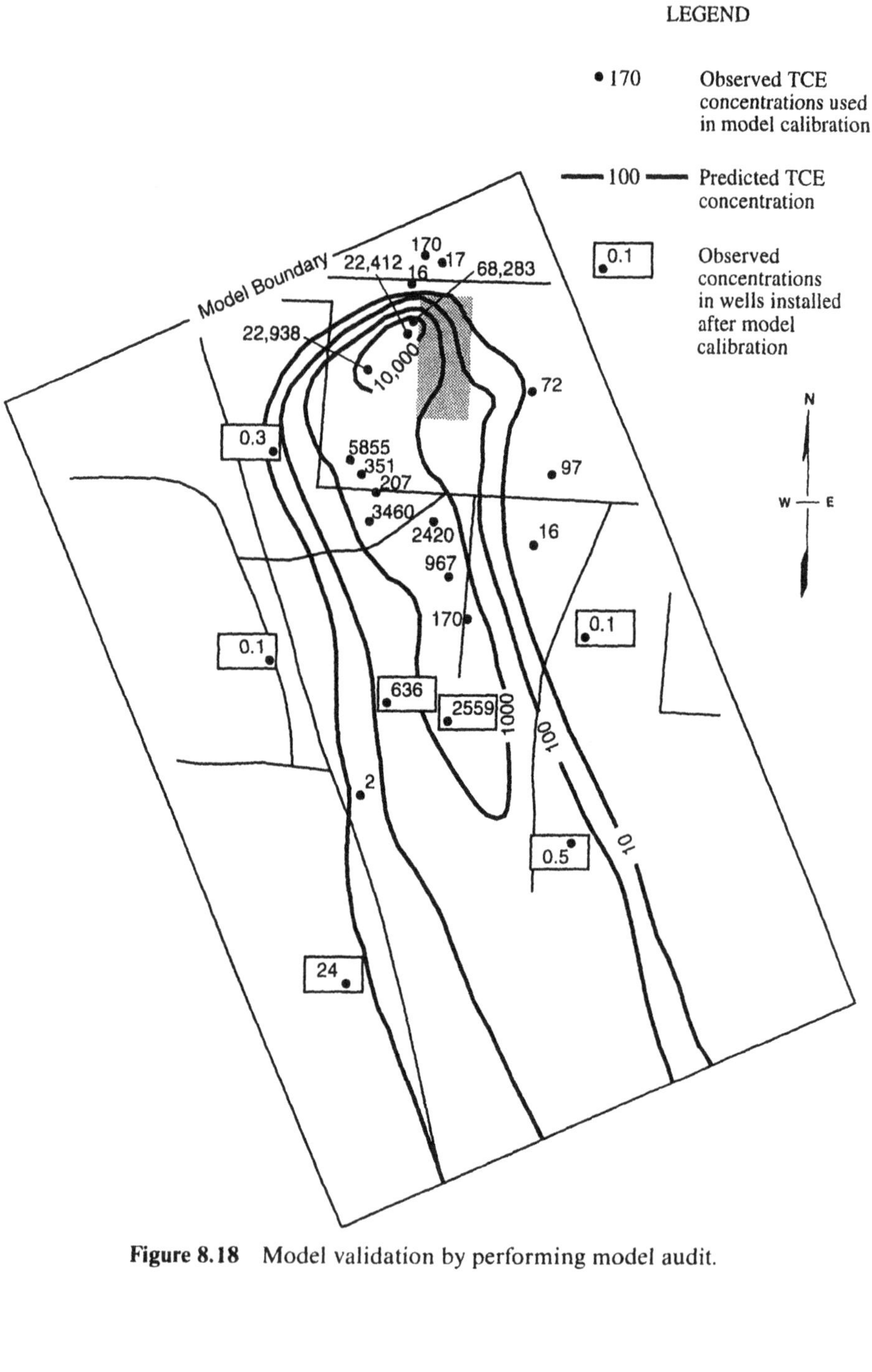

Figure 8.18 Model validation by performing model audit.

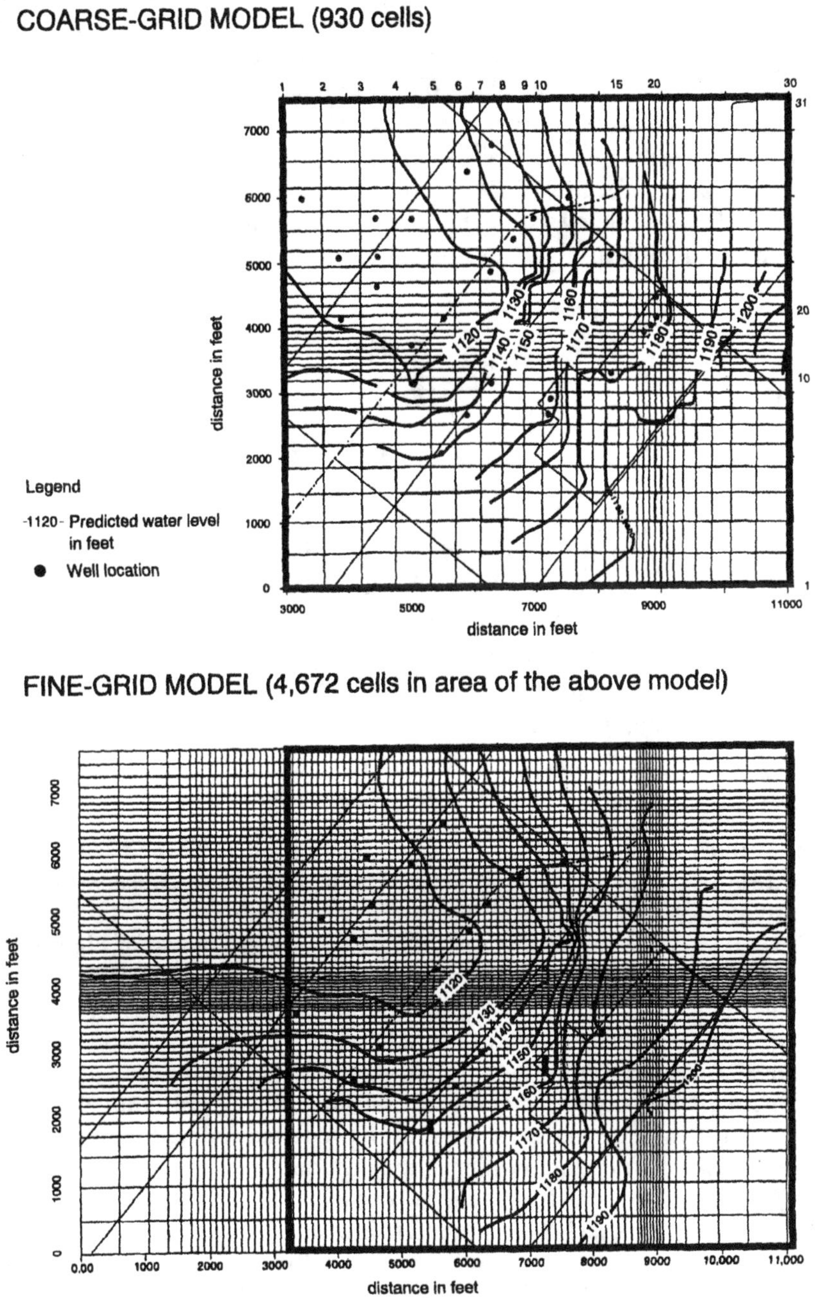

Figure 8.19 Sensitivity of predictions to grid size.

will have been undertaken during the calibration process, but a more rigorous sensitivity analysis is based on the calibrated model. It is not advisable to vary several parameters at once to achieve the "worst case," since this case is usually both highly unlikely and leads to noncalibrated model predictions.

3. Compare the predicted results and interpret in terms of ranking of input data and uncertainty in predictions. The results of sensitivity analyses can be normalized so that the effects of different parameters can be compared quantitatively. For example, a sensitivity index S can be defined

$$S = \frac{|dh|}{(dP/P)} \tag{8.14}$$

where

S = normalized sensitivity index which is a measure of the average change in the predicted variable per fractional change in the input parameter

$|dh|$ = difference in predicted variable, at one or more key locations, between the base case and sensitivity case

dP = change in input parameter value

P = initial input parameter value

The S values so calculated can be numerically ranked. These results will be biased if the ranges of input parameters analyzed are so conservative as to be unrealistic.

Several ideas for presenting sensitivity analyses are shown in Figures 8.20 and 8.21. For example, the results shown in Figure 8.20 suggest that the input parameters should be ranked in importance in this order: dispersivity, hydraulic conductivity, and anisotropy of hydraulic conductivity. The results in Figure 8.21 suggest that ranges in predicted concentrations, for current conditions, of up to an order of magnitude may arise through variation over likely ranges of the input parameters. As a result predicted concentrations are considered to be accurate to an order of magnitude.

The results of the sensitivity analyses occasionally indicate that the calibration parameter data set is not unique. In this case further runs, error analysis, or a search for additional data to help select between the cases is warranted.

Some parameter estimation models (Hill 1993) are also designed to automate the sensitivity analysis process. Viewing interim results sometimes changes the course of a modeling study, but in general model sensitivity analysis is an easily automated step.

PLUME HALF WIDTH AT WATER TABLE

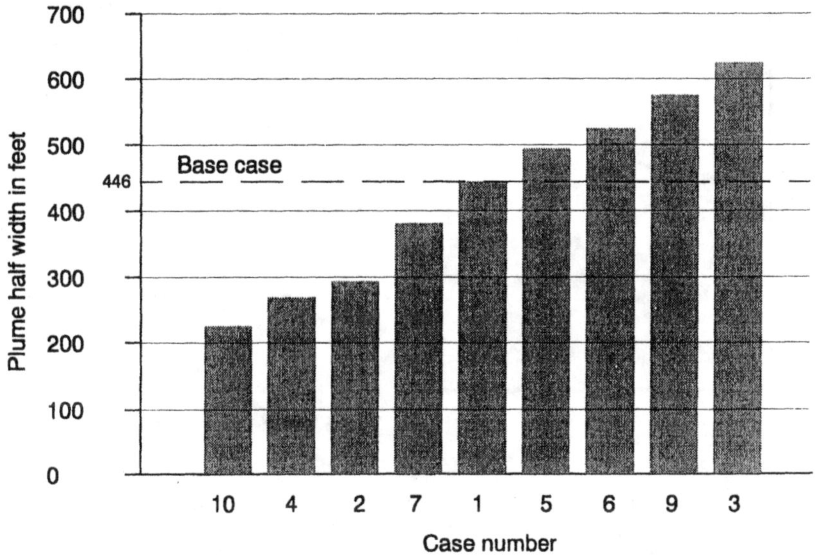

SUMMARY OF CASES EXAMINED					
Case No.	Parameter	Plume half width in feet	Case No.	Parameter	Plume half width in feet
10	Low dispersivity	227	3	High conductivity	628
4	Low vertical to horizontal conductivity	266	9	High dispersivity	577
2	Low conductivity	295	6	1 : 100 Horizontal to vertical conductivity	522
7	Low conductivity interbeds	383	5	1 : 10 Horizontal to vertical conductivity	494

Figure 8.20 Example of summarizing sensitivity analysis.

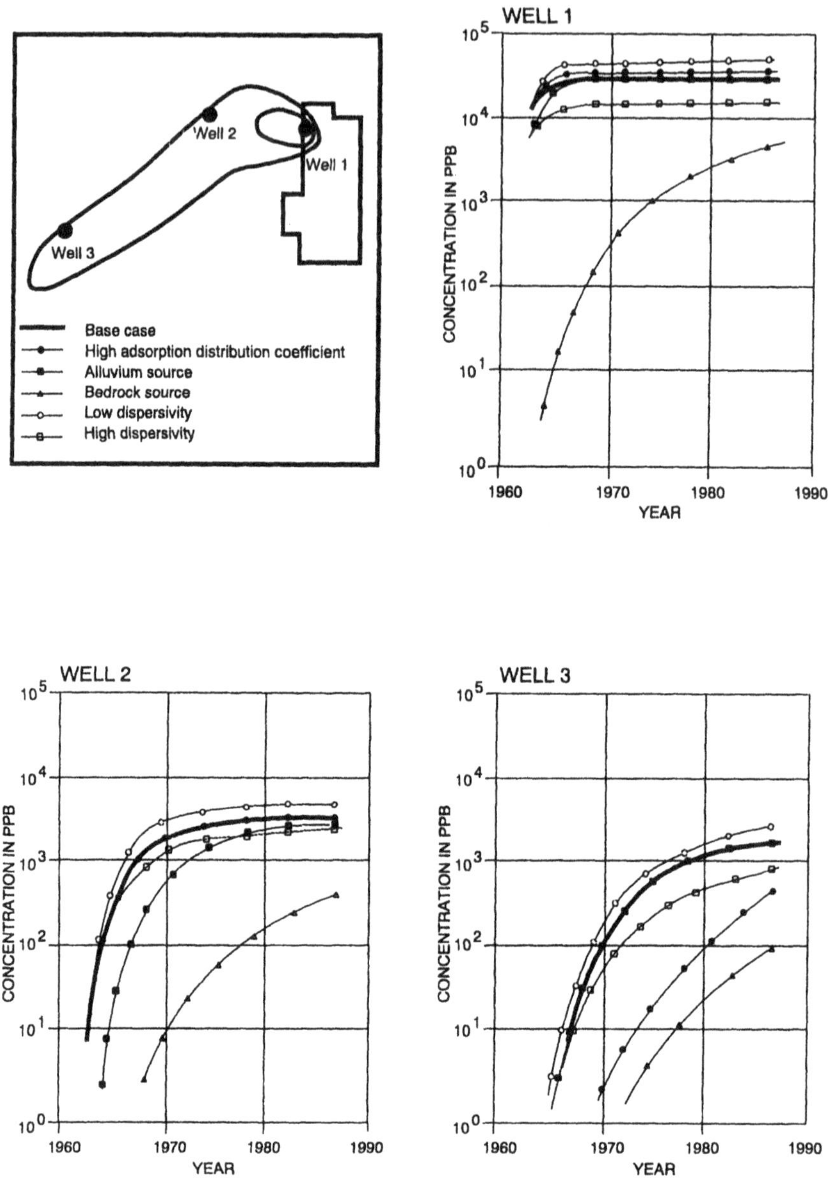

Figure 8.21 Example of illustrating transport parameter sensitivity analysis.

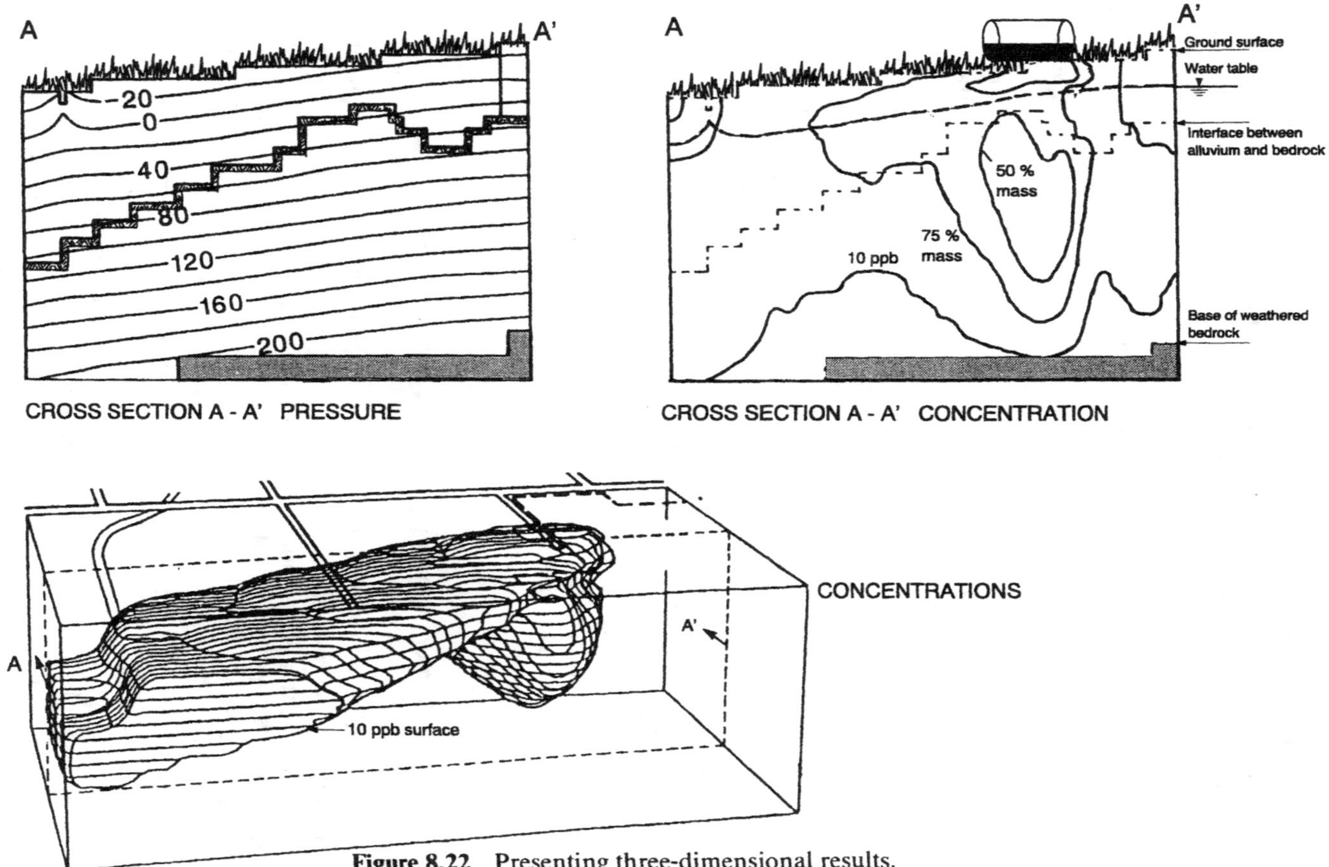

Figure 8.22 Presenting three-dimensional results.

8.9 CONDUCTING MODEL PREDICTIONS

Step nine in the model process is conducting the model predictions. The model predictions are often considered the main purpose of the modeling exercise, although there may be many more purposes to a model than straightforward prediction. The assumptions to be used in the predictive analyses should be well defined. The designer, or end user, of the model results may not understand this aspect of the model. For example, the assumption of a continuing source is not specific enough. Assumptions about the location, concentration, volume, phase(s), accompanying species, and possible depletion of the source also need to be addressed.

If remedial schemes are to be simulated, then the model may be used to predict the capture zones for different schemes, concentrations at any extraction points, time for remediation, concentration time histories at water-supply or compliance points, contaminant mass balances, and ranking of alternate remedial schemes in terms of effectiveness.

Since model predictions are used to generate specific predictive values more often than relative results, concern in the reliability of the predictions is warranted. As model predictions extrapolate over time, the time period of model predictions should be comparable to the period of model calibration. For example, a model that has been calibrated over a 1-year period against water-quality data is not likely to be reliable for 100-year transport predictions. An uncertainty analysis would be helpful in quantifying model reliability under these conditions.

The interpretation of the model predictions should include the modeler's assessment of where the model is more or less accurate and the relative degree of uncertainty in the predictions.

8.10 PRESENTING MODEL RESULTS

Often the last step in the modeling process is writing the report. The presentation of the model results greatly influences the utility of the predictions. It is useful to present tables summarizing the path of development of the model (Table 8.6). The predicted results themselves may be presented in many forms, including three-dimensional graphics (Figure 8.22), concentration time histories (Figure 8.23), vertical cross sections (Figure 8.24), vector illustrations of groundwater flow (Figure 8.25), or comparison of alternate remedial schemes (Figure 8.26). Ideally each figure should be self-explanatory. This means that an explanatory map, site-specific features, and an explanation of the case simulated should be included. Presentation of normalized results (e.g., contours of concentration relative to a source concentration of 1.0) may be used to unify predictions for many contaminants onto one figure.

Due to recent advances in computer hardware and software, there are few restrictions on the model users' imagination for presenting model results. Color plots, as shown in Figures 8.27 to 8.30, may convey the results of complex models in a more informative way for a general audience.

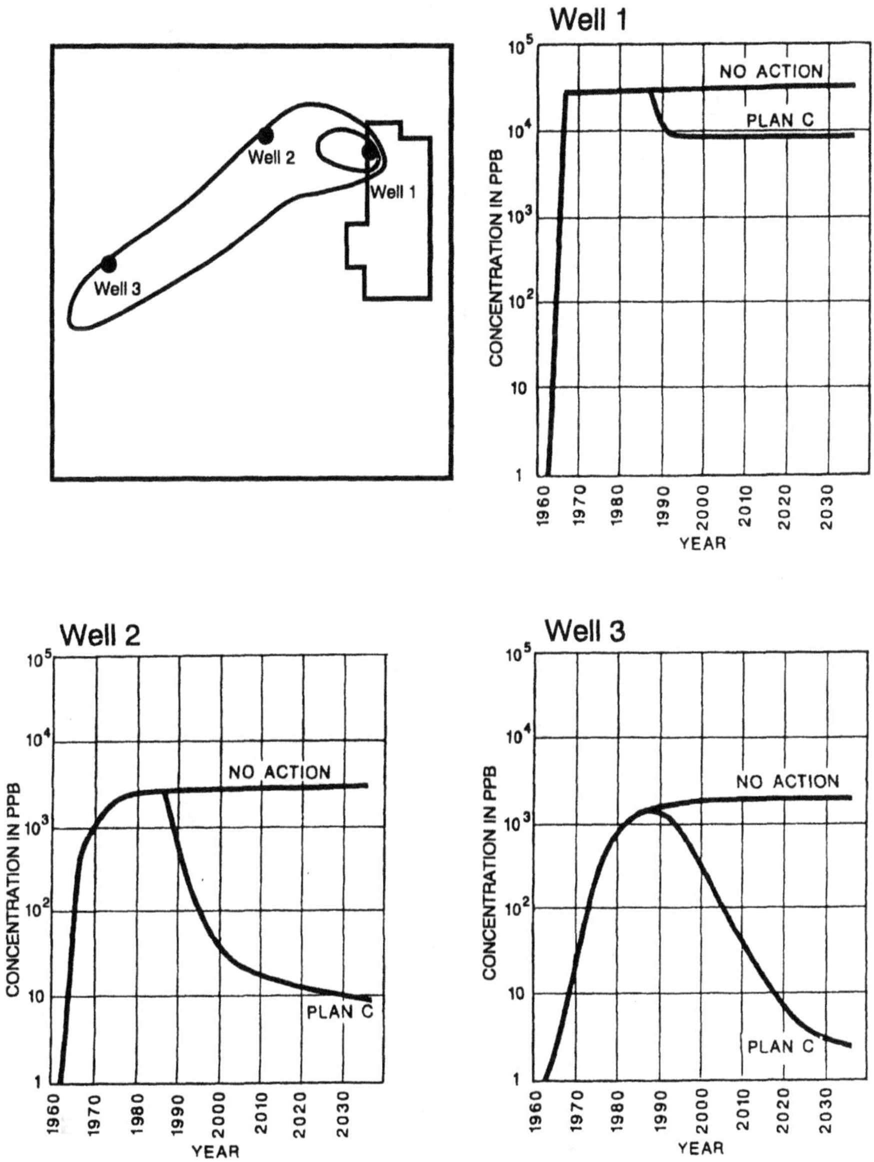

Figure 8.23 Presenting predicted concentration time histories.

TABLE 8.6 Summary of Model Runs

Description and Date	Reason for Performing Calculation	Data Summary[a]								Assumptions Summary	Comments on Results	Suggested Next Stage
		K_x (ft/day)	K_y (ft/day)	K_v (ft/day)	n (1)	S_s (l/ft)	K_d (ml/g)	D_L (ft/day)	D_T (ft/day)			
Case 8 16/10/92	Calibrate flow model	—	—	—	—	—	—	—	—	Zero recharge	Hydraulic heads even lower than case 7. Same conclusions. Mass imbalance 3%.	Revise alluvium hydraulic conductivities to reflect results of new pump tests.
Case 9 17/10/92 flow only	Update model hydraulic properties and investigate influence of hydraulic conductivity on predicted flow patterns	50.0 50.0 0.005 0.05	50.0 50.0 0.005 0.05	0.5 0.5 0.005 0.05	0.28 0.28 0.01 0.01	$5\,E^{-4}$ $5\,E^{-4}$ $1\,E^{-4}$ $1\,E^{-4}$	0.0 0.0 0.0 0.0	100.0 100.0 100.0 100.0	10.0 10.0 10.0 10.0	Recharge assumption as for case 6	Water table drops indicating that surface recharge is not compatible with the revised alluvium K values. Mass balance 1%.	Using best-case hydrodynamics case run transport calculations.
							Not applicable					

Case 10 26/10/92 hydro-dynamics and transport	Using best-case hydro-dynamics to date, predict and compare TCE concen-trations	7.0 5.0 0.005 0.05	5.0 7.0 0.005 0.05	0.05 0.05 0.005 0.05	0.28 0.28 0.01 0.01	$5\ E^{-4}$ $5\ E^{-4}$ $1\ E^{-5}$ $1\ E^{-5}$	0.0 0.0 0.0 0.0	100.0 100.0 10.0 10.0	10.0 10.0 10.0 10.0	Source 2 (for TCE) only: source assumed active 1961 to 1974 incl. Total of 10,000 gals. assumed disposed.	Predictions show deep migration of plume. Influence of alluvium anisotropy and geometry of alluvium/bedrock inter-face can be noted. Depth of plume reflects supplied base head boundary conditions.	Revise bedrock rock-type configuration to reflect latest subcrop map. Also refine base head boundary conditions.

[a] Data listed are presented in the following order: (1) General alluvium, (2) high-conductivity alluvium, (3) tertiary bedrock, (4) Precambrian bedrock.

Note:

K_x = horizontal hydraulic conductivity along x-axis in [ft/day]
K_y = horizontal hydraulic conductivity along y-axis in [ft/day]
K_v = vertical hydraulic conductivity in [ft/day]
n = porosity in [l]
S_s = specific storativity in [l/ft]
K_d = adsorption distribution coefficient in [ml/g]
D_L = longitudinal dispersion coefficient in [ft/day]
D_T = transverse dispersion coefficient in [ft/day]

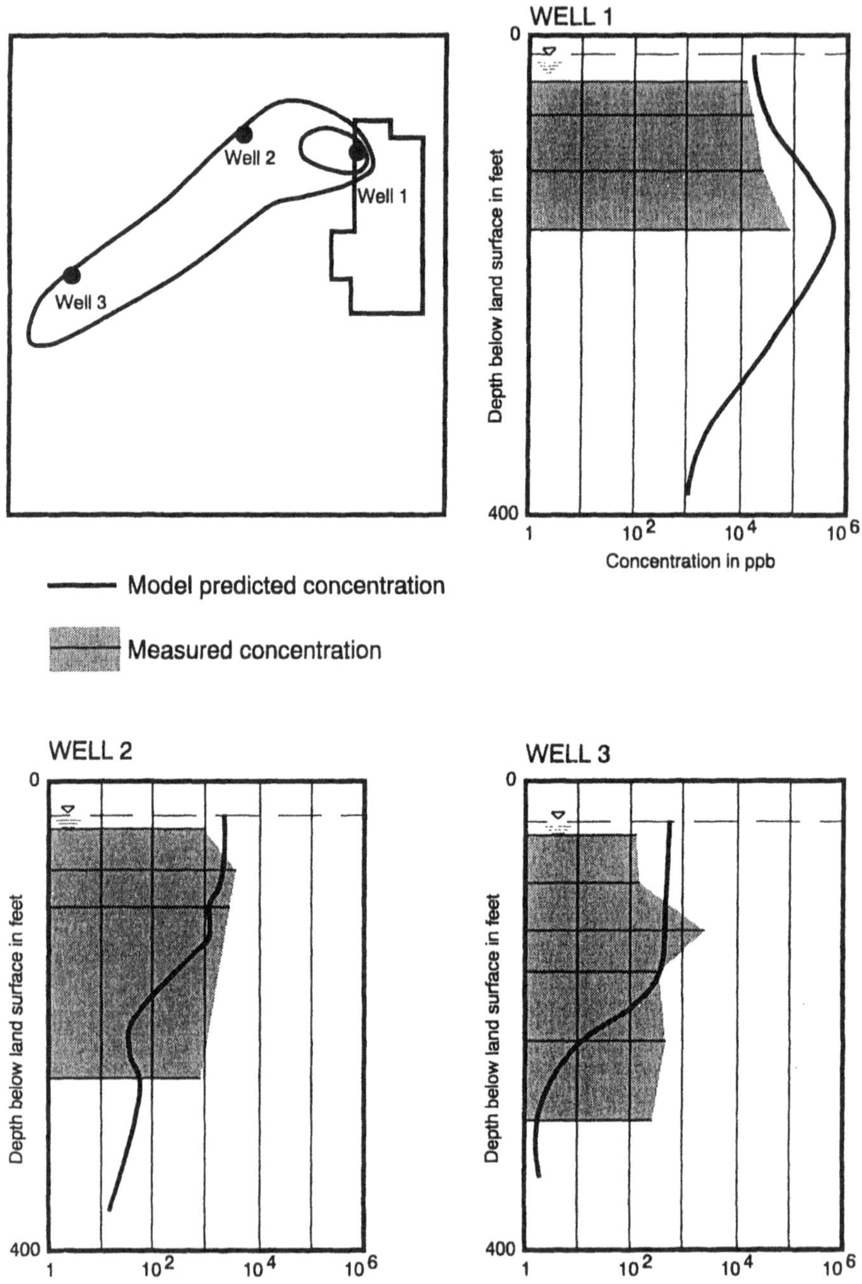

Figure 8.24 Presenting predicted and observed concentrations along plume center line.

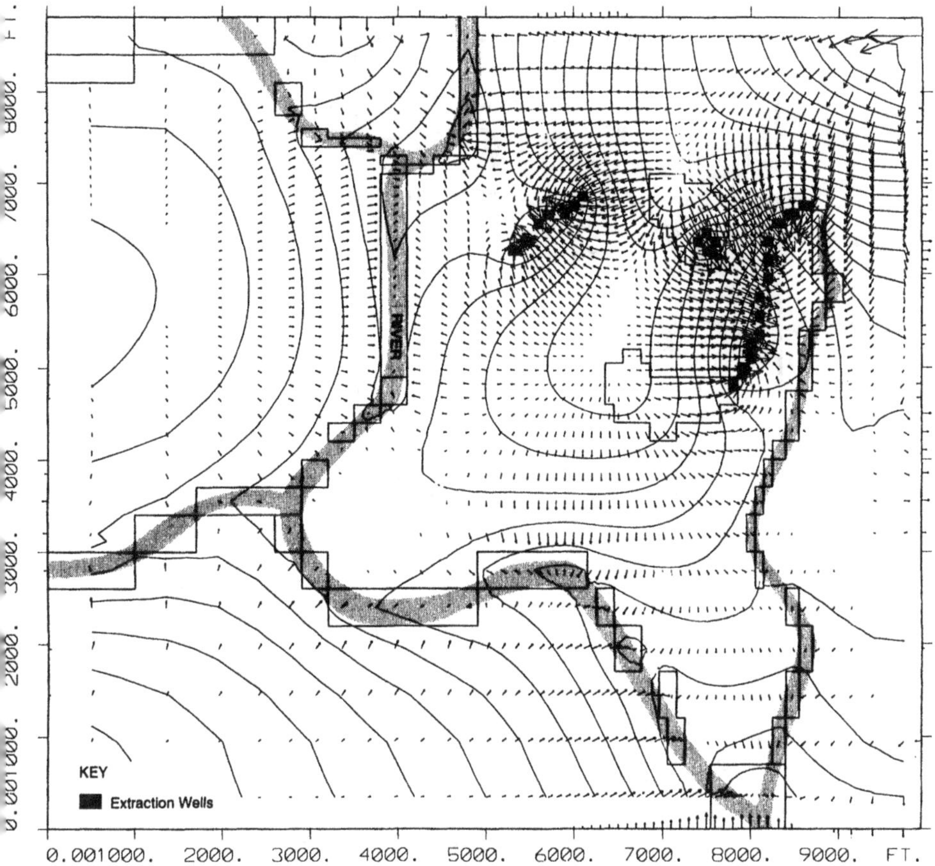

Figure 8.25 Predicted flowfield with remedial wells.

8.11 AUDITING THE MODEL

Occasionally it is possible to conduct a model audit, that is, to compare the model predictions of the future to the actual outcome. Audits can reveal areas of differing model accuracy. Often the actual stresses applied to the system vary from the idealistic stresses simulated in a model; however, a successful audit adds substantially to model credibility. An example of an audit study is presented by Konikow (1986), and another example is illustrated in Figure 8.18.

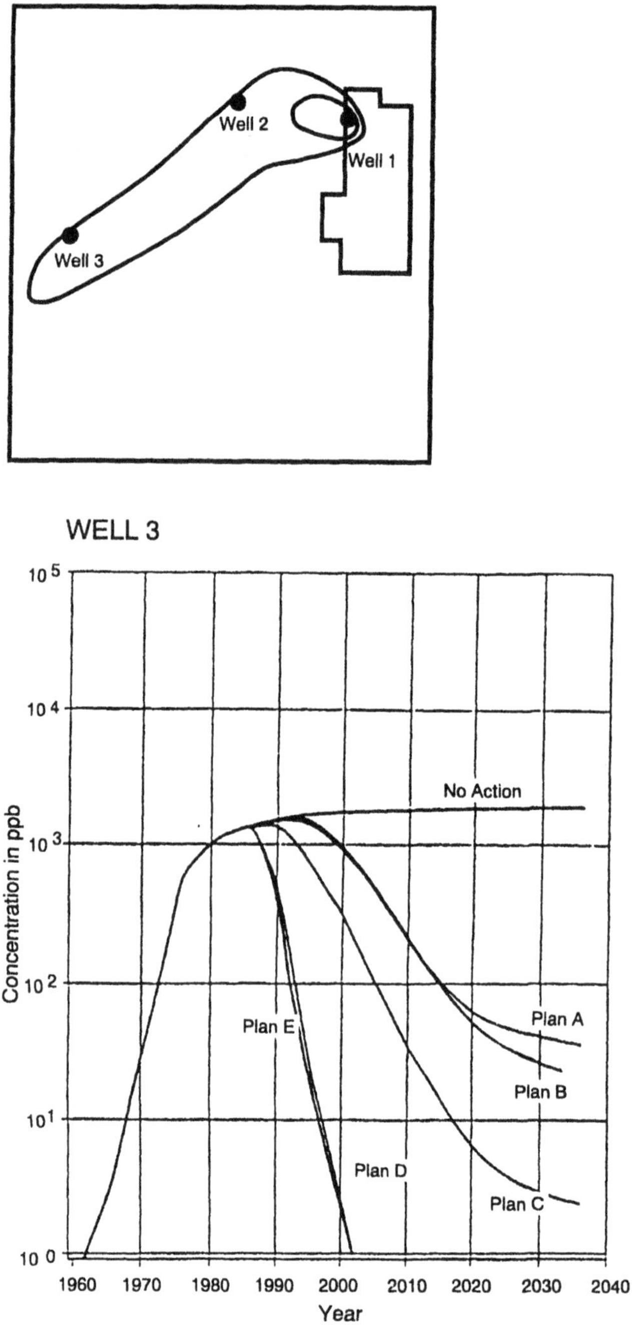

Figure 8.26 Presenting predicted concentrations for alternative remedial plans.

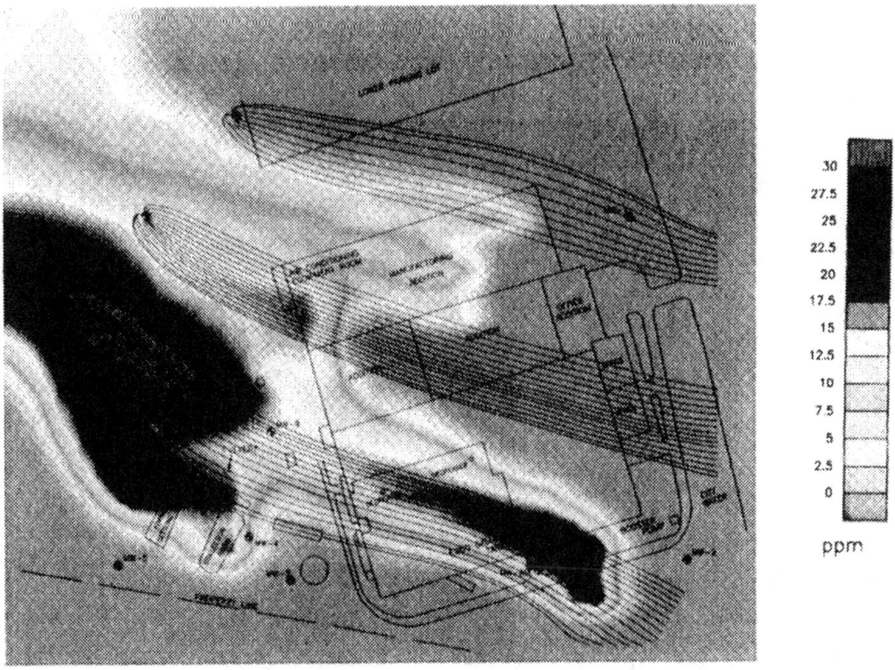

Figure 8.27 Incomplete capture of contaminant plume (provided by T. Franz, Waterloo Hydrogeologic, Inc.).

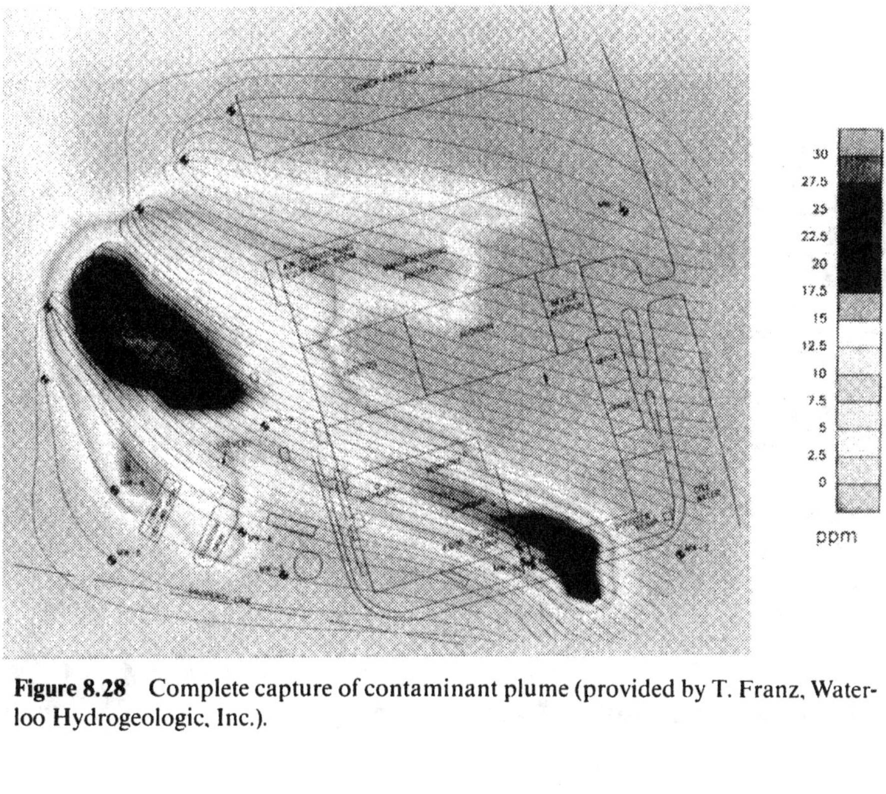

Figure 8.28 Complete capture of contaminant plume (provided by T. Franz, Waterloo Hydrogeologic, Inc.).

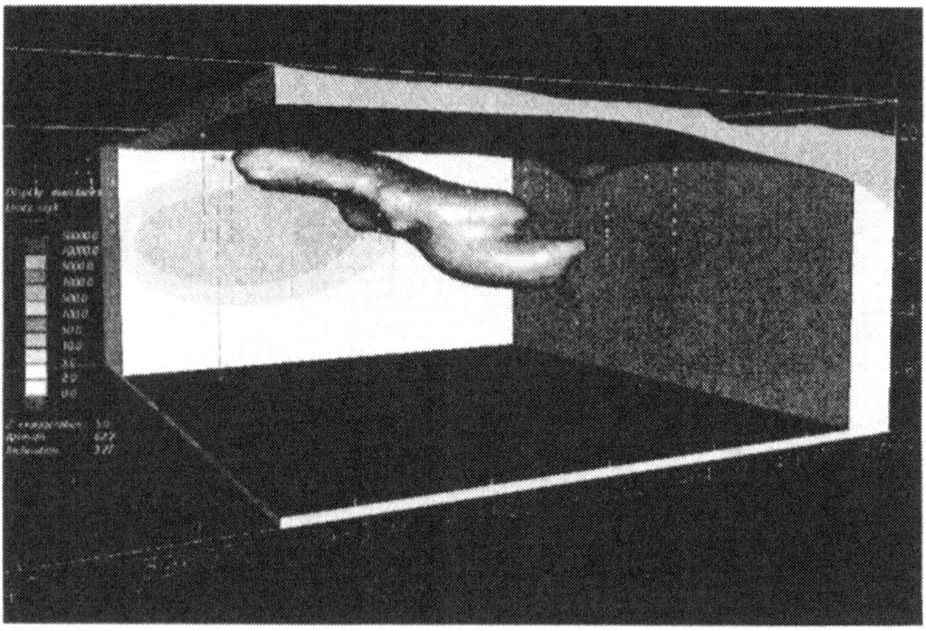

Figure 8.29 Predicted drawdown around oil-spill recovery wells.

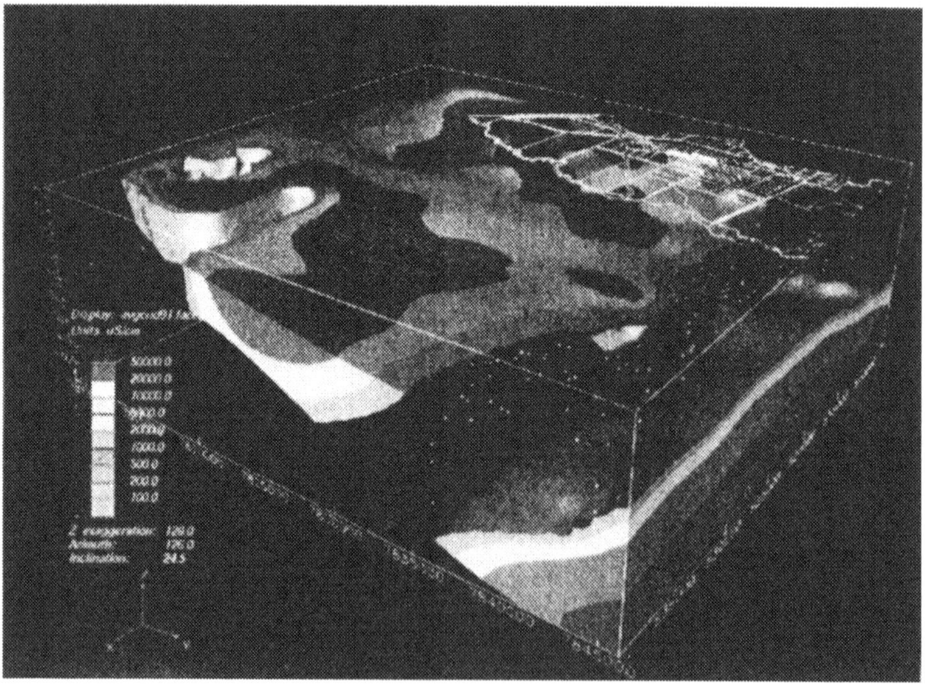

Figure 8.30 Observed concentrations in three dimensions.

TABLE 8.7 Summary of Common Modeling Errors

Error	Typical Result	Remedy
	Data Errors (90% of Errors Fall in This Category)	
Inconsistent units	Infeasible results	Plot all input data and use QA/QC
Misaligned formatted input	Wild results	Check mass balances and plot input and output
Infiltration rate greater than saturated vertical hydraulic conductivity or less than minimum unsaturated hydraulic conductivity in unsaturated flow models	Improbable hydraulic gradients and unstable behavior	Understand physical meaning of specified unsaturated moisture curves
	Conceptual Errors (The Most Difficult Errors to Identify)	
Mismatch between assumptions (e.g., extrapolation of model base to above extrapolated model water table)	Unrealistic predictions	Plot input and output data
Indiscriminate use of fixed value boundaries	Overconstrained model	Replace fixed head with flux boundaries
Misuse of general-head boundaries	These are essentially damped fixed-value boundaries	See above
Specification of all flux boundaries for steady-state run	Nonunique solution	Have at least one boundary cell with a fixed value
Use of fixed-head or flow boundaries in model cells much larger than the feature being simulated	Exaggerated recharge or discharge	Use properly sized model cells, or replace fixed value by fixed-flux boundaries
Unfocused model	Rough predictions	Plan in advance and coordinate with others involved
No simplification, e.g., high degree of heterogeneity	Unnecessarily complex, unwieldy model	Study data for patterns and spend time to develop defensible conceptual model on paper first

TABLE 8.7 (*Continued*)

Error	Typical Result	Remedy
	Code Errors (The Most Difficult Errors to Solve)	
Nondocumented or unrecognized model code limitations	Unrealistic or inconsistent results	Check predictions against expected results: solve by thorough understanding of code, talking to code author or using an alternate code
	Application Errors	
Model boundaries aligned with site boundaries/roads/compass directions, etc.	Potentially unrealistic results	Use natural boundaries
Model boundaries within influence of sources or sinks	Under- or overpredictions	Use appropriate model domain
Model cells and aspect ratios time steps inconsistent	Numerical dispersion	Check Peclet number, Courant number, and stability criteria when setting up advective-dispersive transport models
Model depth = depth of measurements	Biased or incomplete predictions	Use natural model boundaries
Failed model	Inconsistent or no results	Plan alternate approaches from the start
Uncertainty about use of model		Conduct failure analysis (cause/effect/avoidance of failure) or economic analysis (cost of model analysis vs. cost of decisions based on model analyses) or seek assistance

Supplying observed or interpreted concentrations as model input	Unrealistic redistribution due to unbalanced initial conditions	Specify sources rather than observations
Supplying observed or interpreted heads as starting point for transient run	Unrealistic redistribution due to unbalanced initial conditions	Use predicted, steady-state flow pattern as initial conditions
Application beyond range of validation	Poorly defended predictions	Validate model consistent with intended application
Supplying extreme changes in material properties	Unstable results	Use several model cells across interface or define intermediate-value zone
Contaminant plume partially exits model boundary	In predicted remediation the plume may be drawn back through the boundary at unrealistic concentrations, resulting in invalid results	Use larger domain
	Interpretation Errors	
Smoothed contours of predictions hide small-scale problems	Acceptance of incorrect results	Plot unsmoothed data initially
Use of unstable, unconverged or unbalanced results	Acceptance of results containing systematic error	Check mass balances, solution behavior, and plot interim results

8.12 AVOIDING COMMON MODELING ERRORS

The most reliable way of learning the best approach to conduct a modeling analysis is to make many experiments and/or mistakes and learn from them. However, in the interest of saving time and effort, this section provides a compilation of the most commonly noted modeling errors, and ways to avoid them (see Table 8.7). The types of error have leeen divided into five categories for ease of reference:

- Data errors
- Conceptual errors
- Code errors
- Application errors
- Interpretation errors

9 Special Topics

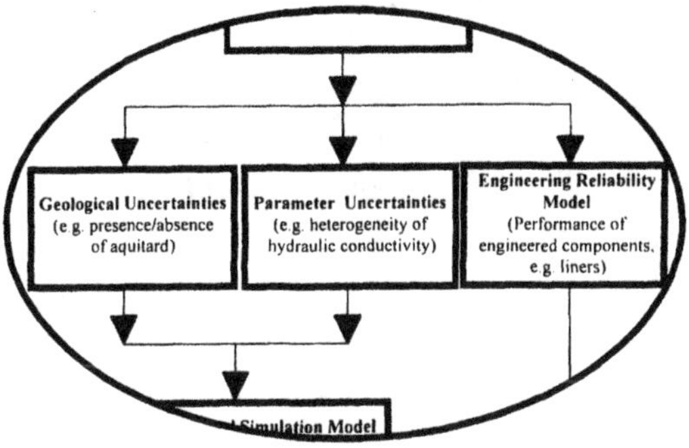

This chapter discusses topics subsidiary to developing a high-quality model but of immediate concern to a model user. These topics are:

- Model quality assurance
- Models in litigation
- Uncertainty and risk in modeling
- Models in a regulatory environment
- Training model users
- Access to groundwater models
- Modeling report
- Reviewing model studies
- Future trends in groundwater modeling

9.1 MODEL QUALITY ASSURANCE

Quality assurance (QA) is the process that is used to ensure that an adequate modeling analysis has been conducted. To a greater or lesser degree it is required in all modeling projects. QA for both code development and code application is discussed in this section.

The most important QA procedures in code development and maintenance are (van der Heijde 1987):

- Verification of model structure and coding
- Validation of the theoretical basis of the model

- Documentation of code development and testing
- Documentation of capabilities and use of code
- Scientific and technical reviews

The quality of a code application rests largely on the expertise of the model users. QA for model application includes the following issues, which are modified from National Research Council (1990):

- Project description and objectives
- Conceptualization used in model
- Description of explicit and implicit assumptions and their justification
- Data acquisition, interpretation, and extrapolation
- Model selection
- Model input data and their uncertainty
- Protocols and criteria for model calibration, validation, and sensitivity analyses
- Analysis and interpretation of results
- Calibration, validation, and sensitivity analysis results
- Presentation and documentation of results
- Evaluation of model utility versus project objectives

In a detailed study involving many parties, an extensive QA program in which these issues are discussed, agreed in advance, documented, and reviewed is appropriate. In a brief study these issues may be documented in the file and not formally reported. At a minimum the elements of model preparation, and the justification for model acceptability, should be presented. The list of instructions presented at the end of subsection 8.4.5 could form the basis for an informal QA procedure. Even labels, which can be attached to model result printouts, that encourage the checking of key items, such as model mass

---

Flow mass imbalance :
Solute mass imbalance :
Cell Peclet number :
Cell Courant number :
Convergence behavior :

Initials :	Date :
Checked :	Date :

Figure 9.1 Sample quality assurance label.

balance for flow and transport, solution behavior, adherence to grid and time step criteria, can provide an element of QA to the shortest study. A sample QA label is shown in Figure 9.1. Some of the ramifications of inadequate QA procedures are addressed in the following section.

9.2 MODELS IN LITIGATION

Litigation often revolves around the liability of various parties. The definition of liability is (Bradley 1983):

$$\text{liability} = \text{legal duty} + \text{breach of duty} + \text{damages caused by breach}$$

where

liability	=	an abstract or monetary measure of failure to fulfill obligations
legal duty	=	current standard practice (not state-of-the-art) and/or contractual obligations
breach of duty	=	failure to fulfill legal duty
damages	=	consequential adverse effects of breach of duty

Any modeling analysis may become evidence in litigation. Being prepared for lawsuits (litigation awareness) is warranted at all times, and this represents sound business practice. The following practices are recommended to limit liability related to modeling analyses:

1. Exercise sound judgment, that is:
   - Use a competent scientific and technical approach.
   - Cover all of the key technical, scientific, and related issues.
   - Document the work.
2. Negotiate any contracts carefully; write proposals or present technical approaches precisely.
3. Work closely with the other parties involved in a modeling analysis:
   - Keep other parties informed of all facts and decisions.
   - Document communications.

If a modeling analysis is being conducted expressly for the purpose of litigation, then the study must be admissible evidence. That is, the modeling analysis must be considered scientifically valid according to generally accepted modeling theories and methods. The validity of the study is measured by whether it would be generally accepted by technical peers. The modeling evidence must also be relevant to the issue under litigation, which is a judicial decision. The value of modeling results as evidence in court can be marginal because the model predictions may not be considered facts but rather opinions.

If the modeling study is to be presented in court, then the principal modeler may appear as an expert witness. In the court's eyes an expert is someone who can provide analysis or opinions beyond the presentation of facts. Expert opinion is expected to be neutral. An expert with experience in analogous situations has greater credibility. Testimony relating to the evidence, the presentation of facts relating to the modeling analysis, is not neutral. The testimony of a witness with firsthand knowledge is given greater weight than others on the modeling team. Preparation for a court case may be extensive. Refer to Bradley (1983) for details.

9.3 MODELING UNCERTAINTY AND RISK

This section presents some of the ideas behind these powerful tools. They enable model users to:

- Make their modeling analyses more useful to designers and decision makers..
- Take more part in the analyses that precede and succeed a traditional modeling study.
- Help to prevent inappropriate use of modeling results.

This section does not present details of the several disciplines and many techniques that are combined in uncertainty and risk analysis.

In a hydrogeological context, uncertainty in the model predictions is largely related to the heterogeneity in the system, as represented in Figure 9.2. The risk associated with a certain design being simulated in a model is linked to its probability of failure, as shown in Figure 9.3. The probability of failure in turn rests on the uncertainty of success of the design and related model predictions. The purpose of a risk-based analysis is to link the modeling analysis and uncertainty analysis with the practical questions of a facility designer. Examples of such questions follow:

- Are 5 or 10 wells needed to dewater or decontaminate this site?
- Will 1 or 2 liners serve to contain this waste?
- Will 100-m or 500-m monitor-well spacing be required for regulatory compliance?
- Is extra data collection worthwhile?

Freeze and others (Freeze et al. 1990; Massmann and Freeze 1989; Massmann et al. 1991) have developed a framework for the practical application of

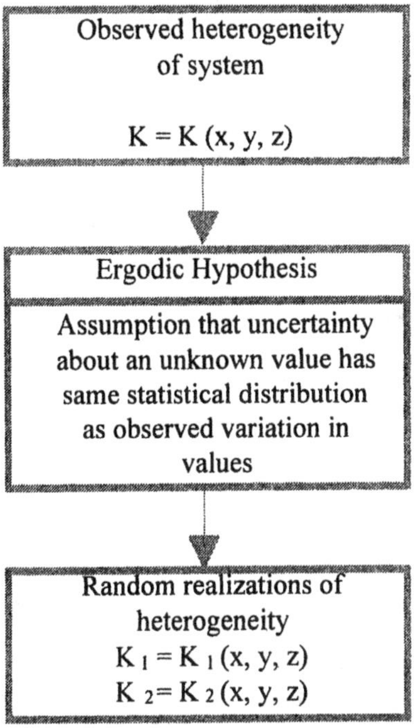

Figure 9.2 Modeling uncertainty.

a risk-based philosophy suited to hydrogeological applications. They combine elements of the disciplines of engineering reliability, Bayesian statistics, mining search theory, decision/risk analysis, and stochastic modeling in their approach. The overall problem that they address is how to achieve the technical objective (e.g., answer one of the questions listed above) subject to technical, legal, and political constraints, as well as the underlying uncertainty of the model predictions. The components of such a design framework are illustrated in Figure 9.4. A detailed analysis of this type is well suited to systems with large uncertainties and high risks (e.g., costly remediation schemes). Even if a full-scale analysis is not performed, this risk-based framework can be used in discussions of modeling studies, as a chart to clarity related interests.

9.4 MODELS IN A REGULATORY ENVIRONMENT

A regulatory agency is generally interested in the long-term value of clean groundwater to human and environmental health. To prevent adverse impacts, a regulatory agency will use performance standards (maximum permissible concentrations after remediation), compliance points (location,

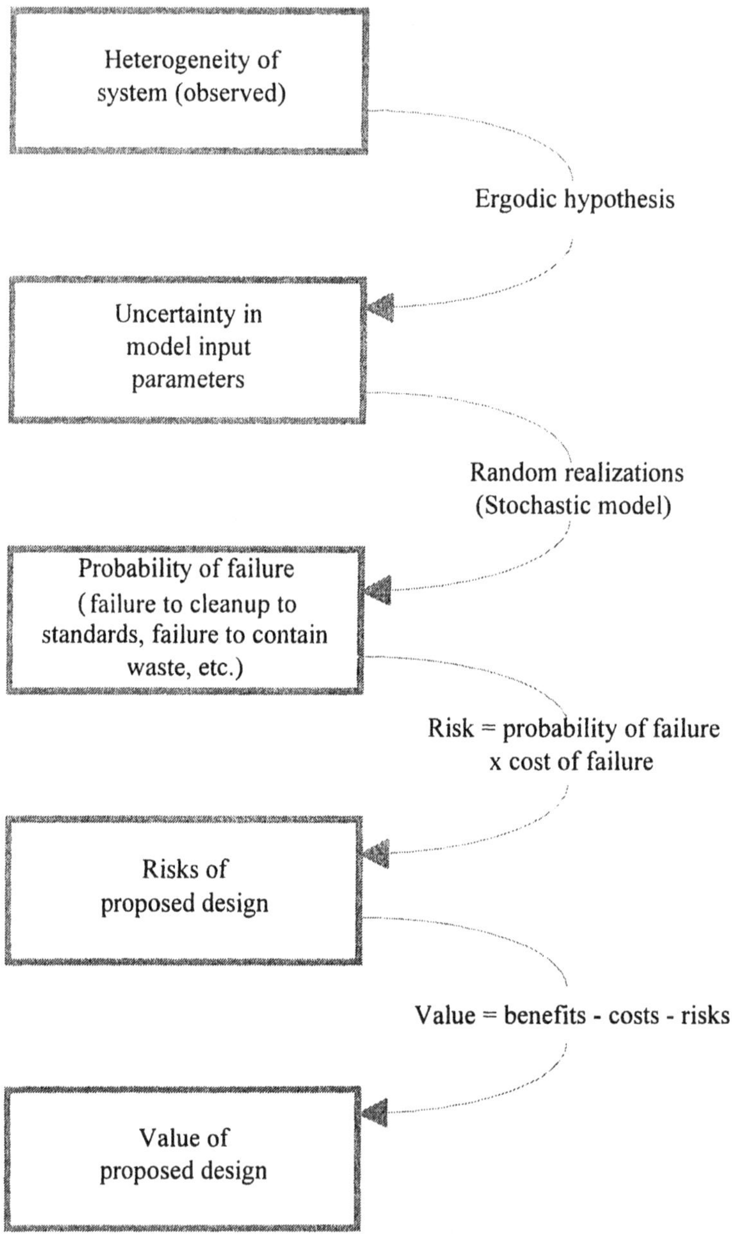

Figure 9.3 Modeling risk.

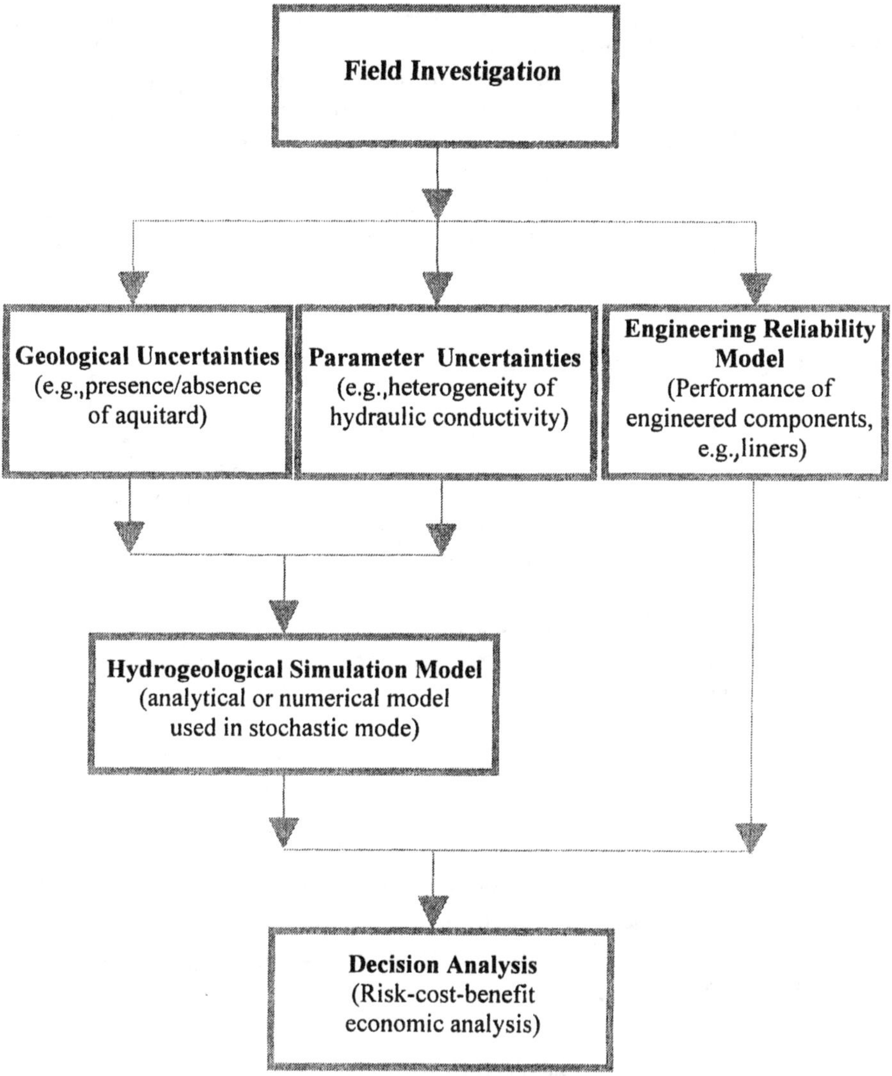

Figure 9.4 Components of hydrogeological design decision analysis (modified from Freeze et al. 1990).

frequency, and chemical-species analyzed) and regulatory penalties (performance bond, fines, and threat of closure) to enforce regulations. In contrast, the perspective of a facility operator may focus on likely revenue and the possible costs of failure. An operator may use the tools of site investigation, containment design, installation of a monitoring network and perhaps modeling analyses to select optimum designs and to demonstrate regulatory compliance.

Most regulatory agencies acknowledge the utility of groundwater modeling in combining diverse data into a tool that can be used to experiment with alternate future courses of action. The attitude of regulatory personnel has changed from one of awe of the sophistication of state-of-the-art models to a wary respect for model results. The regulations and guidance relating to model use in the United States indicate that the use of models is acceptable for a wide range of purposes, but they provide contradictory guidance on whether to use conservative or standard models (based on regulations) or more sophisticated models (based on model selection guidance).

9.5 TRAINING MODEL USERS

The model user is increasingly becoming a generalist who focuses a range of technologies and participates in subjective interpretations and decisions that affect the course of the modeling analysis. Some companies and institutions use designated specialists to oversee appropriate use of groundwater models. This approach may not be practical in every organization. There are many model training courses offered frequently. Some of these courses concentrate on particular groundwater models, while other classes concentrate on specific aspects of the groundwater problem. Model training classes should include the following considerations:

- Address participants key concerns, such as:
  - Which models are best and can I use them?
  - What is a generally accepted modeling approach?
  - How good does a model calibration have to be for it to be acceptable?
  - Where can models and their documentation be obtained?
  - Where can I get expert help if the modeling analysis goes wrong?
  - What effort is usually expended in a modeling study?
  - How do I present modeling results?
- Class length should not exceed three to five days.
- Thirty to 50% of the class should be spent in practical application to give depth to the theory lectures. Examples and problems should be designed to challenge the target audience.
- Two people at a computer terminal is optimum. Examples and problems should not take a long time to run.
- A mixture of presentation approaches (overheads, 35mm slides, video, blackboard, computer screen projected onto the overhead screen, etc.,) will keep the audience alert.
- Practical examples illustrating theoretical approaches or techniques generate enthusiasm.

- Interactive components to all sessions, for example, problems or discussion points, exercises, requests for alternate experiences, involve participants.
- Obtain reviews and suggestions from the training class participants.

Other aspects of lively and informative classes, or presentations of any kind, are described in books by Mambert (1985) and Munson (1984).

9.6 ACCESS TO GROUNDWATER MODELS

There are literally hundreds of groundwater codes in existence (Table 9.1). Some of these are experimental, many have received some practical use, and some have received widespread use and verification in field situations. Several databases maintain information on code capabilities, requirements, utility and history of use, including:

- EPA's IMES (Integrated Model Evaluation System) database
- International Ground Water Modeling Center's (IGWMC's) MARS (Model Annotation Search and Retrieval System) database
- American Geophysical Union monograph by Javandel et al. (1984)

These databases can be used to select an appropriate model (see also Section 8.3 on model selection). The codes themselves can be obtained from various sources listed in Table 9.2. User support for purchased groundwater models is provided to differing degrees. Groundwater modeling courses are provided periodically by IGWMC, NGWA, and TNO-DGV.

9.7 MODELING REPORT

Modeling results may be taken out of context, misused, misinterpreted, or ignored, and it is part of a model user's job to present the results in such a way that their basis is respected and significance appreciated. Model users should write reports to suit the intended audience without biasing interpretations. Possible audiences, and their interests in modeling results, are presented in Table 9.3.

The format for a modeling report varies with current expectations in modeling analyses (e.g., Are sensitivity analyses, error analyses, parameter estimation, and geostatistics standard practice?) and with the intended audience(s). Examples of differing formats are presented in books such as Anderson and Woessner (1992) and in Appendix A.

Academic texts and scientific papers are usually not ideal examples on which to base writing style. Cognitive theory, which is widely applied in

TABLE 9.1 Summary of Popular Groundwater Flow and Transport Models

Model Name	Dimensions	General Capabilities[a]	Code Status/ Availability[b]	Documentation[c]	Field Testing[d]	Experience/ Time Requirement[e]	Notes	Reference/ Author/ Contact	Distributor
Analytical Flow Models									
AQTESOLV									
EPA-WHPA	2	Semianalytical model used to define capture zones in homogeneous 2D steady groundwater flow subject to multiple injection and extraction wells. Monte-Carlo simulation of effect of alternate input parameters included as module.	Latest version 2.11 (4/93)			Moderate/ days			IGWMC
PRINCE		See analytical transport models section							
WALTON		See analytical transport models section							
Analytical Transport Models									
AT123D	3	Reacting solute transport from point, line, area, or volume source in constant parallel uniform flow. Radioactive, chemical, or heat waste can be simulated.	Latest version 1.22 (6/93)			Moderate/ days	Widely used; 3D version is reported to have some problems.	G.T. Yeh, ORNL	IGWMC, GEMS
EPA-WHPA	2	See analytical flow models section. Particle tracking post-processor for 2D flow models included as module.							
MAS3DB	3	Bivariate Gaussian source infinite in y–z plane, uniform flow field		Waterloo Hydrogeologic		Minimal/ hours		Robert Cleary	Waterloo Hydrogeologic
NODK2D3D DK2D3D	2,3	Inverse analytical transport model, with and without decay. Dispersivity, source dimensions, concentration and duration, and decay factor can be solved.	1990	Dale and Domenico (1990)		Moderate/ days	Cannot be applied indiscriminately.		P. Domenico, Texas A&M
PLUME3D	3	Solute transport from point source, uniform flow field		IGWMC		Minimal/ hours			IGWMC

PRINCE	1,2,3	10 analytical models including advective-dispersive transport with decay and retardation	Version 3.0 (1994)	Cleary and Ungs (1994)		Moderate/ hours	User-friendly, menu driven package.	Robert Cleary, Princeton Groundwater, Cleary and Ungs	Waterloo Hydrogeologic, Scientific Software Group
SLUG	2	Solute transport of a slug, uniform flow field		IGWMC		Minimal/ hours			IGWMC
SLUG3D	3	Solute transport of a slug, uniform flow field		IGWMC		Minimal/ hours			IGWMC
WALTON		Flow and solute transport for simple geometries		W.C. Walton		Minimal/ days	Software accompanies book.	W.C. Walton	Scientific Software Group
WMPLUME	2	Solute transport, uniform flow field, point source		IGWMC		Minimal/ hours			IGWMC
				Numerical Saturated Flow Models					
CFEST		See numerical solute transport/ saturated models section							
Golder Groundwater Package		See numerical solsute transport/ saturated models section							
GWTHERM		See numerical solute transport/ saturated models section							
HST3D		See numerical solute transport/ saturated models section							
MODFLOW	quasi-3	MODFLOW (USGS3D-MOD) is a finite difference model that simulates flow in three dimensions. Layers can be simulated as confined, unconfined, or a combination of the two. Flow associated with external stresses, such as wells areal recharge, evapotranspiration, drains, and streams, can also be simulated. The finite-difference equations can be solved using either the strongly implicit procedure (SIP) or slice-successive overrelaxation (SSOR).	Latest version 3.3 (3/91)	McDonald and Harbaugh 1989	Widely used	Extensive/ days	Can be linked to MOC, MODPATH, MT3D or PATH3D for transport calculations.	McDonald and Harbaugh 1989	IGWMC, USGS or Scientific Software Group
PORFLOW		See numerical solute transport/							

TABLE 9.1 (*Continued*)

Model Name	Dimen-sions	General Capabilities[a]	Code Status/ Availability[b]	Documentation[c]	Field Testing[d]	Experience/ Time Requirement[e]	Notes	Reference/ Author/ Contact	Distributor
		saturated models section							
SWENT		See numerical solute transport/ saturated models section							
SWICHA		See numerical solute transport/ saturated models section							
SWIFT II		See numerical solute transport/ saturated models section							
TARGET		See numerical solute transport/ saturated models section							
				Numerical Variably Saturated Flow Models					
CHAMP		See numerical solute transport/ saturated models section							
FEMWATER	2,3	Can be used to delineate wellhead protection areas using the assimilative capacity criterion. Simulates subsurface flows with (1) heterogeneous and anisotropic media consisting of as many geologic formulations as desired (2) considers distributed and point sources/sinks that are spatially and temporally dependent, and (3) accepts four types for boundary conditions (i.e., Dirichlet [fixed-head or concentration], specified-flux, Neumann [specified-pressure-head gradient or specified-dispersive flux], and variable). The variable boundary condition in 3DFEMWATER simulates evaporation/infiltration/ seepage on the soil-air interface.		Yeh and Ward (1979)		Extensive/ months	Can be linked to FEMWASTE for transport calculations.	Yeh and Ward	ORNL

GS3		See numerical solute transport/ unsaturated models section								
MOTIF		See numerical solute transport/ unsaturated models section								
NAMMU		See numerical solute transport/ unsaturated models section								
SUTRA		See numerical solute transport/ unsaturated models section								
SWANFLOW		See numerical solute transport/ unsaturated models section								
TARGET		See numerical solute transport/ unsaturated models section								
TOUGH		See numerical solute transport/ unsaturated models section								
TRACR3D		See numerical solute transport/ unsaturated models section								
TRUST		Precursor of CHAMP: see CHAMP								

Numerical Saturated Flow and Transport Models

AQUA	2	Transient nonisothermal, flow and transport in anisotropic, heterogeneous aquifers. Transport processes are advection, convection, sorption, and decay.		Extensive users manual		Moderate/ weeks	Includes interactive, graphical, data-generation options.			Scientific Software Group
BIOPLUME II	2	Simulates the transport of dissolved hydrocarbons under the influence of oxygen-limited biodegradation and reaeration and anaerobic biodegradation as a first order decay in hydrocarbno concentrations. The model is based on the USGS 2-D solute transport program. Convection, dispersion, mixing, and biodegradation are simulated. Hydrocarbon and oxygen plumes are computed for every time step. Instantaneous reaction between oxygen and hydrocarbon is assumed to	Latest version 1.01 (4/88)	Rifau et al. 1989	Applications reported in literature	Moderate/ hours	Code is hardwired for BTEX compounds.	Rifau		NTIS

TABLE 9.1 (*Continued*)

Model Name	Dimensions	General Capabilities[a]	Code Status/ Availability[b]	Documentation[c]	Field Testing[d]	Experience/ Time Requirement[e]	Notes	Reference/ Author/ Contact	Distributor
		simulate natural biodegradation processes, retarded plumes, and in-situ biorestoration schemes. Also injection wells can be specified as oxygen sources.							
CFEST	3	Transient, nonisothermal flow in anisotropic, heterogeneous, confined, semiconfined, or unconfined aquifers. Includes advection, dispersion, diffusion, sorption, decay, salt dissolution, and leaching.		Gupta et al. 1982	Verified against analytical and semianalytical solutions (Kincaid et al. 1984)	Extensive/ weeks	Predecessor FE3DGW was applied to 2 reported field cases (Thomas et al. 1982; Kincaid and Morrey 1984)	Gupta et al. 1982	BPNWL
CHEMFLO	1	Simulates the movement of water and chemicals in unsaturated soils. Water movement is modeled using Richard's equation. Contaminant transport is modeled using the convective-dispersion equation. Results of the modeling can be displayed as graphs of water content, matrix potential, driving force, conductivity, flux density, or chemical concentration over distance and time. Soil profiles are assumed to be homogeneous.	Latest version 8/89	Nofzinger et al. 1989	Unknown	Moderate/ hours	Three typical examples are presented in accompanying report.	Nofzinger	USEPA/ORD
FLONET: FLOWTRANS	2	Steady-state flow and transient transport in 2D, vertical cross section in anisotropic, heterogeneous, confined, or unconfined aquifers. Biodegradation and sorption may be simulated.	Version 2.2 (1994)	Guiger et al. 1994		Moderate/ days	Includes graphically oriented pre- and post-processors.	Guiger et al. 1994	Waterloo Hydrogeologic, Scientific Software Group

FLOWPATH	2	Steady-state flow and pathlines in anisotropic, heterogeneous, confined, leaky, or unconfined aquifers. Calculates pathlines, travel times, and velocities and capture zones.	Version 5 (1994)	Franz and Guiger 1994	Very widely used. Sample applications included in manual.	Moderate/ days	Includes graphically oriented pre- and postprocessors.	Franz and Guiger 1994	Waterloo Hydrogeologic, Scientific Software Group
FTWORK	3	Steady-state or transient flow and transport in anisotropic, hetero-geneous, confined, or unconfined aquifers. Transport processes are advection, dispersion, sorption, and decay. Parameter estimation available for steady-state flow.	Version 2.2 (1/90)	Faust et al. 1990	Has been applied at low-radio-active waste site. Other applications unknown.	Extensive/ weeks	Can be linked to particle tracking program GEOTRACK	Faust et al. 1990	Geo Trans
Golder Groundwater Package	2	Transient flow and reacting chemical transport in layered aquifer systems	1982	Golder Associates 1983	Widely used in Australia; field problem documented in Grove 1977	Moderate/ days		Marlon-Lambert and Miller 1978	Golder Associates
GWTHERM	2	Transient, nonisothermalf low in anisotropic, heterogeneous, unconfine aquifers. Includes advection, dispersion, diffusion, sorption, decay, and a particle-tracking option.			Used in nuclear waste repository studies (Runchal et al. 1979)	Moderate/ days	PORFLOW derived from this model	Runchal et al. 1979	Dames & Moore
HST3D	3	Simulates groundwater flow and associated heat and solute transport in three dimensions. The three governing equations are coupled through the interstitial pore velocity, the dependence of the fluid density on temperature, pressure and solute-mass fraction, and the dependence of the fluid viscosity on temperature and solute-mass fraction. The solute-transport equation is for a single solute species with linear-equilibrium sorption and linear decay. A complex well flow model may be used to simulate specified	Latest version 1.5 (11/91)	Kipp 1986	Unknown. Has been verified against 8 analytical solutions and com-pared to SUTRA.	Moderate/ days	Only available for VAX. Sometimes requires long computation times.	Kenneth Kipp, USGS, Denver. Kipp 1986	IGWMC

TABLE 9.1 (*Continued*)

Model Name	Dimensions	General Capabilities[a]	Code Status/ Availability[b]	Documentation[c]	Field Testing[d]	Experience/ Time Requirement[e]	Notes	Reference/ Author/ Contact	Distributor
		flow rate and pressure conditions at the land surface or within the aquifer, with or without pressure and flow-rate constraints.							
MOCDENSE	2	Density-coupled version of USGS2D-MOC.	Latest version 2.1 (1/93)	IGWMC		Moderate/ days	Mass balance errors tend to be large at early times, but decrease and stabilize over time.	Sanford and Konikow 1989	IGWMC
MODPATH		Steady-state transport pathline analysis bsed on flow results from MODFLOW.	1989	Pollock 1989		Moderate/ days		Pollock 1989	USGS
MT3D	2,3	Steady-state or transient transport in anisotropic, heterogeneous, confined, semiconfined or unconfined aquifers. Transport processes are: advection, dispersion, nonlinear sorption and decay. Requires results from block-centered finite-difference flow model, such as MODFLOW, as input.	Version 1.8 (10/92)	Zheng 1992		Moderate/ weeks	3D version requires careful use. Small grid sizes and time steps may be needed to ensure useful results. Uses MODFLOW results for flow input. Transient runs require large flow results data storage. Does not accommodate all MODFLOW options.	Zheng 1992	IGWMC or Papadopoulos & Associates
PORFLOW	3	Transient, nonisothermal flow and transport in anisotropic, heterogeneous fractures and confined semiconfined aquifers. Transport processes are advection, convection, dispersion, diffusion.	1981	Runchal 1985	Applications have included aquifer pollution, groundwater resource	Extensive/ weeks	Developed from GWTHERM	Runchal 1985	Analytic and Computer Research or IGWMC

		sorption, decay, multiphase transport (steam/water), and dissolution.			management, nuclear waste and geo-thermal storage (Runchal, 1987; Baca et al., 1988)				
SWENT	3	Transient, nonisothermal, density-coupled flow and transport in anisotropic, heterogeneous, confined aquifers. Transport processes simulated are advection, dispersion, diffusion, sorption, decay, salt dissolution, and leaching.		Intera Environmental Consultants 1983b	3 field cases (Intera, 1983; Sykes et al. 1982a,b)	Extensive/ weeks	Has post-processor capable of steady-state flow sensitivity analysis.	Intera	Intera Environmental Consultants
SWICHA	2,3	Transient, density-coupled flow and transport in anisotropic, heterogeneous, confined aquifers. Transport processes are advection, dispersion, and diffusion.	Version 5.0 (2/91)	Huyakom et al. 1984	Verified, validated and used by public agencies.	Moderate/ weeks		Huyakom et al. 1984	Geotrans
SWIFT II	3	Transient, nonisothermal, density-coupled flow and transport in anisotropic, heterogeneous fractures, confined, semiconfined, or unconfined aquifers. Transport processes are advection, dispersion, diffusion, sorption, decay, salt dissolution, leaching, and dual porosity.		Reeves and Cranwell 1981	Field cases reported: Ward et al. (1984, 1987a, 1987b)	Extensive/ months	Has been applied to heat storage, groundwater contamination, nuclear-waste isolation, and underground deep injection projects.	Reeves et al. 1986; Ward et al. 1984; Reeves and Cranwell 1981	SNL Geotrans
TARGET	2,3	Transient, density-coupled flow and transport in anisotropic, heterogeneous, confined, semiconfined, or unconfined aquifers. Transport processes are advection, dispersion, diffusion, sorption, and decay. A lake-level predicting module is included.	Version 4.42 (5/95)	Dames & Moore 1995	> 200 field applications, including USDOI (1986), Moreno (1989), Sharma et al. (1983).	2D: moderate/ days 3D: extensive/ weeks	Some users report instability at high Peclet numbers.	Selected refs: Moreno and Sinton 1994; Leppert and Bengsston 1990; Sharma 1981	IGWMC

TABLE 9.1 (*Continued*)

Model Name	Dimensions	General Capabilities[a]	Code Status/ Availability[b]	Documentation[c]	Field Testing[d]	Experience/ Time Requirement[e]	Notes	Reference/ Author/ Contact	Distributor
USGS2D-MOC	2	Simulates nonconservative solute transport in saturated groundwater systems. It computes changes in the spatial concentration distribution over time caused by convective transport, hydrodynamic dispersion, mixing, or dilution from recharge, and chemical reactions. The chemical reactions include first-order irreversible rate reaction, reversible equilibrium-controlled sorption with linear, Freundlich or Langmuir isotherms, and reversible equilibrium-controlled ion exchange. Aquifer may be heterogeneous or anisotropic.	Latest version 3.0 (11/89)	IGWMC or USGS		Moderate/ days		Konikow and Bredehoeft 1978	IGWMC
VLEACH	1	Simulates the leaching of a volatile, sorbed contaminant through the vadose zone. It models four main processes: liquid-phase advection, solid-phase sorption, vapor-phase diffusion, and three-phase equilibration. In its current version, VLEACH is subject to a number of major assumptions including steady-state unsaturated-zone flow, no "free product," no degradation.	Version 1.02 (1990)	CH2M Hill 1990		Unknown		CH2M Hill 1990	USEPA
Numerical Variably Saturated Flow and Transport Models									
CHAMP	3	Transient flow and transport in anisotropic, heterogeneous, fractured and deformable porous media. Transport processes are:	1986		Chemical transport in waste disposal	Extensive/ months		Narasimhan and Alavi 1986	LBL

		advection, dispersion, diffusion, sorption, and decay.			(Narasimhan and Alavi, 1986)				
FEMWASTE, 3DLEWASTE	1,2,3	Transient, density/moisture content coupling, transport in anisotropic, heterogeneous, porous media. Transport processes are: advection, dispersion, diffusion, sorption, decay, and 3 sorption models.	Latest version: 1993	Available from NTIS	Have been applied to seepage pond leakage and similar groundwater contamination studies.	Extensive/ months	Can be linked to FEMWATER for flow calculations.	Yeh and Ward 1930, 1931; Yeh 1937, 1939; Duguid and Reeves 1976; Kincaid and Morrey 1934; Morrey et al. 1936	ORNL
GS3	3	Transient flow with free surface evaporation, and transport in anisotropic, heterogeneous, confined or unconfined aquifers. Transport processes are advection, dispersion, and diffusion.		Davis and Segol 1985	2D version has been compared with field experiment (Davis and Segol 1985). Applications to chemical transport in saturated and unsaturated zones (Segol 1976).	Extensive/ weeks		Davis and Segol 1985	Bechtel
MOTIF	3	Transient flow and density-coupled transport in anisotropic, heterogeneous, confined or unconfined, fractured or porous media. Transport processes are advection, dispersion, diffusion, sorption, and decay.			Drawdown and inflow to deep shaft (Guvanasen et al. 1985). Tested under HYDROCOIN (Nea/Ski 1988)	Extensive/ months	Implemented on VAX, IBM, and CRAY computers.	Guvanasen et al. 1985; Reid and Chan 1987; Chan and Scheier 1987; Chan 1988; Chan et al. 1987.	AECL Pinawa, Manitoba, Canada
NAMMU	3	Transient flow and density-coupled transport in anisotropic, heterogeneous, confined or unconfined, fractured or porous media. Transport processes are advection, dispersion, diffusion, sorption, decay, and radionuclide chain decay.	Release 4 (1987)	Need for improved documentation noted under HYDROCOIN.	Sequence of tests through HYDROCOIN established model as state-of-the-art (Nea/Ski 1988)	Extensive/ months	Implemented on 3090, CRAY 2 and VAX 3600.	Herbert et al. 1987; Atkinson et al. 1984; Robinson et al. 1986	UKAEA Harwell Laboratory

TABLE 9.1 (*Continued*)

Model Name	Dimensions	General Capabilities[a]	Code Status/ Availability[b]	Documentation[c]	Field Testing[d]	Experience/ Time Requirement[e]	Notes	Reference/ Author/ Contact	Distributor
SUTRA	2	Transient, density-coupled, nonisothermal flow and transport in anisotropic, heterogeneous, porous media. Transport processes are advection, dispersion, diffusion, sorption, decay, and 3 sorption models.	Latest version 2.0 (11/91)	Voss (1984)	Field experiment (Voss 1984)	Extensive/ weeks	Has been applied to saltwater intrusion, thermal pollution and groundwater contamination studies.	Voss 1984	IGWMC
SWANFLOW	3	Transient, 3-pphase (water, air, and nonaqueous phase liquid) flow and transport in anisotropic, heterogeneous, confined or unconfined aquifers. Transport processes are advection in 3 phases, with no transfer of mass between phases.	Latest version 1.0 (4/91)		Verified by comparison with field experiment.	Moderate/ hours	Available in public-domain and proprietary (newer) versions. Some users report poor convergence.	Faust 1985; Faust et al. 1989	Geotrans, IGWMC
TARGET	2,3	TARGET2DU and TARGET3DU simuklate variably saturated flow and transport. See numerical, saturated flow, and transport models section for other details.					Unsaturated flow characteristics are hard-wired based on Van Genuchten formulation; Brooks/Corey is not supported		
TOUGH	3	Transient, nonisothermal flow and transport in anisotropic, heterogeneous, fractured or porous media. Transport processes are advection, diffusion, dissolution, and multi-phase flow.	Latest version 1986	Pruess (1986)	Nuclear waste isolation in partially saturated, fractured-porous media (Pruess and Wang 1984; Pruess et al. 1985; Rulon et al. 1986)	Extensive/ months	Based on a previous code, MULKOM, which was verified for heat and 2-phase flow (Pruess 1983; Pruess and Narasimhan 1985)	Pruess 1986	LBL

TRACR3D	Transient flow and transport in anisotropic,heterogeneous, deformable, fractured-porous media. Transport processes are advection, dispersion, diffusion, sorption, decay, multiple-species, two-phase flow, with stochastic parameter input.		Travis (1984)	Validated with respect to 3 field experiments (Travis 1984). Aplied to colloid transport (Travis and Nuttall 1987), tracer tests and hydro-chemical transport (Birdsell et al. 1988; Thomas 1988)	Extensive/ months	Travis 1984	LANL
			Selected Geochemical Models				
BALANCE	Chemical equilibrium speciation model used to define and quantify chemical equilibrium reactions between groundwater and minerals. Using (1) the chemical compositions of water samples from two points along a flow path and (2) a set of mineral phases hypothesized to be the reactive constituents in the system, the program calculates the mass transfer necessary to account for the observed changes in composition between the two water samples. Additional constraints can be included in the problem formulation to account for mixing of two end-member waters, redox reactions, and, in a simplified form, isotopic composition.	1982	Plarkhurst et al. 1982		Moderate/ days	Parkhurst, USGS	USGS

TABLE 9.1 (*Continued*)

Model Name	Dimen-sions	General Capabilities[a]	Code Status/ Availability[b]	Documentation[c]	Field Testing[d]	Experience/ Time Requirement[e]	Notes	Reference/ Author/ Contact	Distributor
EQ3/6		Capabilities range from simple solubility calculations to complex kinetic reactions. Can simulate both closed and flow-through systems.	1990	Wolery 1979	Mineral-groundwater interactions (Kerrisk 1984), ruthenium migration (Isherwood 1984) and sedimentary ore formation (Garven and Freeze 1984a,b)	Extensive/ weeks	Related to PATHI, MINEQL, WATEQ	Wolery 1979	LLNL
MINTEQ		Calculates complex chemical equilibria using 6 sorption models	1983	Felmy et al. 1983	Numerous field applications, including clay liners (Peterson et al. 1983), and nuclear waste disposal (Krupka and Morrey 1985)	Moderate/ days	Parent of ECHEM (part of FASTCHEM)	Felmy et al. 1983	EPRI-PNL
PHREEQE		Chemical equilibrium reaction-path model for 19 elements, 120 aqueous species, organics, 3 gases, and 3 redox elements. Simulates ion exchange sorption and 21 precipitation/dissolution minerals.	Version 2.0 (2/92)	Parkhurst et al. 1980	Nuclear waste disposal (Plummer and Parkhurst 1985);	Moderate/ days	Related to MIX	Parkhurst, ESGS Parkhurst et al. 1980; Plummer et al. 1983; Intera Environmental Consultants 1983a,d	USGS

				Seawater evaporation (Plummer and Parkhurst 1990)				
WATEQ4F	Calculates chemical equilibrium in natural waters. It models the thermodynamic speciation of major and important minor inorganic ions and complex species in solution for a given water analysis and in situ measurements of temperature, pH and redox potential. From this model the states of reaction for the water with solid and gaseous phases are calculated. The examination of reaction states may suggest the origin of the dissolved constituents and assist in the prediction of the chemical effects of groundwater production, recharge, irrigation, and inorganic pollution.	1987	Ball et al. 1987	Used extensively in USGS field investigations (Ball et al. 1981; Krupka and Jenne 1982)	Moderate/ days	Related to WATEQ series. Thermodynamic database is probably the best available.	Ball et al. 1987 Nordstrom et al. 1990	USGS
Selected Hydrogeologic and Geochemical (Hydrochemical) Models								
CPT	Coupled chemical equilibrium speciation and groundwater flow with precipitation/dissolution	1983		Validated with respect to several field experiments (Tsang and Doughty 1985)	Extensive/ months	Based on PT	Mangold and Tsang 1983	PNL
DYNAMIX	Couples TRUMP (flow and transport) and PHREEQE (equilibrium chemical reactions). Accounts for aqueous speciation, ion-exchange, and precipitation/ dissolution.	1989		Compared to field data (Narasimhan et al. 1985, 1986)	Extensive/ months		Narasimhan et al. 1985	LBL

TABLE 9.1 (*Continued*)

Model Name	Dimen-sions	General Capabilities[a]	Code Status/ Availability[b]	Documentation[c]	Field Testing[d]	Experience/ Time Requirement[e]	Notes	Reference/ Author/ Contact	Distributor
FASTCHEM		Package of codes for flow, geochemistry and transport. Geochemistry model is ECHEM (based on MINTEQ). Transport model is based on SATURN. Accounts for aqueous speciation, ion-exchange, surface complexion, and precipitation/dissolution.	1989	Hostetler and Erikson 1989	Shares validation of parent codes.	Extensive/ months	Intended to be used to study waste disposal from electric utility fossil fuel plants	Hostetler and Erikson 1989	EPRI-PNL

Note:

AECL	Atomic Energy of Canada Limited
BPNWL	Battelle Pacific Northwest Laboratories
EPR-PNL	Electric Power Research Instsitute, Pacific Northwest Laboratory
GEMS	Graphical Exposure Modeling System
IGWMC	International Groundwater Modeling Center
LANL	Los Alamos National Laboratory
LBL	Lawrence Berkeley Laboratory
LLNL	Lawrence Livermore National Laboratories
ORD	Office of Research and Development
ORNL	Oak Ridge National Laboratory
SNL	Sandia National Laboratories
UKAEA	United Kingdom Atomic Energy Authority
USDOI	United States Department of the Interior
USEPA	United States Environmental Protection Agency
USGS	United States Geological Survey

[a] These descriptions are derived from Users Guides, the EPA Integrated Model Evaluation System Database, Mangold and Tsang (1991) and Hall et al. (1990).
[b] Latest version is assumed to be the date of the Users Guide unless other versions are reported in the literature.
[c] Documentation is highly variable in content and standard.
[d] Validation with respect to field experiments, field audits and reviewed field applications are listed.
[e] Model author estimates are presented where available (about 50% of models listed). Remainder are user estimates, or estimates based on use of similar codes.

TABLE 9.2 Model Code Sources

Code Source	Location	Code Description
International Ground Water Modeling Center (IGWMC)	Colorado	Codes of widespread use and relatively low cost. Also custodian for archived codes
Scientific Software Publishers	Virginia	Codes with pre- and postprocessors and higher-priced proprietary codes
National Ground Water Association (NGWA)	Ohio	User-friendly codes and codes with books
Rockware International	Colorado	Codes with geological application
United States Geological Survey (USGS)	Virginia	USGS codes
TNO-DGV Institute of Applied Geoscience	Delft, the Netherlands	Same codes as IGWMC
Waterloo Center for Ground Water Research, University of Waterloo	Ontario, Canada	Some practical and some research codes
Code authors	Consultants, universities, and publicly funded government laboratories	Proprietary, research codes

TABLE 9.3 Modeling Report Audience

Audience	Interest
Managers of facility, facility operators	General conclusions, costs, and duration of engineering solution
Engineers	Specific conclusions, detailed concentrations and hydraulic responses through time and at specific locations, and accuracy of predictions
Regulators	Specific conclusions, degree of calibration, validation, and reliability of model, comparisons of alternate future actions in terms of performance standards and compliance points.

marketing literature but rarely applied in scientific writing, demonstrates that writing that directly addresses the readers viewpoint and is formatted for ease of comprehension is more readable. For example, the use of interspersed lists instead of continuous text will focus the reader on important points and be less tiring to read. Awkward passages can be dramatically improved by verbally explaining the section to an imaginary audience and then writing down the spoken version. Since the modeling report is not likely to be read several times in its entirety, the main points need to be reiterated to ensure comprehension. Important conclusions can be restated in the introduction, main text, and conclusions. For example, in this book model boundary conditions, which may be the single most important decision in a modeling analysis, are discussed three times with different emphasis.

The figures and tables are the key to a usable report. They should be as simple as possible while focusing on the main points to be illustrated. Background information and maps are better relegated to appendixes or a separate technical volume. Specific suggestions for modeling illustrations are given in Section 8.8.

The most difficult part of a modeling report is the discussion of how well the model represents reality. The problem is that the conclusions on this are a matter of opinion and this opinion, will be based on the experience and training of the reviewer or reader. It is easy for the modeling results to be misrepresented or misunderstood. The facts need to be clearly presented and inferences explained. Here are some examples of what may occur when comparing model predictions with site observations:

- The observed data may be so variable in time and/or space, and variable in accuracy, that the model cannot be expected to well represent the entire data set. There are scales of heterogeneity that cannot be incorporated in the model. Subsets or averaged data based on the scale or duration of interest, may be used to generate a more usable data set. The variability of the observed data, over model-cell distances and times, may be used as a measure of the expected correlation between predicted and observed results. Sensitivity analyses and stochastic analyses will allow you to take a more objective approach to quantifying the accuracy of results, and may point out the need for additional definitive data.
- The purpose of the model defines the required absolute accuracy, and hence reality, of the model predictions. For example, a model used to compare the effectiveness of alternate actions may not need to be as accurate as a model used to predict influent concentrations for a water treatment plant.
- The site-specific data used as model input may be insufficient to define a realistic model. There are various ways to interpolate and extrapolate data meaningfully and make the best use of the available information (see subsection 8.4.5). Alternatively you can define zones of more or less confidence in the model predictions based on a combination of correla-

tion between observations and predictions and degree of definition provided by site-specific input data.

- The model results may be so hedged with uncertainty as to prevent direct application of the predictions. This model may be used as a basis for general conclusions about site conditions, or to generate recommendations for improving site characterization.
- The model predictions may be, or appear to be, inconsistent with the observations. This situation can be dealt with by model error analysis, as described in Section 8.5 on model calibration.

Your familiarity with the site and setting, gained by drawing together the data and preparing the model, will make you the best judge of the utility of your model and its predictions.

9.8 REVIEWING MODEL STUDIES

A constructive model review can provide focus, balance, and depth to a model report. The model user simulates phenomena that cannot be directly observed and makes predictions for a system that is not fully understandable. In modeling projects there is no absolute right or wrong, but some model predictions are more realistic than others. In reviewing groundwater models, you search for the plausibilty of model results, such as:

- Model conclusions must follow logically from reasonable hypotheses. Model input data must be reasonable, and the conceptual model must be appropriate and supported by data.
- Model data and results must be presented comprehensively and honestly; evidence that contradicts model conclusions or that impose restrictions on model accuracy should also be provided.
- Model conclusions must be well supported. In many cases sensitivity analyses strengthen model conclusions and can provide a safeguard against the possibility of errors, coincidences, unrealistic or nonunique solutions, and misuse of predictions.
- The modeling study should focus on the geology and hydrology of the site rather than the numerics of the model.
- The model should not serve "to artfully turn garbage to gospel by three-dimensional, color-coordinated graphics" (Lehr 1990).

Stochastic modeling establishes confidence intervals (i.e., that a particular result is more or less likely) for groundwater predictions. You might conclude that stochastic modeling is the safe and correct modeling approach. However, stochastic analyses are bound to fail if the underlying conceptual model is not appropriate or if there are insufficient data to support the analysis. No matter what model is used, a model reviewer must check, step by step, the plausibility of the modeling approach. Table 9.4 lists relevant questions.

Perfectly accurate models do not exist. Therefore the model purpose should be considered in reviewing a model study. Models designed for general analyses may lack detailed calibration or sensitivity analyses and still produce useful results. Models applied to the initial design of a remedial measure may use part of the field data, ignore unrelated questions, and still solve the problem at hand. However, models designed for detailed applications that claim accurate predictions at key locations or times require accurate, plentiful input data, calibration, and sensitivity analyses. In summary, the model purpose and model effort should balance.

TABLE 9.4 Questions to Use in Reviewing a Model Study

Natural System

- Is the natural system understood?
- Are there sufficient data to support a model?
- Are model input data reasonable?
- Are relevant model input data missing?
- Are natural boundaries appropriately defined?
- Is groundwater flow density-dependent?

Investigated Problem

- Does the purpose of the model address the needs of project?
- Does the model clarify the investigated problem?
- Is modeling cost-effective?
- Does an alternative to modeling exist?
- Is the level of sophistication appropriate?

Conceptual Model

- Is the simplification of the aquifer system appropriate?
- Are model boundaries plausible and unrestrictive?
- Is the model domain appropriate?
- How many dimensions are needed? Which ones?
- Are the governing processes known and incorporated?
- Is the problem steady or transient?

Model Calibration and Validation

- Are time and space discretization consistent?
- How is calibration achieved?
- Are calibration criteria appropriate to the model purpose and amount of data?
- Is a complex distribution of model input data the result of:
    - Insight into the groundwater system?
    - Experience in trial-and-error calibration?

TABLE 9.4 ***(Continued)***

Automated calibration using inverse models?
Combination of the above?

- How much data are available for model calibration and verification?
- Is the time period of calibration comparable to the predictive time frame? If not, are the uncertainties in predictions discussed?
- Are discrepancies between predicted and observed values sufficiently discussed?

Model Sensitivity Analysis

- Are key input data varied over reasonable range?
- Are sensitivity results discussed in relation to model prediction accuracy, applicability, and use?

Model Application

- Are model results plausible?
- Are model simulations affected by arbitrary boundary conditions?
- Are model results supported in general by independent simplified analytical results?
- What is known about the accuracy of the model results?
- What is known about the predictive capability of the model?
- Does the calibration support the model application (e.g., use of transient calibration to support transient prediction)?
- In transport modeling, is numerical dispersion discussed or checked by reducing cell sizes and comparing predictions?
- Are mass balances calculated appropriately and checked?
- Are results presented to significant figures supported by the model predictive accuracy?
- Do model conclusions step beyond the limits set by model resolution in time and space?
- Are model-based conclusions reasonable and well supported?

9.9 FUTURE TRENDS IN GROUNDWATER MODELING

This section briefly discusses some of the research trends and application trends in groundwater modeling. Research into groundwater flow processes is focusing on fracture flow (Smith et al. 1990), unsaturated flow (Custodio et al. 1988), multiphase flow (Abriola 1988), nonaqueous phase flow (Faust et al. 1989), gas flow (Thorstenson and Pollock 1989), and unsaturated flow in fractures (Wang and Narasimhan 1990). Since simplified solutions to unsaturated flow problems are currently available, research of immediate practical interest is that on flow in fractures and nonaqueous phase flow. Nonaqueous phase-flow modeling benefits from data and research from the oil and gas industry and is already leading to practically applicable models (Faust et al. 1989; Mercer and Cohen 1990). However, despite the efforts expended on fracture flow, models currently available are inadequately validated due to lack of data (Evans and Nicholson 1987; Pruess and Wang 1987).

Areas of transport modeling research include solute transport in fractured media (Tsang and Tsang 1989; Wels and Smith 1994), the proper mathematical description of dispersion (Gelhar 1992; Neuman 1990), chemical reactions due to biological processes (Baveye and Valocchi 1989), unsaturated transport (Wierenga and Bachelet 1988; Pruess and Wang 1987; Nielsen, van Genuchten, and Biggar 1986), gas-phase transport, and chemical reactions in the subsurface (Melchior and Bassett 1990). Of immediate practical interest are transport in fractured media (which await better fracture-flow models), description of dispersion, and chemical reactions. Despite the volume of data on observed or modeled dispersion, no unifying, implementable mathematical description is available. Transport models that simulate chemical reactions due biological processes are evolving with experience gained in bioremediation projects. Models of nonbiological reactions are developing along various paths. Sorption models based on equilibrium partition coefficients are commonly used in groundwater transport models. This approach is limited by the fact that the partition coefficient varies in space and time due to the underlying mechanisms of sorption, and the effects of pH or other geochemical conditions. Alternate, more realistic, approaches are offered by the ion-exchange or surface-complexation models invoked in hydrochemical models (geochemical models embedded in flow and transport models), which are starting to be applied to practical cases. Due to the computer resources, databases, and model user experience required, these models cannot yet be classified as practical models.

Trends in the practical application of models are listed and briefly discussed below:

- Stochastic models. Probabilistic simulation will be in demand for strongly heterogeneous sites, long-term predictions, and where model predictions cannot be readily verified by sampling.
- More efficient numerical solutions to complex problems. A range of solution algorithms each suited to a particular suite of problems is evolving, also adapting algorithms to parallel processing and adaptation of mainframe codes to personal computers is adding more codes to the list of practical models.
- Interactive models with data pre- and postprocessing. Interface codes and/or interfaces supplied with codes are allowing more rapid, although perhaps not more accurate, application of models. Such codes do not supplant the need for model user training and experience.
- Parameter-estimation and parameter-identification models. Embedded routines or wraparound codes increasingly provide more options to check or estimate input data distributions and their uncertainty.
- Expert decision-support systems. Use of groundwater modeling results in combination with related considerations including: risk factors, local

regulations, aversion to or acceptance of risk, cost, benefit, and multiple interdisciplinary models. This trend is similar to that of surface-water decision-support systems.

- GIS-based groundwater models. GIS systems often use a similar database to groundwater models, and so some interwoven systems are emerging.
- Regulatory-driven changes. There are trends toward using publicly funded codes, toward developing ASTM procedures to define expected standards of groundwater modeling practice, and toward requiring the use of models to define the expected duration of groundwater contamination.
- Industry-driven changes. As end users become more sophisticated, there are trends toward more direct use of models by end users, toward higher expectations from modeling projects, and toward the use of models as long-term management tools.

APPENDIX A
Groundwater Modeling Case Studies

This appendix contains five case studies illustrating a wide range of applications of groundwater models. Most of these case studies are arranged with sections in the following order:

- Specifying the modeling problem
- Evaluating the real world conditions
- Developing the conceptual model
- Selecting model input data
- Selecting the model and discretization
- Calibrating the model
- Performing model simulations
- Model study summary

While these case studies also cover some of the areas included in a typical model report, the description of each case study is too brief to be complete. It is not intended to lead the reader through a detail analysis covering all aspects of model selection, calibration, validation, modification, and application. For more complete descriptions of case studies, refer to Bredehoeft et al. (USGS Water Supply Paper 2237), three case studies presented in Mercer and Faust (1981) Chapters 4 and 5, and case studies presented in National Research Council (1990), Chapter 5. The selected model studies presented here are:

A.1 Applying a Black Box Model to Design a Groundwater Management Plan for the Südhessische Ried, Germany (single-well model).

A.2 Modeling the Groundwater System of the Semarang Basin, Indonesia (two-dimensional depth-integrated model).

A.3 Design of Monitoring Network (combination of two models).

A.4 Saltwater Intrusion Modeling for Coastal Aquifer System in Florida (two-dimensional cross-sectional and three-dimensional model).

A.5 Site Remediation (phased two- and three-dimensional models).

A.1 APPLYING A BLACK BOX MODEL TO DESIGN A GROUNDWATER MANAGEMENT PLAN FOR THE SÜDHESSISCHE RIED, GERMANY

Groundwater in the area of interest, the Südhessische Ried located south of Frankfurt, Germany, is extensively tapped for drinking water supply and for industrial and agricultural use. Natural groundwater replenishment is limited. In periods with poor rainfall, groundwater demand has to be matched by release of stored water in the aquifer. The decline of the groundwater, however, is restricted in areas with vegetation sensitive to significant changes in the natural water table. Groundwater shortage in the entire region was experienced in 1976 due to two consecutive years with low annual rainfall.

To allow effective groundwater management, Rhine River water injection into the aquifer started in 1989 for controlled groundwater replenishment. In comparison to the uncontrollable natural groundwater recharge in the region, the artificial injection of water helps limit groundwater fluctuation in defined areas, increase the total groundwater volume, and decrease groundwater stress for the region.

A.1.1 Specifying the Modeling Problem

Groundwater management in the region must satisfy the following restrictions:

- Groundwater discharge and groundwater recharge must average in the long term, although annual discrepancies are acceptable.
- Strict limits for groundwater fluctuation for three zones protect local vegetation.

The purpose of the model study was to design a groundwater management plan that demonstrates possibilities and limits of groundwater management based on controlled water injection. In particular, the study was to produce guidelines about how to relate injection rates to natural groundwater replenishment and to actual groundwater production to ensure stable groundwater levels in ecologically sensitive areas.

A.1.2 Evaluating the Real World Conditions

The area of interest is part of the Upper Rhine River Valley. To the west and to the east, mountain ranges border the alluvial plain. The overall thickness of the unconsolidated deposits in the Rhine River Valley reaches more than 3000 m. Only the upper part of approximately 120-m thickness, consisting of fluvial sand and gravel, yields significant amounts of water. Deeper groundwater becomes increasingly salty.

Hydrogeological parameters of the aquifer are well-known. The hydraulic conductivity with a mean value of 3×10^{-4} m/s is in the typical range for comparable deposits. The specific yield of 0.15 dominates release/storage of groundwater.

It is estimated that recharge due to precipitation contributes roughly two-thirds of the natural groundwater replenishment in the area of interest. Rainfall from November to April averages to 270 mm or 40% of the total mean annual rainfall of 660 mm. Fluctuation of the annual rainfall from a low of 360 mm to a high of 1020 mm was observed. One-third of the annual rainfall reaches the water table, most of it in the period from November to April. Groundwater recharge is delayed after the rainfall event due to the passage through the unsaturated zone, which measures up to 10 m in the investigated area.

A fault running from the south to the north separates the Hessisches Ried from the mountain range in the east. Subsurface flow in the crystalline rock is small and is limited to flow in fractures.

Numerous creeks, originating in the mountains to the east, flow westward through the investigated area and empty into the dominating Rhine River. While the Rhine River communicates directly with the aquifer system (prescribed head boundary), the magnitude of water exchange between surface water and aquifer has to rely on estimates.

Groundwater production in the area is well-known and documented. The main groundwater user in the area using approximately 20 million m^3 per year is a waterworks serving approximately six million people. Agricultural, commercial, and industrial activities consume an additional five million m^3 per year.

Artificial groundwater recharge was installed in 1989. Infiltration is induced by nine infiltration ditches and several wells. Planning is to inject up to 12 million m^3 per year depending on the natural groundwater recharge and the actual discharge figures.

A.1.3 Developing the Conceptual Model

Key elements of the conceptual model were defining the model boundaries, simplifying the aquifer system, and determining the required model dimensions to produce meaningful conclusions: A description of each of these elements follows.

- Selecting model boundaries. The fault in the east represented a natural boundary (flux boundary) of the investigated aquifer system. Other natural model boundaries did not exist. The Rhine River is too far away from the area of interest to be useful. Meaningful boundaries were found by estimating the total catchment area of the investigated groundwater stress area, as shown in Figure A.1.1. The border of the catchment area represented a groundwater divide and served as a no flow boundary. The estimate of the catchment area was based on field measurements of

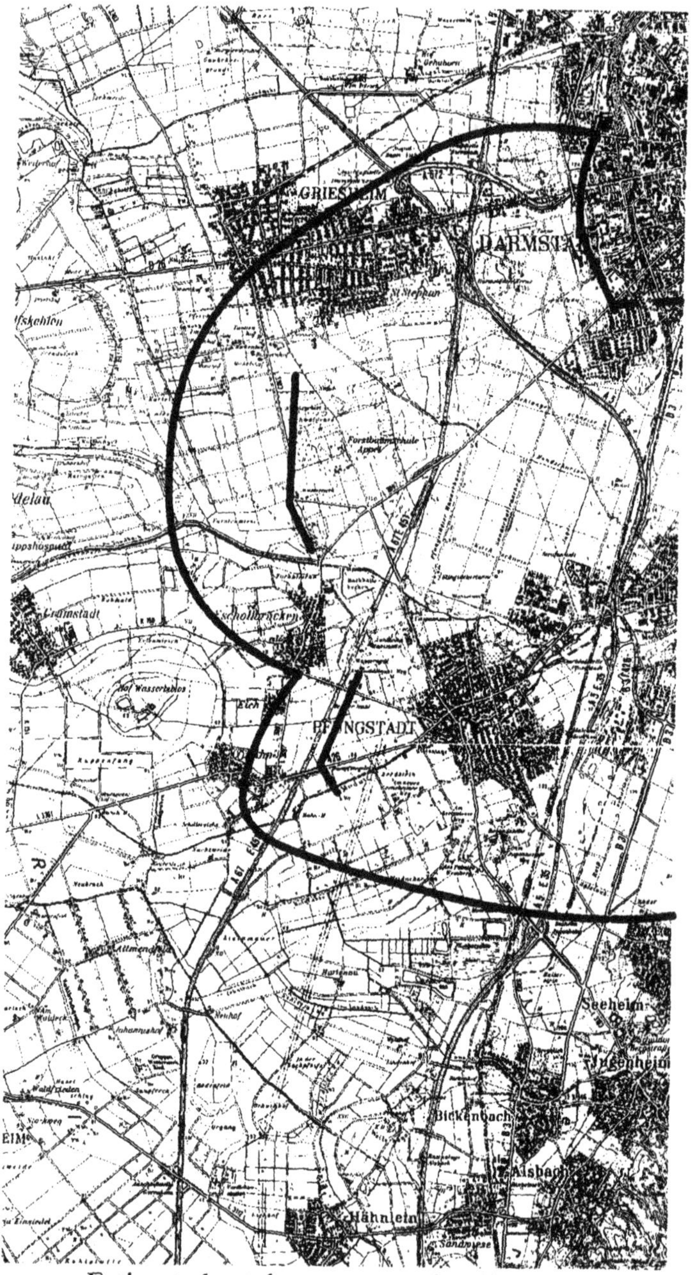

– – – Estimated catchment area of wells

—— Water wells

Figure A.1.1 Site location.

groundwater levels and an approximate regional water balance. It was accepted that in reality the size of the catchment area will vary depending on the difference of groundwater inflow and outflow.

- Simplifying the aquifer system. Partly clay layers separate the water carrying deposits into an upper and lower part. These layers are, however, believed to be discontinuous, so the aquifer is seen as one continuous hydrogeologic unit.
- Determining required model dimensions. The groundwater management plan was designed using the most basic groundwater model, the global water balance or black box model, as shown in Figure A.1.2. Black box models represent the aquifer as one single control volume and represent a zero-dimensional modeling approach. Groundwater levels are not calculated as a function of space but as regional averages.

A.1.4 Selecting Model Input Data

Black box models do not require information about hydraulic conductivity. They focus on the regional water balance components including stored and released water. All existing data pertinent to groundwater inflow and outflow to the system are compiled. Several simplifications of the investigated groundwater system allow the specification of the model input data. The main assumptions are:

- The catchment area is constant.
- In the year of the study, groundwater production and natural groundwater replenishment balance.
- Mean annual groundwater recharge is due to precipitation.
- Groundwater recharge due to precipitation and groundwater production are the only time-dependent components of the regional water balance. Time series for groundwater production rates and rainfall events exist. Since most of the groundwater is produced to match drinking water supply, fluctuations of groundwater production are small in comparison to fluctuations of natural groundwater recharge.

A.1.5 Selecting the Model and Discretization

For the purpose of the study, a black box model was selected. The chosen approach illustrates in sufficient accuracy the effect of controlled water infiltration on the regional groundwater system. Once the regional groundwater balance was reached, the reaction of the system due to various stress situations could be forecasted. Local groundwater level information for defined areas was found by analytical approximations.

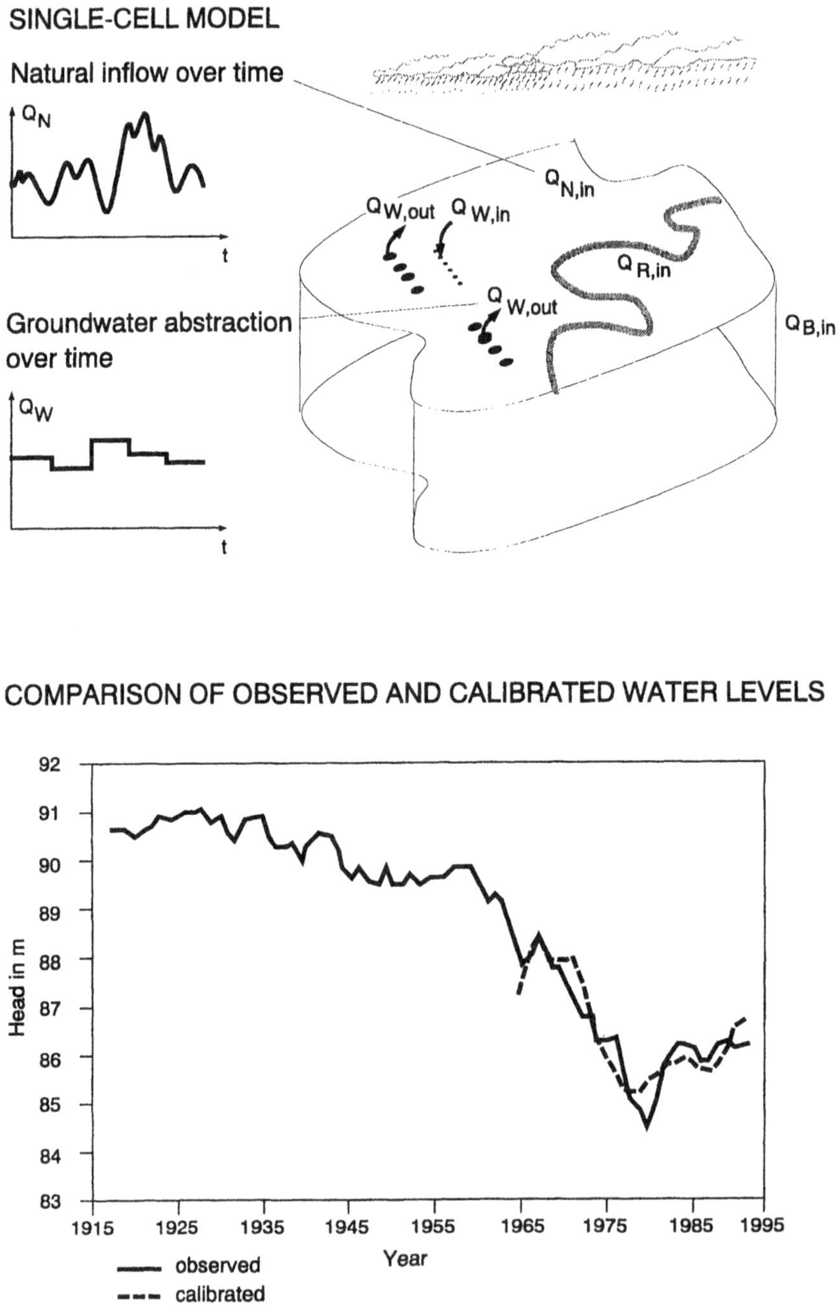

Figure A.1.2 Conceptual model and comparison of observed and calibration groundwater level for a selected monitoring well.

The governing equation for the global mass balance is given in Equation 5.2. Calculation of the mean water table elevation is reduced to balancing inflow and outflow. Groundwater fluctuation is related to changes in mass by the following relationship:

$$[(h(t + \Delta t) - h(t)]n_e\text{area} = Q_{\text{in}} - Q_{\text{out}} \tag{A.1.1}$$

In this case study, inflow and outflow were expressed in time series. With a known groundwater level at the beginning of the calculation, step by step calculation of the mean groundwater elevation was possible. A time step of one month was chosen.

The delayed response of groundwater recharge to a rainfall event was approximated using the linear storage theory (Equation 5.5). It was assumed that in the winter period from November to April, 80% of the rainfall would contribute to the groundwater recharge and that in the summer evapotranspiration would balance precipitation and natural water consumption. The coefficients in the linear storage model were calibrated using observations of natural groundwater recharge.

The effect of artificial infiltration on the water table in the close vicinity of the infiltration locations was approximated analytically using the relevant well equation.

A.1.6 Calibrating the Model

Calibrating a black box model is possible by either comparing calculated aquifer responses with average groundwater heads gained from the field data, or by using groundwater heads in selected representative observation wells. The result of the second approach is shown in Figure A.1.2.

Groundwater observations can be traced to 1916 for the selected monitoring well. Rainfall events have been available from 1964. Figure A.1.2 compares measured values with calculated values. A similarly good match between observed and simulated groundwater heads was found at several other observation locations representative of the region. However, significant differences between observations and calculations existed at other observation wells. Generally, it is impossible to simulate local groundwater elevations accurately with a model concept based on averages.

In summary, the black box model could be calibrated to approximate the average response of the aquifer system due to various stress situations.

A.1.7 Performing Model Simulation

Model simulations were performed assuming an increased groundwater production of 12.7 million m^3 per year and assuming that the additional water production is compensated by artificial groundwater recharge. Besides natural groundwater recharge, water injection depended on the following restrictions:

1. Infiltration rate will be constant.
2. Infiltration rate will depend exclusively on rainfall event.
3. Infiltration rate will be technically limited to a defined figure.
4. Infiltration will vary between a specified minimum and maximum.
5. No infiltration will be allowed in selected months.

Figure A.1.3 illustrates the predictive model application for cases 1 and 2 over a period of 20 years. While the effect of controlled groundwater injection on the average water table for cases 1 and 2 is basically the same (Figure A.1.3), the operation of the Rhine water treatment plant and the infiltration units would have been different. Since the storage capacity of the aquifer ensures that fluctuation in groundwater recharge over several months are compensated by release or storage of water without significant effect on the water table (Figure A.1.3), the infiltration scheme can focus on technical requirements as long as one guarantees a long-term balance of infiltration and increased groundwater production. Dry periods over several years without infiltration or with a constant infiltration (case 1), however, would result in a drastic decline of the water table.

A.1.8 Model Study Summary

This case study illustrates how meaningful results can be achieved without sophisticated modeling. An understanding of the investigated problem, together with an appreciation of mass balances allows the model user to address the demanding problem in a straightforward manner. However, there were severe restrictions. Local water-table fluctuations remained unknown. No differentiation of discharge and recharge locations was possible using this approach. Analytical approximations had to complement the black box simulation as required to estimate the effect of injection locally. A numerical model (multiple-cell model) may prove to be justified as limitations on groundwater management increase.

A.2 MODELING THE GROUNDWATER SYSTEM OF THE SEMARANG BASIN, INDONESIA

The Semarang basin is part of the coastal plain between the volcanic mountains in North Java and the Java Sea. In 1989, the year of this case study (Spitz 1989), the population of the city of Semarang was around 1.2 million. The surrounding flat area is intensively used for agriculture, but commercial and industrial activities occur close to the city.

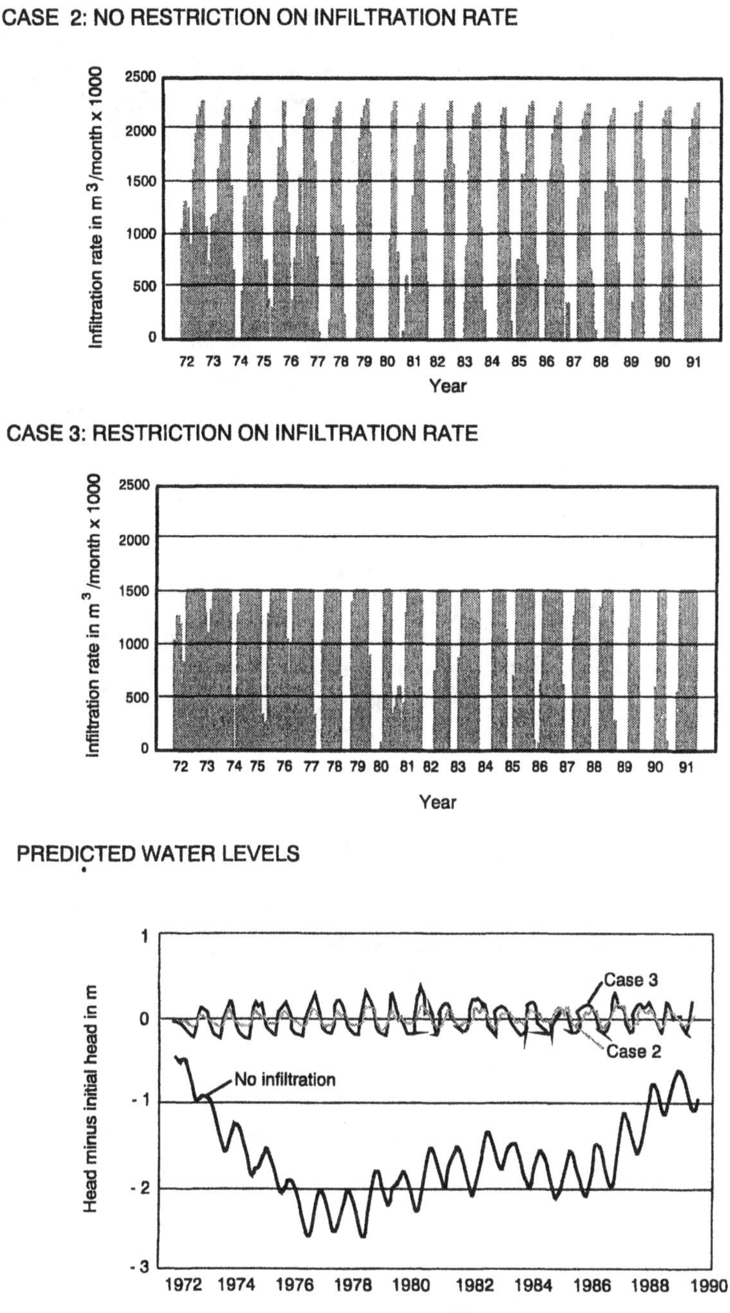

Figure A.1.3 Calculated infiltration schemes and predicted water fluctuation.

The water demand for domestic and industrial consumption is only partly supplied by public water supply systems. Domestic water supply relies, to an extent, on dug wells tapping the shallow aquifer. In recent years a rapidly increasing number of deep wells have been drilled to match the required industrial water demand. The actual number of deep wells and their total discharge is unknown since some are still unregistered.

A.2.1 Specifying the Modeling Problem

In 1989 there were no rigorous groundwater management strategies in place for control of groundwater extraction in the Semarang basin. The level of groundwater extraction exceeded the natural groundwater replenishment. Increasing saltwater intrusion due to the decrease in the heads in the coastal plain was one of the undesirable consequences of this overextraction. In Semarang the lowest head dropped as much as 13 m below sea level.

The purpose of the model study was to support the efforts to formulate groundwater management strategies for this groundwater stress area. The model was developed to combine all currently available field data, to guide new field investigation and to forecast the aquifer system's response to various management scenarios such as maintaining the present level of water production or by doubling it. Simulations focused on the prediction of the water table. Consequently, the magnitude of saltwater intrusion could be estimated.

A.2.2 Evaluating the Real World Conditions

The investigated area is centered on the City of Semarang, Figure A.2.1. The terrain is flat with an average surface elevation of about 5 m above sea level, becoming gently undulating south of the city. To the north the Java Sea limits the investigated area. The southern end is formed by the abruptly rising steep foothills of the volcanic mountain range. To the east and west the coastal plain extends beyond the boundaries chosen for the model study. The coastal plain in the region of interest has an average width of approximately 5 km with increasing width eastwards. East of Semarang the mountain range gently trends inland while the coastline bends to the north.

The area has the distinctive climatic features of a tropical oceanic zone. Each year has a cycle of a wet and a dry season, with the latter continuing from June to September. In 1988 the main annual precipitation in Semarang was 2644 mm. Numerous rivers originating in the mountains to the south, flow almost parallel northward through the investigated area.

The industries that use most of the groundwater produced from deep wells generally flank the main east–west road. Groundwater monitoring and investigation of the aquifer system at Semarang has been performed since 1970. The oldest Dutch monitoring records, however, are dated 1842, when one of the first wells in Indonesia was drilled at Fort Wilhem I, Semarang.

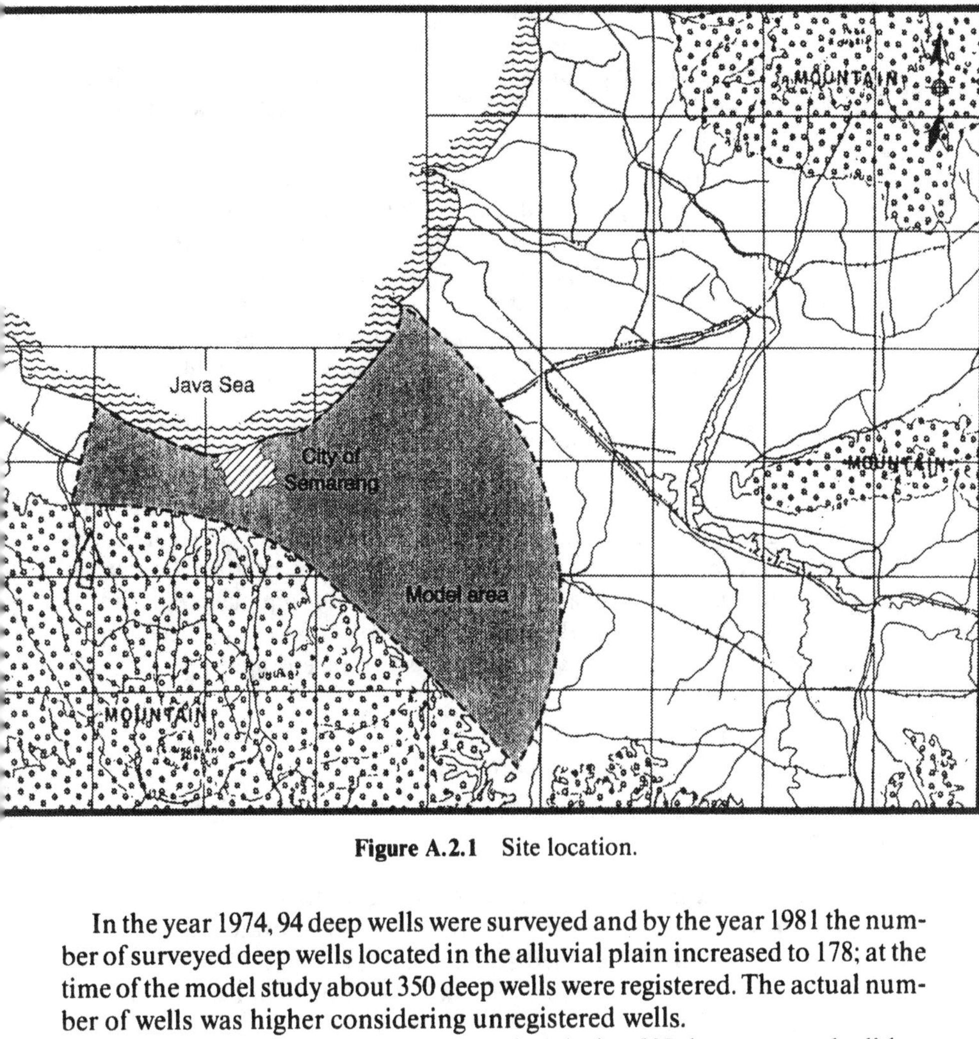

Figure A.2.1 Site location.

In the year 1974, 94 deep wells were surveyed and by the year 1981 the number of surveyed deep wells located in the alluvial plain increased to 178; at the time of the model study about 350 deep wells were registered. The actual number of wells was higher considering unregistered wells.

Most marine sediments at the coastal plain is of Holocene age; the lithological unit is clayey materials with a high content of mollusk fragments. Due to the low hydraulic conductivity of the materials, saline water entrapped during marine deposition is still present. The alluvial deposits show a structure of alternating layers of medium grained sands with gravel and clayey sands and clayey materials. In layers with high hydraulic conductivity, the entrapped marine water has been flushed out by fresh water originating from the recharge areas.

Precise identification of the main aquifers is rather difficult. The interpretation of drilling records allows the geological units to be classified into the following aquifer systems, ranking the aquifers from ground surface downward:

- Shallow aquifer. The depth of the phreatic aquifer is about 4 to 10 m below the ground surface with increasing depth towards the hilly areas in the south. Groundwater is extracted by numerous dug wells, mainly for domestic water supply. Generally the water-table pattern does not suggest that groundwater extractions are significant locally although the total extraction figure is significant. The hydraulic gradient is nearly constant from south to north.

- Aquifers formed dominantly by marine sediments. These aquifers are about 30 to 90 m in depth and are difficult to identify. Aquifers are comprised of sand and clayey layers that are interbedded with massive marine clay sediments of low permeability. The depositional near-shore environment indicates that the sandy, silty, and clayey materials are lenticular and do not form continuous uniform layers. Most of the groundwater is brackish or salty and is generally not suitable for most uses.
- Aquifers formed dominantly by alluvial sediments. These aquifers consist of sand layers and are referred to as the *deep aquifer system.* This system is found at an average depth of 60 m, with an average thickness of 10 m. The permeable layers are superior to all other aquifers in terms of the amount and the quality of groundwater that can be produced by wells. Water is transmitted freely and seawater has been completely replaced by fresh water. The alluvial deposits range from clayey sand to gravel. Groundwater exploitation via deep wells is concentrated on this deep aquifer system.
- Aquifers formed dominantly by volcanic sediments. These aquifers are formed by volcanic sediments which are deposited in a marine environment. Groundwater in these low-permeability aquifers is salty, and it is not used.

Field data indicate that the natural subsurface flow in the Semarang basin basically follows the surface drainage system. Groundwater in the deep aquifer system was mostly confined and artesian prior to extensive groundwater production. The groundwater system in the investigated coastal area interacts with the aquifer system in the hilly area.

A.2.3 Developing the Conceptual Model

Key elements of the conceptual model were defining the model boundaries, simplifying the aquifer system, and determining the required model dimension to produce meaningful conclusions. A description of each of these elements follows.

- Selecting model boundaries. In the north, the Java Sea forms a natural boundary to the investigated groundwater basin prescribing the heads. A zero-flux boundary separated the model area from the neighboring coastal area to the west. The boundary was elongated along the estimated streamline under natural groundwater flow conditions. It had to be verified at the end of the performed analysis that the chosen streamline remained practically unaffected. This was done by repeating the analysis with a boundary located at larger distance from the region of pumping. The simulations remained practically the same, which indicated that the chosen streamline boundary was a good one. Otherwise, the boundary would have to be moved further away until any changes in the boundary did not significantly affect the water levels in the region of interest.

Only a small amount of field data was available to support a choice for an appropriate boundary in the eastern part of the study area. A choice of prescribed heads along rivers seemed to be inappropriate. The observed heads in the deep aquifer system do not correspond to the water table in the rivers. Rivers and deep aquifer systems in the investigated area are, in most cases, separated by thick layers of low permeability, which still contain saltwater entrapped during marine deposition.

A flux boundary that followed the main river in this area was selected, as shown in Figure A.2.2. Although there was no evidence of a strong interrelationship between the river and the deep aquifer, the river represented a natural boundary being far enough away from the main area of interest at Semarang. In addition the assumed natural groundwater flow direction coincided approximately with the river pattern. Close to the sea the flow direction of both river and groundwater were identical (boundary flux is equal to zero). Further to the south increasing lateral subsurface inflow into the model area from the neighhoring plain in the south–east was allowed.

The southern boundary of the model area coincided with the border line of the hilly area. A flux boundary simulated recharge coming from the mountain area. Altogether the boundaries formed a model area which was narrow in the west and wide in the east. The total area was 749 km^2. The boundary conditions are summarized in Figure A.2.2.

- Simplifying the aquifer system. The model study focused on the flow conditions or the head distribution in the deep aquifer system. Due to its superior transmissivity, groundwater exploration is concentrated on this hydraulic unit causing significant groundwater depressions. The interaction between the deep aquifer system and the shallow aquifer was thought to be relatively small.

  The layers forming the deep aquifer system were grouped in the model study into one deep confined aquifer. Leakage at the top due to vertical head gradients was allowed. The base of the aquifer formed by the very low permeability formation was considered to be an impermeable boundary, as shown in Figure A.2.2.
- Determining required model dimensions. The average thickness of the deep aquifer system is very small in comparison to the horizontal extent of the model area. Overall flow is dominantly horizontal allowing the use

MODEL BOUNDARIES

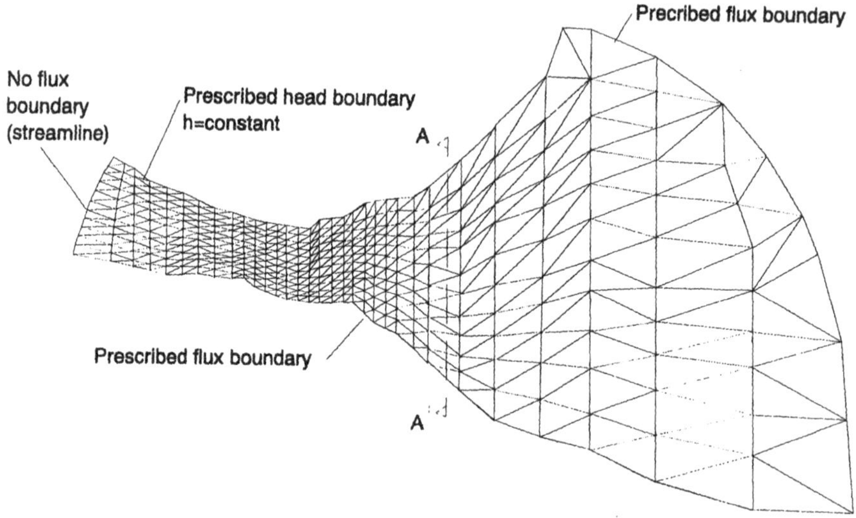

SIMPLIFIED AQUIFER SYSTEM
(Cross section A - A')

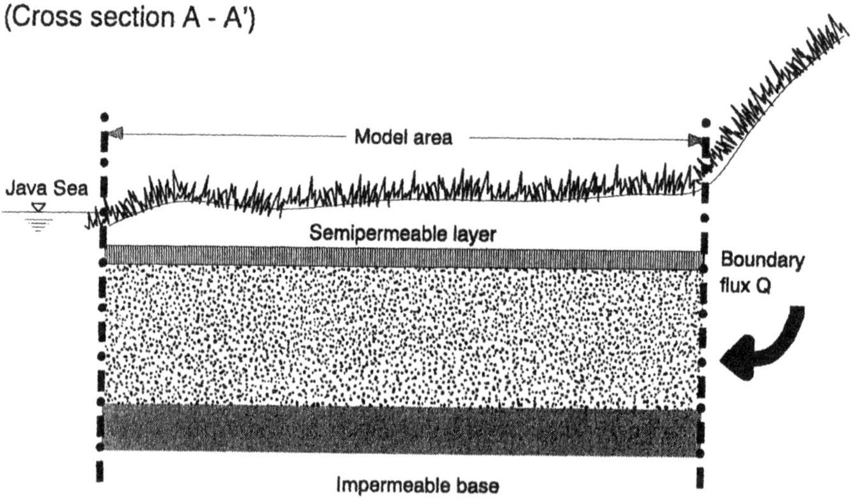

Figure A.2.2 Boundary conditions and simplified aquifer system.

of a two-dimensional horizontal flow model. A more detailed representation of the deep aquifer system (e.g., with a multi-layer model) seemed unnecessary due to the lack of data.

A.2.4 Selecting Model Input Data

Field data provided estimates of hydrogeological and boundary conditions. The model input data have been selected as follows:

- Aquifer parameters. Values for the thickness of the investigated hydraulic aquifer system were obtained from the drilling records and from the geophysical survey. The data indicate an average thickness of the high-permeability strata ranging between 1 and 20 m. The heads were always above the top of the aquifer and the flow remained confined. Therefore the actual aquifer thickness was less important than the transmissivity.

  Only a few investigations were performed to measure reliable values of the transmissivity and the storativity of the confined aquifer system. A hydraulic conductivity of 12 m/day (about 10^{-4} m/s) was initially chosen that gave a transmissivity of 120 m^2/day in the western and 180 m^2/day in the eastern part of the study area. These initial estimates compared with data found in other regions of the coastal plain. The storativity figures for the aquifer system had to rely on estimates.

  The leakage factor was kept small throughout all performed simulations. Field observations indicated that the amount of leakage was very small.
- Boundary flux. The lateral groundwater inflow into the model area along the southern and eastern part directly influenced the average elevation of the heads. Estimates of the boundary flux were obtained by:

  A simple, but reliable calculation according to Darcy's law assuming that the hydraulic gradient and the transmissivity are known along the boundary.

  A separate water balance of the adjacent watershed system that required additional intensive field investigations.

  Comparison with other areas with similar flow conditions.

  Model calibration.
- Groundwater extraction. The historical development the groundwater extraction was estimated using the data obtained from the registered wells. A linear increase of the extraction rate between two field surveys proportional to the increase of registered wells was assumed.

A.2.5 Selection of the Model and Discretization

A finite element model was used, since it had advantages over the finite difference model in regard to the shape of the model area. Groundwater flow in the confined aquifer was assumed to flow in two dimensions and to depend on the prescribed boundary conditions, the amount of external recharge and discharge, the leakage through the top of the aquifer and the aquifer parameters, transmissivity and storativity.

Small elements were chosen in areas with strong groundwater extraction, yielding strong changes in the heads, and they were bigger in areas where the changes are smoother, as shown in Figure A.2.2. The total element number was 694 by 388 node points. The grid system covered a total area of 749 square kilometers, yielding an average element area of about 1 km^2. Element values such as transmissivities and nodal values such as well discharge were assigned using the preliminary estimates.

A.2.6 Calibrating the Model

The items to be altered by calibration were initially chosen as boundary flux and transmissivity of the confined aquifer. A variation of the storage coefficient and of the leakage factor was not considered because they affect the flow behavior in the investigated area less than boundary flux or transmissivity.

Data for the following three time periods were used in the model calibration and, by comparison of the simulations for each period, in the model verification.

- Data prior to 1960. The collected data prior to 1960 provided an estimate of the head distribution under relatively undisturbed conditions. Since all data were concentrated around the old city center, a flow calibration for the entire model area was not possible.
- Data from the year 1981. The head distribution measured in 1981 already showed the effect of the increasing groundwater exploitation. Field measurements were performed mostly in the same area for which historical data were available.
- Data from the year 1989. This data set allowed a more complete picture of the overall conditions, including data not only in the vicinity of the center but also in the western and eastern parts of the model area.

A.2.7 Performing Model Simulation

Model simulations were performed for the following three scenarios estimating the resulting head distribution for the year 2000:

- Prediction A. Prediction A was based on the assumption that the total extraction rate does not increase. Although this scenario seemed unlikely, it provided the steady-state picture for the groundwater extraction rate at the time of the study. No significant differences to the measured head distribution was found for the year 2000. The reason for the similarity in the head distributions is the fast response of the confined aquifer to changes in the well production rates.

- Prediction B. It was assumed that the annual increase in the extraction rate that was observed in the last 10 years would increase by a factor of 2. Simulations revealed that one has to expect maximal drawdowns of about 25 m below sea level.
- Prediction C. For this prediction the observed increase of the total groundwater extraction was extrapolated until the year 2000. The exploitation will force the heads to fall in large areas below sea level with a minimum of the head distribution around 35 m below sea level.

Model results for predictions A, B, and C are shown in Figure A.2.3.

A.2.8 Model Study Summary

The model study of the Semarang basin addresses the changes of groundwater heads due to groundwater extraction. Hence, it represents the most common model application. The finite-element model allows the model user to adjust the discretization to the irregular shape of the model boundary. The shortage of model input data makes it necessary to rely on estimates when defining the boundary, boundary conditions, and aquifer properties. An extensive review of existing records such as drilling records that go back more than 100 years yields helpful information about the groundwater system. Plausibility of model simulations are checked by mass balances, by comparison of model results with investigations of groundwater systems in North Java, which compare with the Semarang basin, and by a sensitivity analysis of model input data. The main constraint of the model study is that when groundwater extraction rates increase, unnatural boundaries become more and more affected by model simulations.

A.3 DESIGN OF MONITORING NETWORK

The increase of groundwater contamination in Germany has initiated a number of state programs for the design of groundwater monitoring systems aiming to locate sources of groundwater contamination. To minimize efforts the design of groundwater monitoring networks should take advantage of existing data and existing groundwater models.

A.3.1 Specifying the Modeling Problem

Due to the various sources of groundwater contamination, interpreting groundwater quality data and designing monitoring systems is difficult. The relevant area of influence of each monitoring location has to be estimated in

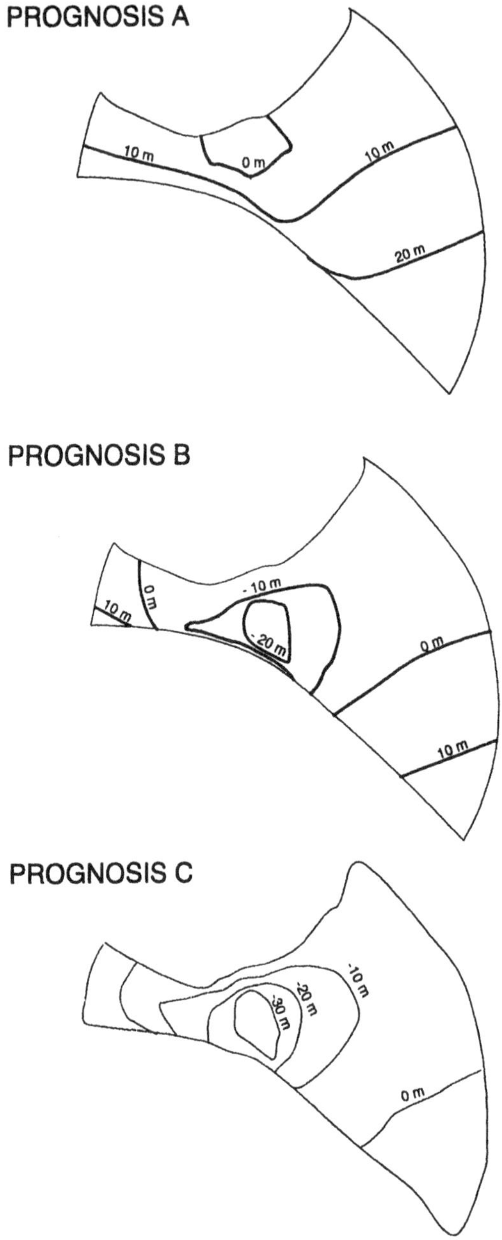

Figure A.2.3 Model results for three scenarios.

order to relate the groundwater sample to groundwater origin. The groundwater sample taken several meters below groundwater level does not reflect the site-specific conditions of the sampling location. Analogous to the catchment area of pumping wells, the relevant area of influence of a monitoring location defines the area where the sampled groundwater originated. Solutes entering the groundwater system in this area, together with the corresponding soil and aquifer type, affect the quality of the groundwater at the monitoring location. In this model study the model user was asked to determine the areas of influence of existing monitoring wells for an alluvial groundwater system to identify major factors influencing the observed groundwater quality. Additionally the model study had to guide the design of new monitoring wells.

A.3.2 Evaluating the Real World Conditions

The investigated groundwater system is part of the upper Rhine valley and is shown in Figure A.3.1. To the right, the groundwater system is bounded by a mountain range. Subsurface inflow exists. The aquifer is structured in interwoven layers of sand and gravels. The total thickness of the aquifer system is several hundreds of meters. Two hydrologic units with different hydraulic conductivities are identified. A fault results in an abrupt change of aquifer thickness in the investigated model area. Natural groundwater recharge exists but differs depending on land use and soil type. A significant part of the model area is urban. The remaining part is forest or used for agriculture. Groundwater flow is predominantly in the direction of the River Rhine, which is represented by a prescribed head boundary. Groundwater flow is affected by numerous pumping wells used for drinking water supply. The catchment area of one group of pumping wells is shown in Figure A.3.2.

A.3.3 Developing the Conceptual Model

Sampling depth, that is, the estimation of the relevant area of influence for a monitoring well, depends to a great extent on vertical flow components. Hence groundwater flow has to be represented three-dimensionally. This is illustrated in Figure A.3.2. Figure A.3.2 shows pathlines along the main flow direction. The relevant area of influence of a depth-specific monitoring location is calculated by defining the relevant three-dimensional stream tube. The intersection of the stream tube with the boundaries of the groundwater system (water-table or model inflow boundary) approximates the relevant area of influence. The main flow direction in the area shown in Figure A.3.1 is known from a regional groundwater flow model covering a total area of 17 $\times$ 25 km. The existing multi-layer model is based on a grid with a constant cell size of 250 m in both horizontal directions. In the vertical direction the aquifer is represented by three layers. The upper two layers of the groundwater model correspond to the two upper geological units, the third layer represents the deeper

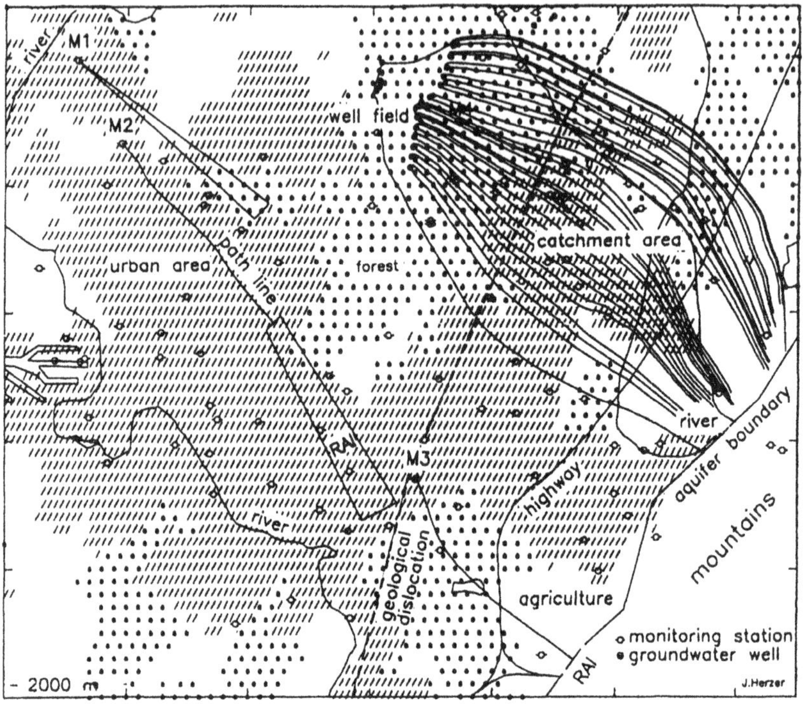

Figure A.3.1 Groundwater flow on the horizontal plane.

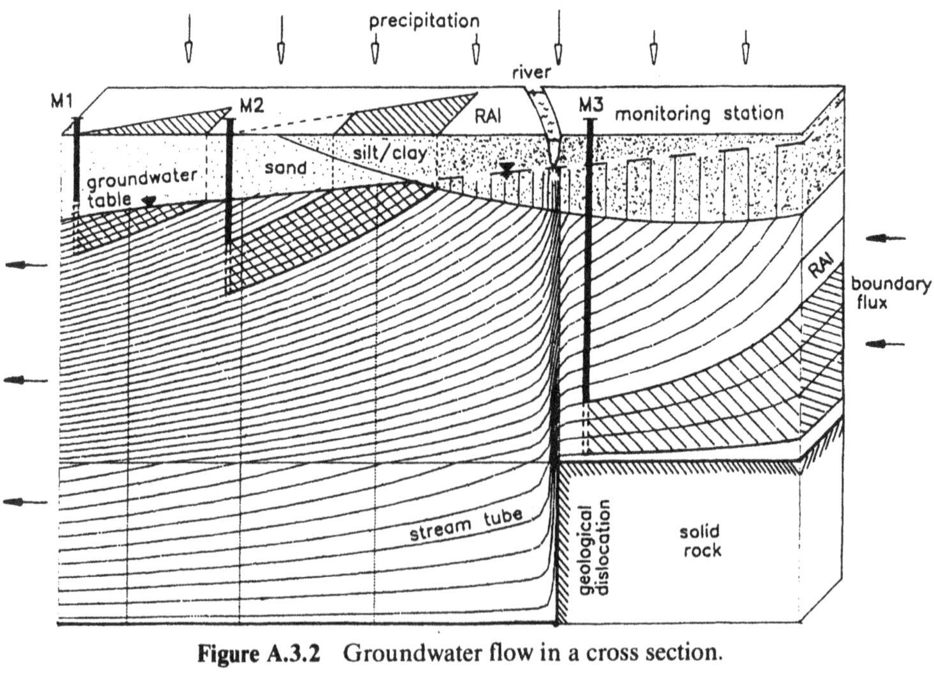

Figure A.3.2 Groundwater flow in a cross section.

aquifers not considered in the current modeling approach. The vertical discretization of the existing model does not allow calculation of three-dimensional streamlines within the upper two layers.

A.3.4 Selecting the Model

Due to the limitation of the existing multi-layer FD model in representing vertical flow components, modeling was carried out using a combination of numerical and analytical techniques. Horizontal groundwater flow was simulated using the existing multi-layer flow model. Main flow direction streamlines in a horizontal direction were calculated. Figure A.3.1 shows streamlines near the group of wells. In addition streamlines toward the three monitoring wells M1, M2, and M3 are shown. Vertical flow in cross sections oriented along these horizontal streamlines are approximated analytically.

A.3.5 Performing Model Simulations

Stream tubes are calculated for steady groundwater flow. Deviation from simplified conditions have been accounted for by additional concepts. Fluctuation of natural groundwater flow and dispersive spreading results in an increase to the width and length of the calculated relevant area of influence. The simplest estimation is to increase the geometry of the area of influence linearly with distance to the monitoring location. Such additional concepts have only been applied to modify the width of the well catchment area. It was assumed that the length of the catchment area corresponds to the length of the well filter.

For the monitoring wells M1, M2, and M3, the relevant areas of influence are calculated and shown in Figure A.3.1. The sampling interval of monitoring well M1 transects the water table. Hence the site-specific conditions are monitored as well as a thin strip of groundwater upstream. Monitoring well M2 samples groundwater that originated at a distance from the monitoring location. Monitoring well M3 does not sample groundwater that has originated in the Rhine valley. Hence this monitoring well is not suitable to detect contamination sources in the Rhine valley.

A.4 SALTWATER INTRUSION MODELING FOR COASTAL AQUIFER SYSTEM IN FLORIDA

The study area is adjacent to Tampa, on the central western coast of Florida. Saltwater intrusion in this region has been a concern since the 1920s due to population growth and the resultant increases in groundwater pumpage inducing saline degradation of groundwater. Groundwater supplies the bulk of the agricultural, domestic, and industrial water needs of the area, since surface-water supplies are typically brackish. There are three major well fields, and numerous individual wells, in the area of concern. A water supply authority was formed in the 1970s to manage the area water supply, but encroaching saline water may limit future well development.

A.4.1 Specifying the Model Problem

The purpose of the model study is to develop a computational tool suitable for evaluating the safe yield of the western well fields and predicting the impacts of future well field development on the groundwater resources. Specific aims of the modeling study include assessment of the impact of deepening selected wells and predicting the recent transient migration of chloride in three dimensions. The resulting calibrated and verified model is being used as a groundwater management tool to evaluate well permit applications and develop regional groundwater strategies.

A.4.2 Evaluating the Real World

The terrain of the study area is generally flat and low lying with the highest elevation being about 70 feet (ft) above mean sea level. Ancient shorelines resulting in linear terraces, and numerous sinkholes surrounded by swamps are the most prominent physiographic features of the area.

A 10- to 50-ft layer of surficial sand overlies several thousand feet of carbonate rocks. The carbonate rocks of the study area were deposited in a horizontal position and subsequently tilted downward to the southwest, resulting in beds that dip and increase in thickness to the southwest. The uppermost three aquifers and three confining units in this system were incorporated in the model.

The surficial aquifer system is the sand layer described in the previous paragraph. The upper confining unit (Hawthorn Formation) is irregular in thickness, varying from zero near the coast to 70 ft in the southwest. The upper Floridan aquifer (Tampa Limestone and Suwannee Limestone) ranges in thickness from 110 ft in the northwest to 480 ft in the southwest. The semi-confining unit (Ocala Limestone) has low porosity and is approximately 250 ft thick. The base of the lower Floridan aquifer (Avon Park and Lake City Limestones) ranges from 900 to more than 1200 ft below ground surface. The base of the Lake City Limestone contains layers of gypsum and anhydrite that retard the flow of water; these layers form the base of the model.

The primary source of groundwater recharge is precipitation, which is about 52 inches/year (in./yr). Average runoff is 11 in./yr, with half of the runoff occurring as overland flow and half as base flow. Average evapotranspiration has been estimated to be 35 to 40 in./yr. In addition the hydrologic balance may be significantly altered locally due to lowering of the potentiometric surface under pumping conditions.

Groundwater discharge occurs as pumpage, groundwater outflow to the ocean, and upward leakage. In May 1987, 129 million gallons per day were withdrawn as permitted pumpage from the study area. This is equivalent to a discharge of 8.4 in./yr throughout the area.

The elevation of the water table fluctuates by several feet seasonally in response to varying precipitation. Groundwater flows from the northeast toward the west (Gulf of Mexico) and southwest (Old Tampa Bay); see Figure A.4.1. Depressions in the water table occur around all major well fields due to downward leakage from the surficial sands to underlying aquifers. The flow pattern in the Floridan aquifer system follows a similar distribution.

A.4.3 Developing the Conceptual Model

The groundwater system in this area is well-known because of the hundreds of wells, long history of water development, and numerous studies that have been conducted. The technical issues of concern were the prediction of the location of the saltwater front subject to alternate groundwater management practices and development of a usable management tool. The key physical processes to be simulated included the density-driven flow of the saltwater and the numerous pumping wells in the area.

The modeling approach selected involved:

- Developing several two-dimensional cross-sectional models to test boundary conditions, calibrate some material properties, and run sensitivity analyses.
- Developing a three-dimensional model to incorporate the effects of pumping wells and realistically represent saltwater movement in all three aquifers.

The size of the model domain was selected based on the existence of natural boundaries (the ocean and Old Tampa Bay to the west and southwest), the extent of site-specific data (the base of the model), interference with external well fields (to the north and east), and the area covered by other models (to the north and east).

The model was visualized as an essentially horizontally layered three-aquifer system with a saltwater wedge underlying fresh groundwater to the west and southwest, and eight well fields extracting water from the upper and lower parts of the Floridan aquifer. A combination of specified values and specified fluxes were used to define the model boundary conditions (Figure A.4.2). Observed head and chloride values, which varied with depth, were supplied along the west boundary. Along the seaward boundary both saltwater inflow (at depth) and freshwater outflow (near-surface) occurs. So a zero-gradient concentration boundary condition was assumed where outflow from the model domain occurs (according to the predicted equivalent freshwater head gradients). Along the model's base, outside the saltwater wedge, a sup-

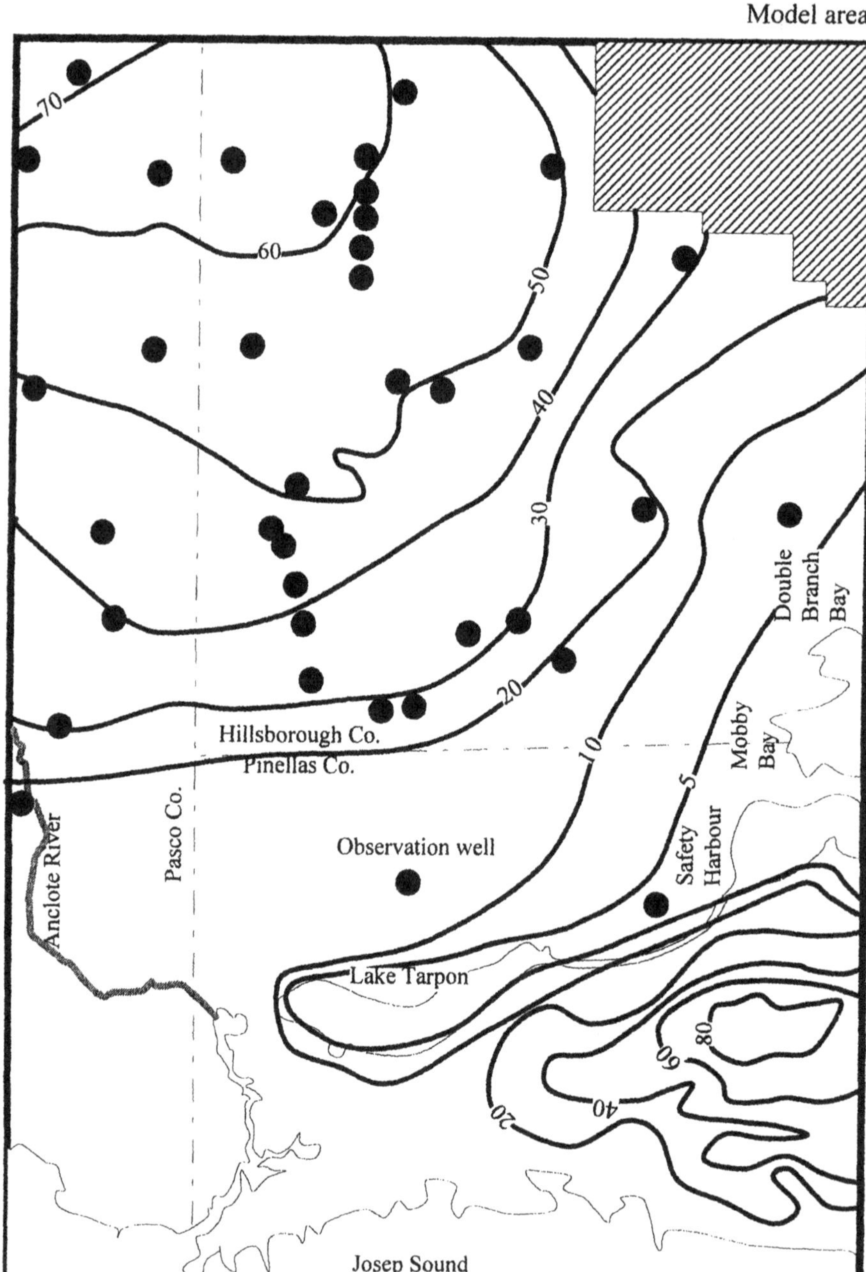

Figure A.4.1 Site location.

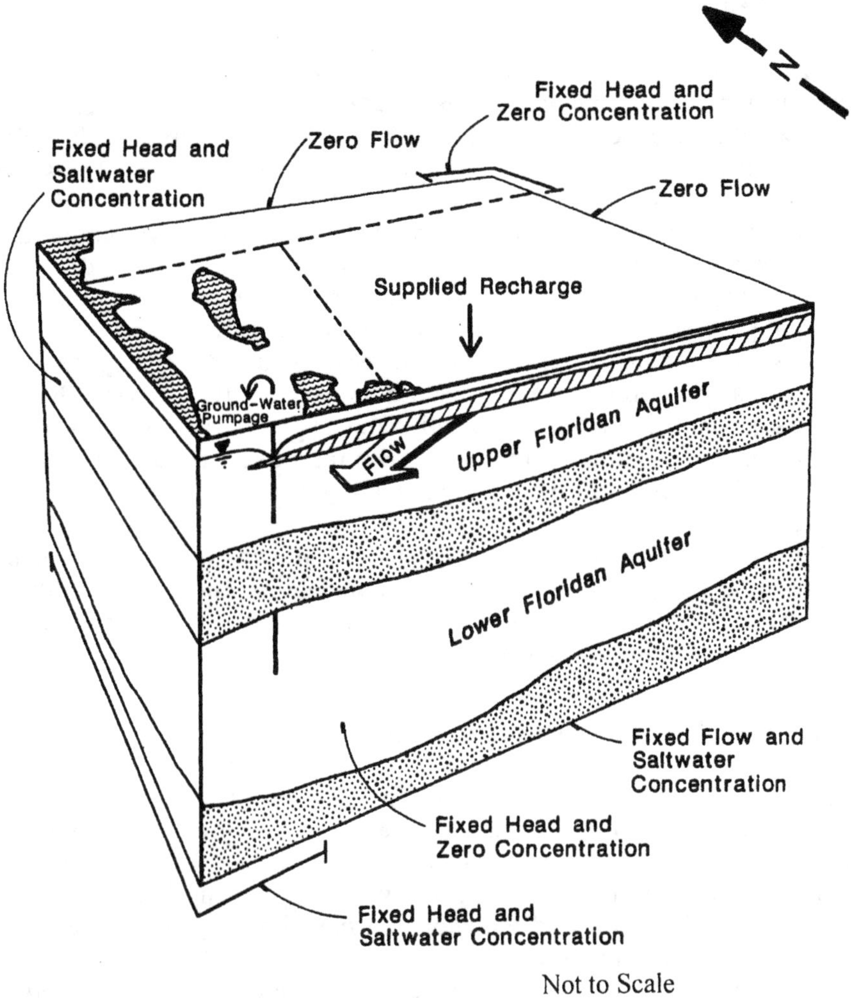

Figure A.4.2 Conceptual model and boundary conditions.

plied flux condition was specified. This flux represents saline recharge from underlying aquifers. Since only one or two wells penetrate to these depths, the concentration of this saline recharge is unknown and was a parameter varied during model calibration. The north and east boundaries of the model, with the exception of the northeast corner, coincide with groundwater divides. These model boundaries were simulated as zero-flow boundaries, with recharge occurring from the northeast corner. Recharge and discharge corresponding to groundwater/surface water interaction, infiltration from precipitation, and groundwater pumpage vary both spatially and with time and were simulated as such.

A.4.4 Selecting Model Input Data

Six hydraulically distinct units, identified on the basis of numerous hydraulic conductivity measurements, were specified in the model and aquifer parameters were assumed to be constant in each unit. The thickness of each aquifer and confining unit was discretized in the three-dimensional model as well as possible (with particular care in the vicinity of the well fields), and variations in hydraulic conductivity (for accurate overall transmissivity) were used in places where the thickness variations could not be well represented.

Groundwater pumpage for more than 800 wells was known, and agricultural pumpage was estimated based on permitted rates and crop coefficients. The pumpages were distributed vertically over the open interval of each well at rates proportional to the local transmissivity. Pumpages that fell in the same model cell were combined. A model preprocessor was developed to combine the disparate pumping databases, map the data onto the model domain, and provide data in a format readily accepted by the numerical model.

Since the initial model runs were for steady-state conditions, the model initial conditions serve only to expedite the calculations. The initial conditions used were a simplified hydrostatic head pattern based on observed water levels and zero chloride concentrations.

A.4.5 Selecting the Model and Discretization

The selection of a numerical model for application was based on three main considerations: the three-dimensional nature of the problem, the density-driven saltwater flow, and the implementation of the model as a groundwater management tool on available computer systems. A small number of codes fitted these requirements, and the choice of model was agreed between the groundwater agency and the modelers.

The three-dimensional model extends about 22 miles (east–west), 15 miles north–south) and 1500 ft (vertically). The orientation of the domain was selected to coincide with the predominant flow direction in the vicinity of the major well fields. The dimensions of the cells were designed to avoid the numerical problems associated with extreme cell aspect ratios or Peclet numbers, and they were selected to be finer near well fields and coarser near the inland boundaries. The cell sizes ranged from 1000 by 2000 by 20 ft thick to 6000 by 12,000 by 280 ft thick, for a total of nearly 25,000 cells.

A.4.6 Calibrating the Model

Generally, flow calibration precedes transport calibration. However in this case the flow and transport calculations are strongly coupled by density effects, so flow and transport calibration were carried out simultaneously.

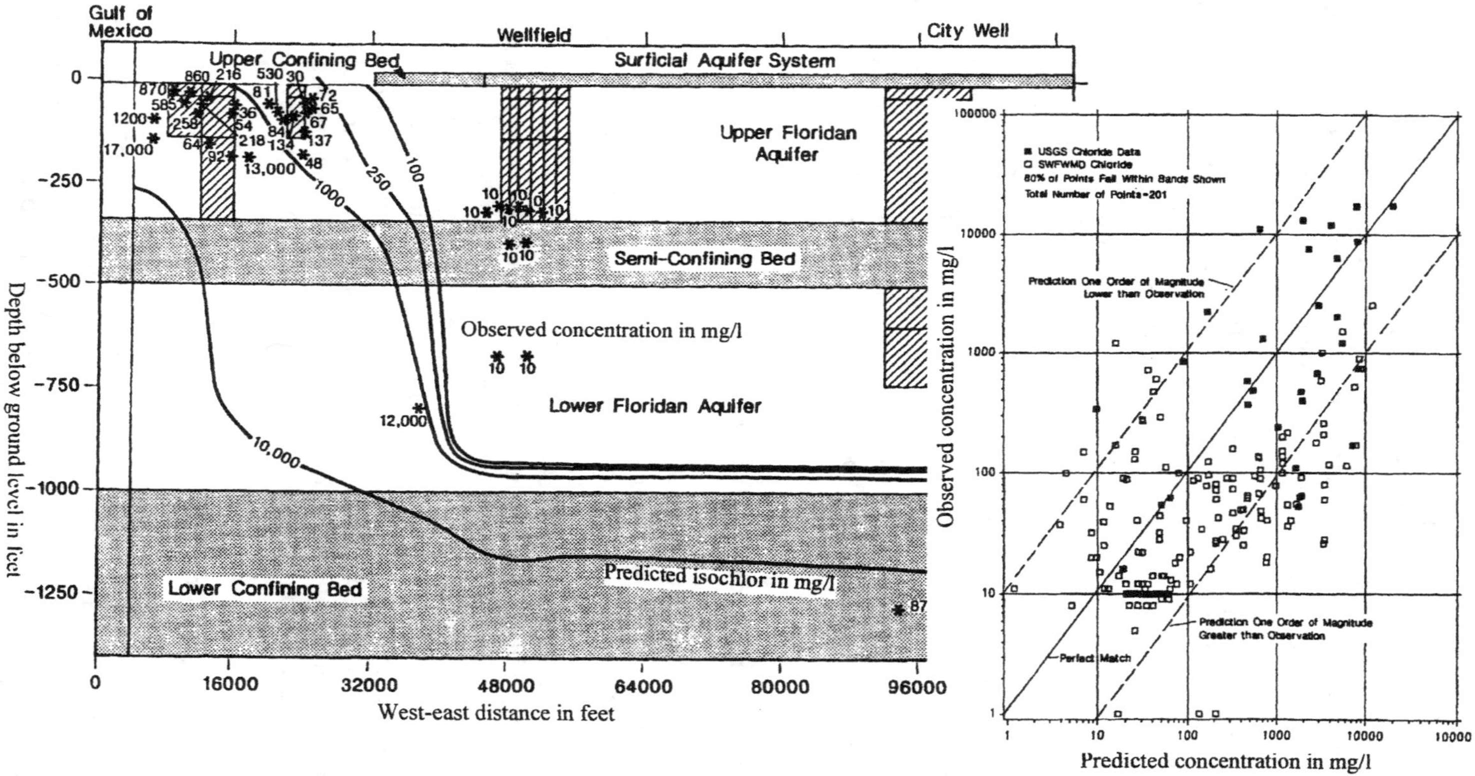

Figure A.4.3 Predicted chloride distribution.

May 1979 data were chosen for the steady-state calibration because it was an unusually wet period with extensive data, and at a time before the rapid increases in pumping rates of the 1980s that could be approximated as an equilibrium period. Fourteen runs were carried out in which uncertain boundary conditions were adjusted, hydraulic conductivity and anisotropy of the Hawthorn Formation were varied, and dispersivities altered by up to a factor of four. The resulting predicted potentiometric surface and water-table elevations were within a few feet of those observed. The predicted chloride distributions were compared with more than 200 observations.

The observations are highly variable and of differing quality. Therefore the expectations for calibration rested on comparisons at the locations of greatest concern; at the well fields. A sample cross section is shown in Figure A.4.3 Overall the predictions are within one order of magnitude of the observations 80% of the time, with best agreement at high concentrations near the three major well fields. The resulting calibrated model was then used to run a steady-state and a transient verification run (i.e., with different pumpage and rainfall, but with all other parameters remaining the same). The steady-state verification was not entirely satisfactory, but the transient verification resulted in realistic predictions for conditions during the period May 1979 to May 1981.

A.4.7 Performing Model Simulations

Two situations were simulated with the calibrated and verified model. First the effect of deepening 20 wells in the largest well field was predicted. The results showed worsening water quality in the lower aquifer and improvement in the upper aquifer as the zone of higher pumpage altered. Second, a comparison of steady-state and transient predictions was made in order to check whether steady-state conditions were being approached (Figure A.4.4). Also the effect on predicted results of varying the recharge and pumpage monthly versus annually was checked. Because of the differences between the steady-state and transient predictions it was concluded that the transient calculations of chloride movement were more realistic and that annual time steps for varying pumpage and recharge were adequate, since the movement of chloride occurs on a time scale of decades rather than months. In addition recommendations for further use of the model and for future data collection were developed.

A.5 SITE REMEDIATION

This case study describes some aspects of a groundwater modeling analyses of the migration and cleanup of several solvents at a site in an arid climate. Data presented in this case study represent interpretations at one period in time; current interpretations may differ from the conclusions presented here. The solvents leaked from an underground tank and were disposed of into a dry

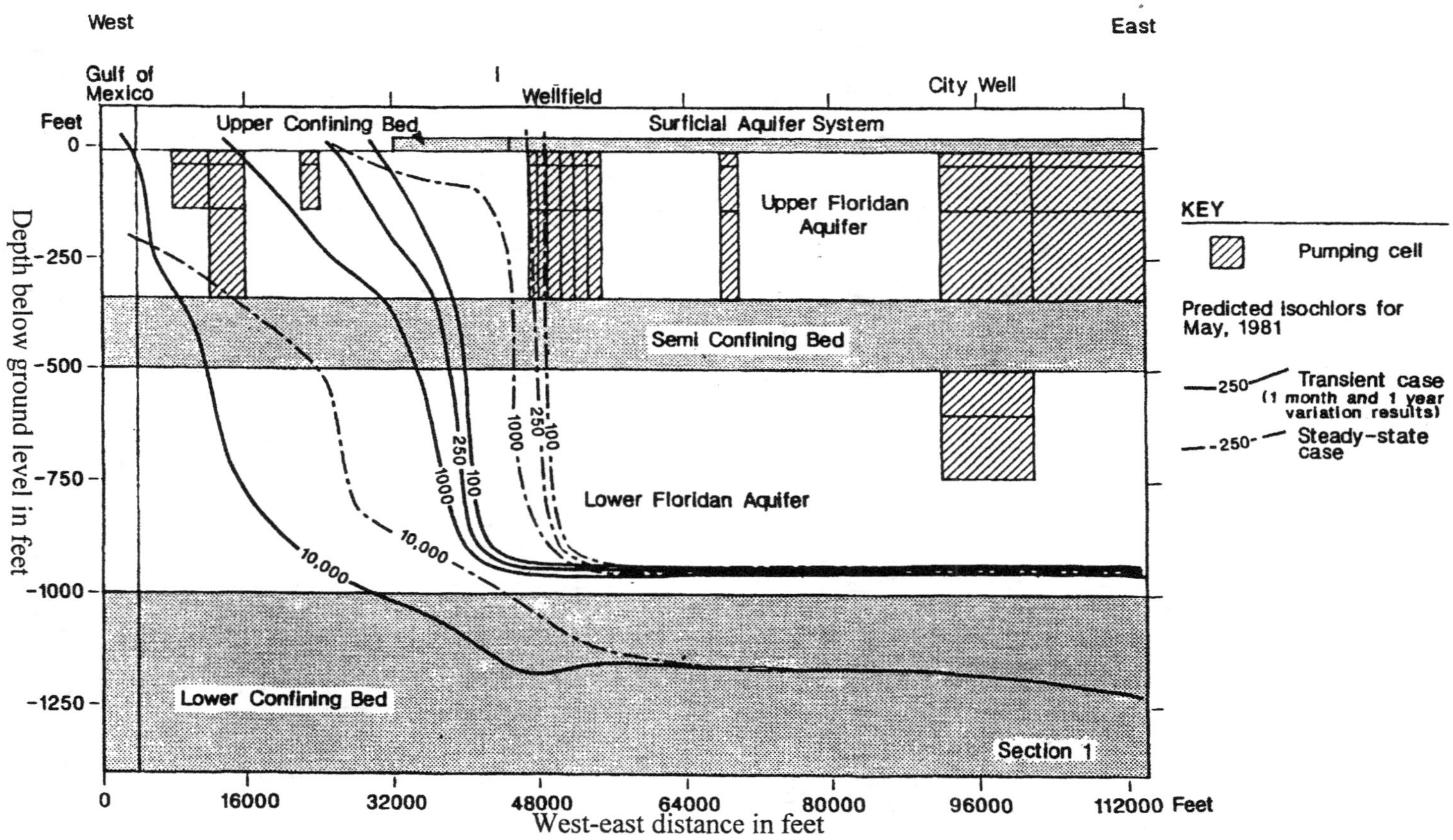

Figure A.4.4 Comparison of chloride distributions for transient and steady-state simulations.

well for several years. The surrounding land was originally farmed as orchards but has developed into a mixture of industrial and residential properties. There are a couple of domestic irrigation wells within a mile of the site, but the public supply wells are located in the more transmissive aquifer to the south.

A.5.1 Specifying the Modeling Problem

The major constituent of the release was trichloroethylene (TCE). TCE is denser than groundwater and travels faster through soil and rock because it is less viscous than groundwater. TCE gradually biodegrades in the subsurface, and it may also become adsorbed or vaporize. Prior to the 1970s there were few regulatory controls on disposal of this type of waste. However at the time of writing, the applicable aquifer water quality standard for TCE in this region is 5 parts per billion (ppb) (see also Figure 3.2).

One of the main purposes of this modeling study was to help to design an effective way to capture the contaminated groundwater. However, since the hydrogeologic regime and behavior and location of the contaminant plume were initially unknown, the first purposes of the model were to develop a representative conceptual model, help in selecting well locations for site characterization, and predict the existing extent of contamination.

A.5.2 Evaluating the Real World Conditions

The plant site lies near the margin of a major river basin. The bedrock outcrops to the west of the plant, and the main valley and aquifers lie to the south. Drill hole investigations indicate the presence of six geologic units ,the surficial unit being silty sand or sandy gravel alluvium. The alluvium thickness ranges from 20 ft on the plant site to 240 ft thick two miles to the south. Five bedrock units were encountered. The bedrock surface is irregular, with a bedrock high to the south of the plant site and an overall slope to the southwest. The bedrock contains southwest trending faults and exhibits differing degrees of weathering and fracturing.

The water table lies at depths of 100 to 130 ft below the ground surface; the saturated thickness of the alluvium varies from 10 ft to 140 ft (Figure A.5.1). Aquifer tests of the alluvium and bedrock indicate heterogeneity and anisotropy in each unit. The overall flow system comprises an upgradient recharge area to the west (the outcrops) and a downgradient discharge area (the main river valley), subject to varying stresses due to changing pumping patterns.

A.5.3 Developing the Conceptual Model

At the time this modeling effort was undertaken, the hydrogeologic regime of the alluvium was generally well understood. However, details of flow in the

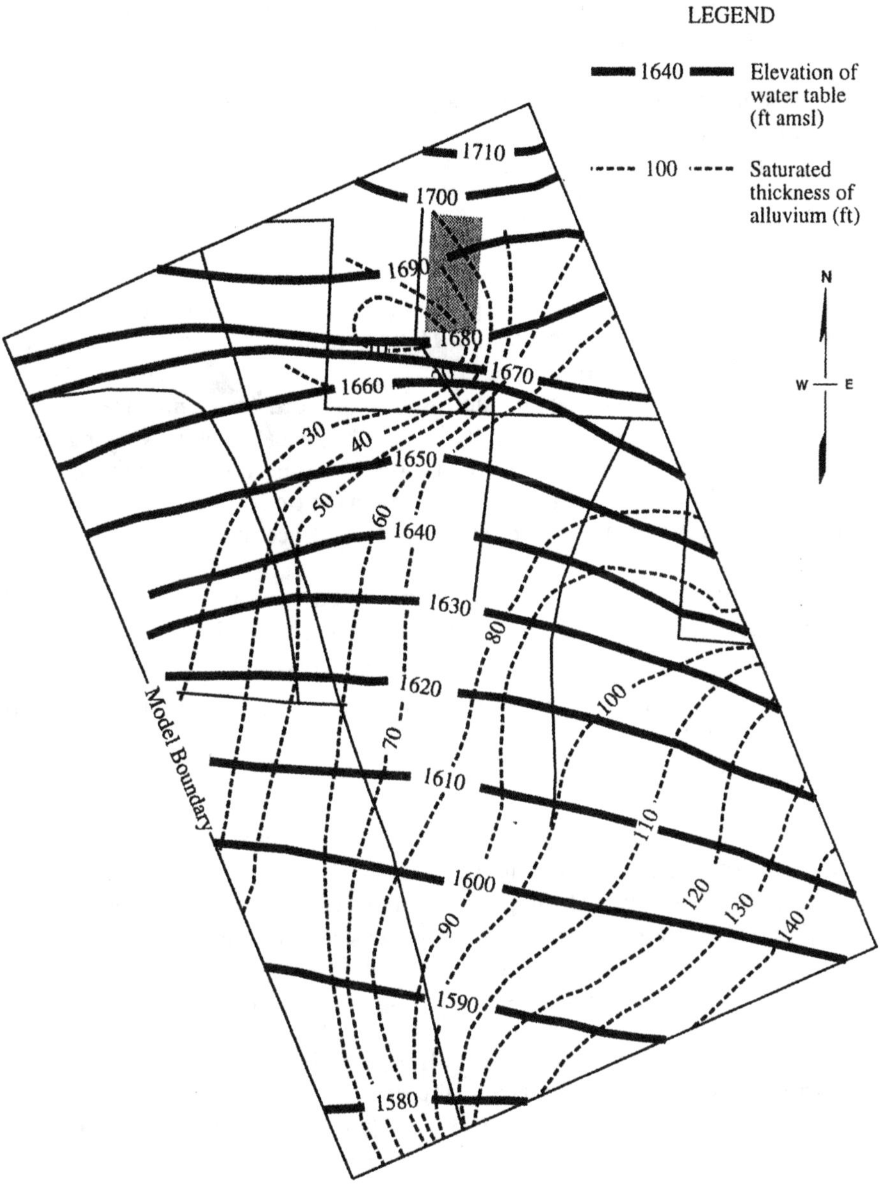

Figure A.5.1 Water table and saturated thickness of the alluvium.

bedrock were unknown, and the lack of full details about historical and source data led to uncertainties in the overall conceptual model. The technical issues to be addressed by the modeling analysis included:

- Design of the capture system (location, depth, pumping rates of extraction wells, etc.).
- Prediction of long-term (30 years in the past and 50 years into the future) concentration distributions.
- Explanation of the contaminant plume behavior observed.

The key chemical and physical processes to be simulated included density-driven flow of TCE, the dissolution of TCE from pure phase to solute, and flow and transport in the bedrock.

A three-stage modeling approach involving both two- and three-dimensional modeling analyses was used (Figure A.5.2) because of the varying levels of data available and purposes of the model at different stages of the six-year project. Two-dimensional models were used in the preliminary investigations. Three-dimensional models were eventually used because:

- Groundwater contamination was observed in both the alluvium and bedrock, so both units were included in the model domain.
- The solvent sources were mostly point sources.
- Vertical hydraulic gradients (upward and downward) influence contaminant migration.
- Simulation of cleanup wells screened at various depths was required.

The general model was envisioned to be a two-layer system (alluvium and bedrock) with vastly differing rates of groundwater flow between the two units, but similar rates of solvent transport in both units, and upward and downward interchange between the two units across the model domain. The conceptual model for the solvent source was based on a variety of indirect data. For example, undissolved TCE initially remained in pores above the water table, undissolved TCE is observed in the bedrock (near the primary area of release), the highest TCE concentrations are still observed close to the source area 20 years after TCE was no longer used at the site, and observed groundwater concentrations appear to account for only a few percent of the total TCE volumes potentially released. Based on these data, it was assumed that most of the TCE migrated downward, due to its specific gravity of 1.46, into the fractured bedrock to depths of 150 ft below ground. The resulting zone of higher-concentration TCE remained stationary and will dissolve gradually over a lengthy period. These assumptions were tested in the model.

The size of the model domain was primarily chosen to encompass the future extent of the contaminant plume, as estimated by analytical and two-dimensional numerical models. The base of the model was taken at the interface with less permeable bedrock (about 400 ft below ground surface), and this

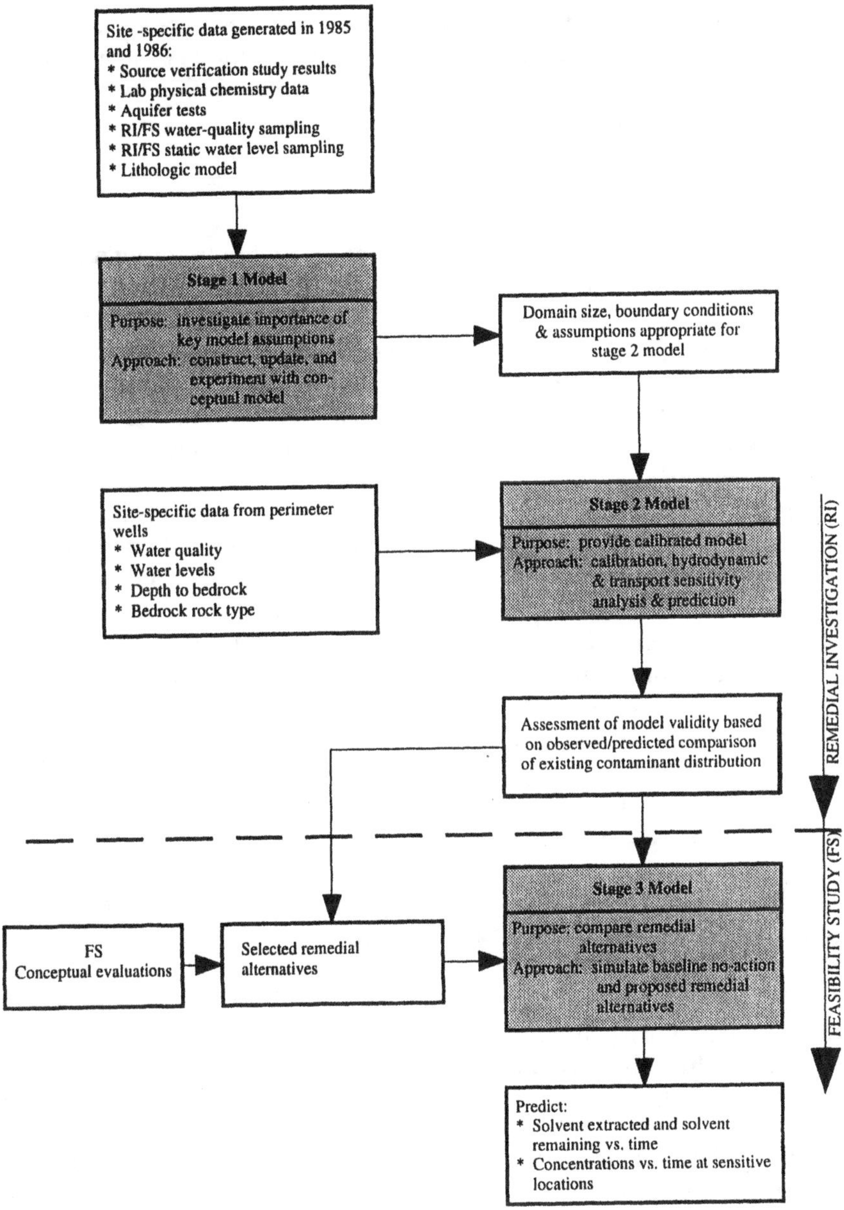

Figure A.5.2 Three-stage groundwater modeling approach.

boundary was initially treated as a flux boundary based on observed hydraulic gradients. The vertical boundaries were assumed to be specified head boundaries, with some vertical flow. The surface boundary incorporated a patchwork of infiltration rates based on land use, irrigation practices, and leakage from canals.

A.5.4 Selecting Model Input Data

The field program at this site involved installing and monitoring more than 200 well points (up to 20 points vertically in some multipoint wells), soil-gas studies, and source verification studies. The laboratory program included physico-chemical analyses, and biodegradation studies and adsorption studies. This model had to accommodate uncertain historical flow regime in the bedrock and lack of detail on potential sources. The latter parameters were varied during model calibration and the impact of their uncertainty assessed by model sensitivity analyses.

The alluvium is heterogeneous; its hydraulic properties were tested by means of aquifer pumping tests. The alluvial layer varies in thickness and a correlation between the saturated thickness and hydraulic conductivity was found. This relationship was used to interpolate hydraulic conductivities between measurement locations, resulting in a distribution of conductivities varying over an order of magnitude. The bedrock hydraulic conductivities were measured by means of packer tests, and varied over two orders of magnitude. The hydraulic conductivity within each of the five bedrock units was assumed to be constant.

Steady-state flow analyses were used to develop the initial flow field. Initial concentrations were assumed to be at background levels throughout the domain.

A.5.5 Selecting the Model and Discretization

The selection of the model was initially based on the need to simulate two-dimensional flow of a dense contaminant. However, later model selection was based on the three-dimensional nature of the problem, the need to simulate multipoint cleanup wells, and requirements that model runs should be practical on microcomputers.

The model domain, as discussed previously, encompasses the present contaminant plume, and extends approximately 2 miles (along the plume centerline) by 1.5 miles (perpendicular to the direction of travel) by 400 ft vertically. A larger domain would have entailed extrapolation of field data and incorporation of other solvent plumes.

The orientation of the domain was selected to coincide with the dip of the bedrock surface (to the southwest near the plant site) and trend of the faults (to the southeast). This orientation was chosen because a dense plume might follow the bedrock topography rather than the direction of groundwater flow. The dimensions of the model cells were chosen to be finest close to the sources, in order to capture density effects. However, relatively fine cells were also needed wherever pumping wells might be placed. The model discretization was tested by running the same transport case with a finer grid to demonstrate that numerical dispersion did not distort the predictions. The calculation cells varied from 100 by 100 by 10 ft to 1000 by 1200 by 60 ft, for a total of 32,000 cells.

A.5.6 Calibrating the Model

Model calibration proceeded in several stages because the field program expanded as modeling progressed. Five site models of differing complexity were applied over a six-year period of study. Calibration of the three-dimensional model involved running 54 cases and sensitivity analyses for the major input parameters, including time step and calculation grid. Two aspects of the field program helped to generate confidence in the model predictions.

First, observed plume center line concentrations were difficult to explain; the highest observed concentrations were found 0.5 miles to the south of the plant, although the groundwater flow direction is to the southeast. The model predictions indicated that denser than TCE water would tend to follow the bedrock topography to the southwest initially and then (at lower concentrations) travel to the southeast, following the groundwater flow. This dogleg in the predicted plume is consistent with field observations (see wells with concentrations exceeding 20,000 ppb in Figure A.5.3). Second, model predictions suggested that groundwater contamination existed a mile farther offsite than the initial monitoring network. Monitor wells subsequently installed to locate the fringe of the plume (wells shown in boxes in Figure A.5.3) showed that the plume did extend farther than estimated based on data from the earlier monitoring network. These comparisons between model predictions and subsequently collected monitoring data were considered a form of validation of the model predictions.

The calibrated flow model resulted in predicted hydraulic heads in the alluvium within 1 ft of those observed and reasonable agreement among predicted and observed heads in the bedrock, saturated thicknesses, and vertical hydraulic gradients. Predicted and observed TCE concentrations, in the alluvium, are shown in Figure A.5.3. The sensitivity analyses showed that ranges in predicted concentrations of up to an order of magnitude may arise through variations, over likely ranges, of uncertain input parameters. The standard deviation of the observed concentrations is, at many wells, of the same order of magnitude as the mean observed value. As a result predicted concentrations were considered to be accurate to an order of magnitude, and discrepancies

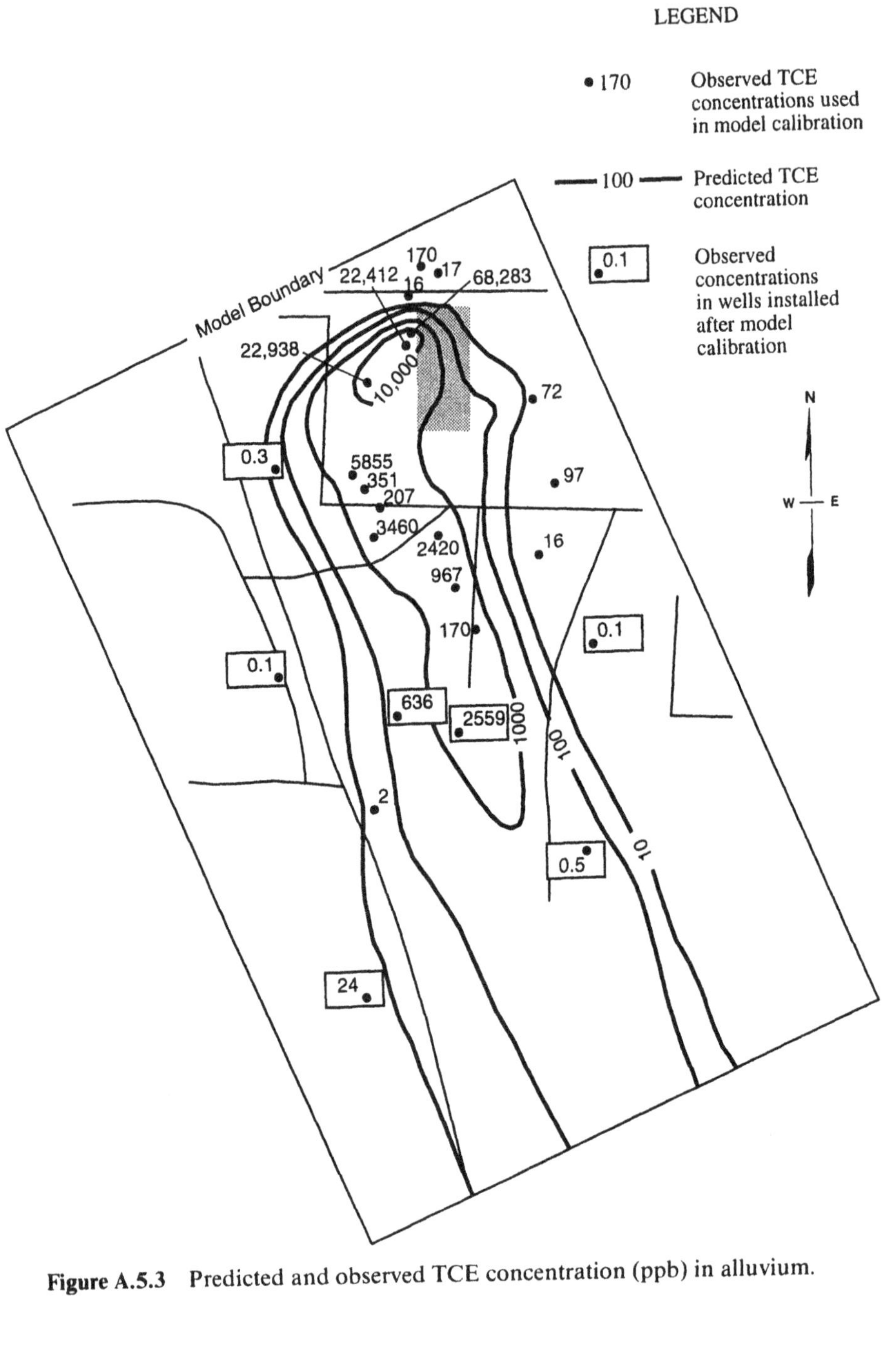

Figure A.5.3 Predicted and observed TCE concentration (ppb) in alluvium.

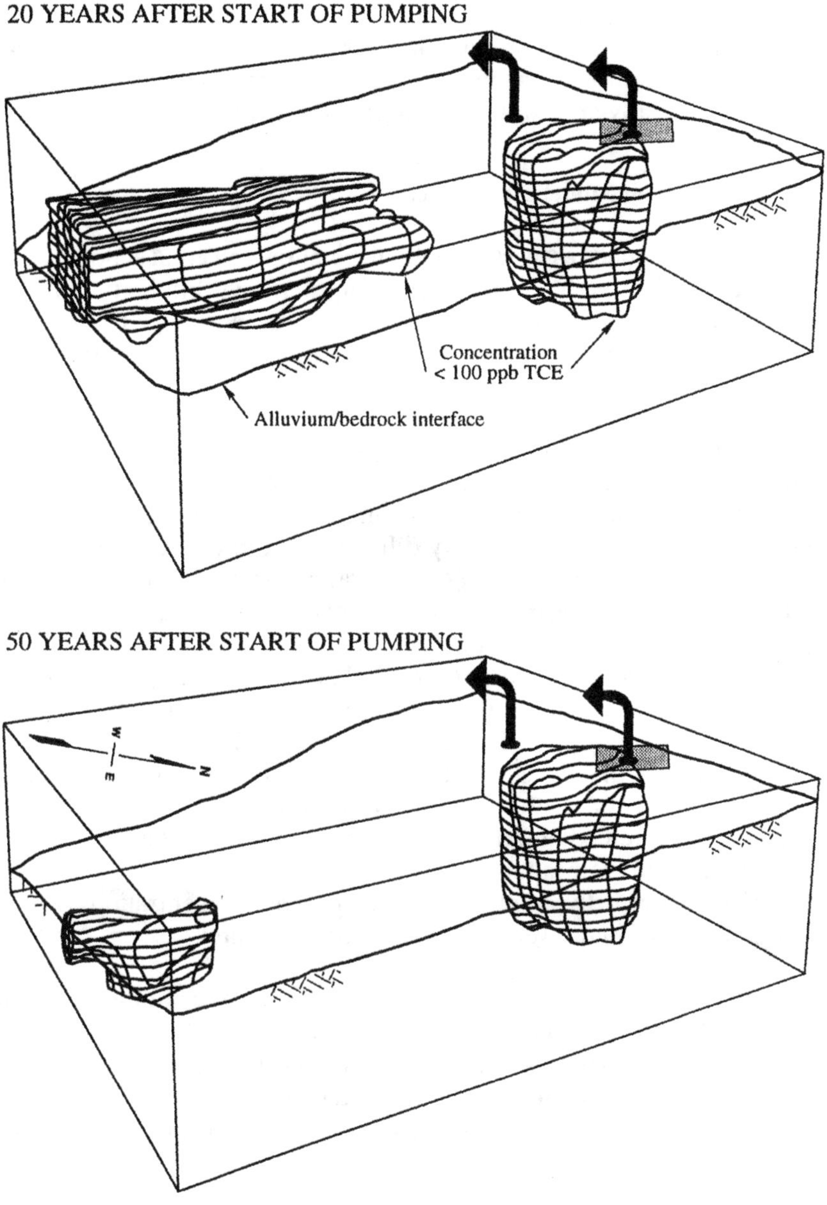

Figure A.5.4 Predicted TCE plume.

between predicted and observed values of less than an order of magnitude were considered insignificant. On this basis the transport predictions correlate well with observations at more than 80% of monitor locations (of which there are more than 200). The model predictions are less realistic outside the hydrogeologic study area and upgradient of the sources.

A.5.7 Performing Model Simulations

The calibrated model was used to simulate potential remedial actions to compare their effectiveness in cleaning up the aquifer. Five different networks of pumping wells, together with assorted barriers, groundwater flushing, and source treatment, were evaluated. The assumption, developed during model calibration, that a continuing source of TCE exists as a higher-concentration zone in the bedrock implies that removal of this source is necessary to terminate potential contamination of the aquifer. Predictions for a 50-year period from 1986 show that TCE migrating from the plant site would be contained but that some outlying TCE beyond the interim remedy would continue to migrate, in an increasingly dilute plume, to the southeast (Figure A.5.4). Predicted time histories of concentrations in the remedial wells and predicted mass remaining in the contaminant plumes were used to assess the likely effectiveness of cleanup wells over time.

Several years after the modeled remedial actions were implemented, predicted and observed concentrations in remedial wells were compared. A formal audit was not performed. However, it was generally found that the model predictions at remedial wells near the source significantly overpredicted (by more than an order of magnitude) the future concentrations. Predicted concentrations in the remedial wells midway along the contaminant plume were well approximated (within factor of two) by the model.

The discrepancy between predicted and observed concentrations near the source could be viewed several ways, three of which are:

- Assumptions about the longevity, volume, and concentration of future sources could have been too conservative.
- Remedial wells near the source could have been more effective at collecting source contamination than the model predicted.
- The model could inadequately describe site conditions near the source.

APPENDIX B
Compilation of Typical Values of Hydrogeologic Parameters

TABLE B.1 Porosity

Material	Porosity in [l]	Reference
Anhydrite	$(5 \text{ to } 50) \times 10^{-3}$	J
Chalk	0.05 to 0.20	J
Basalt	0.03 to 0.35	A
Basaltic lava and sediments	0.10	L
Chalk	0.023	L
Chalk, fractured	0.05	L
Clay	0.34 to 0.57	A
Clay	0.45 to 0.55	B
Clay, lacustrine	0.40	D
Clay, lacustrine	0.44	D
Clay, marine	0.48	C
Clay, glacial, till	0.3 to 0.35	E
Clay silty	0.38	F
Clay, glaciomarine	0.64	G
Clay, silty	0.41	D
Dolomite	0.05 to 0.15	J
Dolomite	0.034	M
Dolomite, fractured	0.18	W
Dolomite, fractured	0.12	N
Dolomite, fractured	0.024	O
Dolomite, fractured	0.07 to 0.11	P
Dolomite and limestone, fractured	0.06 to 0.6	Q
Gabbro, weathered	0.42 to 0.45	A
Granite, weathered	0.34 to 0.57	A
Granite	$1 \times 10\ 10^{-3}$	K
Granite, Ossian	$(3.1 \text{ to } 22) \times 10^{-3}$	H
Granite, Scottish Lowland	$(2.2 \text{ to } 2.7) \times 10^{-3}$	H
Granite, Skene Complex	$(1.2 \text{ to } 4.6) \times 10^{-3}$	H
Granite, Cornish Carnmenellis	$(6 \text{ to } 8.8) \times 10^{-4}$	H
Granite, Sweden	$(5 \text{ to } 700) \times 10^{-4}$	I
Granite, fractured	0.02 to 0.08	V
Gravel	0.30 to 0.40	B
Gravel, fine	0.25 to 0.38	A
Gravel, coarse	0.24 to 0.36	A
Gravel and sand	0.20 to 0.35	B
Limestone	07 to 0.56	A
Limestone	0.01 to 0.10	B
Limestone	0.05 to 0.15	J
Limestone	0.35	R
Limestone	0.23	S
Limestone	0.12	O
Limestone, fractured and calcareous sandstone	0.25	T
Limestone, fractured	0.01	U
Salt	1×10^{-3}	J
Sand, fine	0.26 to 0.53	A
Sand, fine to medium	0.30 to 0.35	B

TABLE B.1 ***(Continued)***

Material	Porosity in [1]	Reference
Sand, uniform	0.30 to 0.40	B
Sand, medium to coarse	0.35 to 0.40	B
Sand, coarse	0.31 to 0.46	A
Sand, coarse alluvium	0.41	D
Sand, medium alluvium	0.43	D
Sand, fine alluvium	0.51	C
Sand, marine	0.41	C
Sandstone	0.14 to 0.49	A
Sandstone	0.10 to 0.20	B
Sandstone	0.05 to 0.15	J
Sandstone	0.32 to 0.48	X
Schist	0.04 to 0.49	A
Shale	0.01 to 0.10	B
Shale	0.01 to 0.1	J
Silt	0.34 to 0.61	A
Silt	0.40 to 0.50	B
Silt, sandy	0.39	D
Silt, clayey	0.34	D
Silt, loess	0.50	C
Siltstone	0.21 to 0.41	A
Soils	0.50 to 0.60	B

References:

(A) Morris, D. A., and A. I. Johnson, 1967. Summary of hydrological and physical properties of rock and soil materials as analyzed by the hydrologic laboratory of the U.S. Geological Survey. USGS Water Supply Paper 1839-D.

(B) Todd, D. K., 1980. *Ground Water Hydrology*, 2d ed., New York: Wiley, p. 535.

(C) MacCary, L. M., and T. W. Lambert, 1962. Reconnaissance of ground-water resources of the Jackson Purchase region, Kentucky. USGS Hydrol. Inv. Atlas HA-13.

(D) Johnson, A. I., and D. A. Morris, 1962. Physical and hydrologic properties of water bearing deposits from core holes in the Las Banos-Kettleman City area, California. Denver, CO, USGS Open File Report.

(E) Grisak, G. E., and J. F. Pickens, 1980. Solute transport through fractured media 1 & 2. *Water Resources Res.* 16(4), 719–739.

(F) Crooks, V. E., and R. M. Quicgley, 1984. Saline leachate migration through clay: A comparative laboratory and field investigation. *Can. Geotech, J.* 21, 349–362.

(G) Desaulniers, D. E., 1986. Groundwater origin, geochemistry, and solute transport in three major glacial clay plains of east-central North America. Ph.D. thesis. Dept. of Earth Sciencies, University of Waterloo, Ontario, p. 445.

(H) Lever, D. A., and M. H. Bradbury, 1975. Rock-matrix diffusion and its implications for radionuclide migration. *Mineralogical Magazine* 49, 245–254.

(I) Skagius, K., and I. Neretnieks, 1986. Porosities and diffusivities of some nonsorbing species in crystalline rocks. *Water Resources Res.* 22(3), 389–398.

(J) Cross, A. G., T. F. Lomenick, R. S. Lowrie, and S. H. Stow, 1985. Evaluation of five sedimentary rocks other than salt for high level waste repository sitting purposes. Oak Ridge, Tenn., Oak Ridge Nat. Lab. ORNL/CF-85/2/V2.

(K) Norton, D., and R. Knapp, 1977. Transport phenomena in hydrothermal systems: Nature of porosity. *Amer. J. Sci.* 27, 913–936.

TABLE B.1 ***(Continued)***

(L) Grove, D. B., 1977. The use of Galerkin finite-element methods to solve mass transport equations, Rep. USGS/WRD/WRI-78/011. USGS, Denver, CO. (Available as NTIS PB 277-532 from Natl. Tech. Inf. Serv., Springfield, VA)

(M) Halevy, E., and A. Nir, 1962. Determination of aquifer parameters with the aid of radioactive tracers. *J. Geophys. Res.* 67(5), 2403–2409.

(N) Grove, D. B., and W. A. Beetem, 1971. Porosity and dispersion constant calculations for a fractured carbonate aquifer using the two-well tracer method. *Water Resources Res.* 7(1), 128–134.

(O) Kreft, A., A. Lenda, B. Turek, A. Zuber, and K. Czauderna, 1974. Determination of effective porosities by the two-well pulse method. Isot. Tech. Groundwater Hydrol., Proc. Symp., vol. 2, pp. 295–312.

(P) Walter, G. B., 1983. Convergent flow tracer test at H-6: Waste isolation pilot plant (WIPP), southeast New Mexico (draft). Hydro Geochem. Inc., Tucson, AZ.

(Q) Claasen, H. C., and E. H. Cordes, 1975. Two-well recirculating tracer test in fractured carbonate rock. *Nevada Hydrol. Sci. Bull.* 20(3), 367–382.

(R) Bredehoeft, J. D., and G. F. Pinder, 1973. Mass transport in flowing groundwater. *Water Resources Res.* 9(1), 144–210.

(S) Fenske, P. R., 1973. Hydrology and radionuclide transport, monitoring well HT-2m, Tatum Dome, Mississippi. Proj. Rep. 25, Tech. Rep. NVD-1253-6. Center for Water Resour. Res., Desert Res. Inst., Univ. of Nev. Syst., Reno.

(T) Segol, G., and G. F. Pinder, 1976. Transient simulation of saltwater intrusion in southeastern Florida. *Water Resources Res.* 12(1), 65–70.

(U) Rabinowitz, D. D., and G. W. Gross, 1972. Environmental tritium as a hydrometeorologic tool in the Roswell Basin, New Mexico. Tech. Completion Rep. OWRR:A-037-NMEX, New Mexico. *Water Resources Res. Inst.*, Las Cruces.

(V) Dieulin, A., 1980. Propagation de pollution dans un aquifere alluvial: L'effet de parcours. Doctoral dissertation. Univ. Pierre et Marie Curie-Paris VI and l'Ecole Natl. Super. des Mines de Paris, Fontainebleau, France.

(W) Bentley, H. W., and G. R. Walter, 1983. Two-well recirculating tracer tests at H-2: Waste Isolation Pilot Plant (WIPP), southwest New Mexico. Draft paper, Hydro Geochem., Inc., Tucscon, AZ.

(X) Oakes, D. B., and Edworthy, D. J., 1977. Field measurement of dispersion coefficients in the United Kingdom. *Ground Water Quality, Measurement, Prediction, and Protection.* Reading, England: Water Research Centre, pp. 327–340.

TABLE B.2 Specific Yield

Material	Specific Yield in [l]	Reference
Clay	0.01 to 0.18	A
Gravel, fine	0.13 to 0.40	A
Gravel, medium	0.17 to 0.44	A
Gravel, coarse	0.18 to 0.43	A
Limestone	0.00 to 0.36	A
Loess	0.14 to 0.22	A
Sand, fine	0.01 to 0.46	A
Sand, medium	0.16 to 0.46	A
Sand, coarse	0.18 to 0.43	A
Sand, Eolian	0.32 to 0.47	A
Sandstone, fine	0.02 to 0.40	A
Sandstone, medium	0.12 to 0.41	A
Schist	0.22 to 0.33	A
Schist, weathered	0.06 to 0.21	B
Silt	0.01 to 0.39	A
Siltstone	0.01 to 0.33	A
Tuff	0.02 to 0.47	A

References:

(A) Morris, D. A., and A. I. Johnson, 1967. Summary of hydrological and physical properties of rock and soil materials as analyzed by the Hydrologic Laboratory of the U.S. Geological Survey. USGS Water Supply Paper 1839-D.

(B) Stewart, J. W., 1964. Infiltration and permeability of weathered crystalline rocks, Georgia Nuclear Laboratory, Dawson County, Georgia. USGS Bull. 1133-D, p. 57.

TABLE B.3 Horizontal Hydraulic Conductivity

Material	Hydraulic Conductivity in [m/s]	Reference
Basalt	$(1.89 \text{ to } 47{,}200) \times 10^{-11}$	A
Basalt	1.16×10^{-7}	B
Basalt	$(0.2 \text{ to } 4{,}250) \times 10^{-10}$	C
Basalt, permeable	$(4.75 \text{ to } 475{,}000) \times 10^{-7}$	B
Basaltic lava and sediments	$(1.8 \text{ to } 180) \times 10^{-3}$	Q
Beach sand	$(4.72 \text{ to } 18.9) \times 10^{-5}$	A
Beach sand	$(8.10 \text{ to } 19.7) \times 10^{-4}$	B
Calcium kaolinite	$(2.08 \text{ to } 11.6) \times 10^{-8}$	B
Caliche, compacted	$(4.98 \text{ to } 99.5) \times 10^{-11}$	B
Carbonate rocks, augmented by tubes, tunnels, cavities	$(4.76 \text{ to } 11{,}400{,}000) \times 10^{-10}$	F
Chalk	3.6×10^{-4}	P
Chalk, fractured	2.2×10^{-4}	P
Clay	$(4.72 \text{ to } 4{,}720) \times 10^{-10}$	A
Clay	$(? \text{ to } 9.95) \times 10^{-10}$	B
Clay	$(0.1 \text{ to } 47) \times 10^{-10}$	D
Clay, montmorillonite	1.00×10^{-12}	C
Clay, kaolinite	1.00×10^{-10}	C
Clay, unweathered marine	$(4.98 \text{ to } 12{,}700) \times 10^{-13}$	B
Clay, silty	$(4.75 \text{ to } 9{,}380) \times 10^{-10}$	B
Clay, sandy	$(2.55 \text{ to } 2{,}550{,}000) \times 10^{-12}$	B
Clay, lean	$(1.97 \text{ to } 27.8) \times 10^{-11}$	B
Clay, sodium Boston blue	$(1.62 \text{ to } 995) \times 10^{-12}$	B
Clay, Vicksburg buckshot	$(3.01 \text{ to } 11.0) \times 10^{-12}$	B
Clay, compacted Boston blue	$(3.59 \text{ to } 30.1) \times 10^{-11}$	B
Clay, loess	$(4.05 \text{ to } 4.98) \times 10^{-11}$	B
Cobbles	$>1.89 \times 10^{-2}$	A
Dolomite	$(4.25 \text{ to } 9.43) \times 10^{-11}$	A
Dolomite	1.16×10^{-8}	B
Dolomite and limestone, fractured	$(3 \text{ to } 7) \times 10^{-3}$	H
Dolomite, fractured	$(25 \text{ to } 5) \times 10^{-4}$	I
Dolomite, fractured	1.14×10^{-5}	J
Dune sand	$(9.43 \text{ to } 28.3) \times 10^{-5}$	A
Dune sand	$(2.31 \text{ to } 30.1) \times 10^{-4}$	B
Gabbro, weathered	$2.31 \text{ to } 10^{-6}$	B
Gabbro, weathered	$(0.5 \text{ to } 3.8) \times 10^{-6}$	C
Glacial till	$(9.95 \text{ to } 11{,}600{,}000) \times 10^{-13}$	B
Glacial till, mostly sand	5.67×10^{-6}	B
Glacial till, mostly gravel	3.47×10^{-4}	B
Glacial till, NE Ohio	$(3.82 \text{ to } 4{,}280) \times 10^{-10}$	B
Glacial till, surficial, Montgomery Co., Ohio	$(4.75 \text{ to } 231) \times 10^{-9}$	B
Glacial till, buried, Rohrers Island, Ohio	$(1.39 \text{ to } 6.37) \times 10^{-8}$	B
Glacial till, S. Illinois	$(3.85 \text{ to } 28.9) \times 10^{-8}$	B
Glacial till, S. Dakota	$(1.39 \text{ to } 2{,}310) \times 10^{-10}$	B

TABLE B.3 (*Continued*)

Material	Hydraulic Conductivity in [m/s]	Reference
Glacial deposit, outwash	$(4.98 \text{ to } ?) \times 10^{-4}$	B
Glacial deposit, esker, Westfield, Massachusetts	$(9.95 \text{ to } 127) \times 10^{-5}$	B
Glacial deposit, delta, Chicopee, Massachusetts	$(9.95 \text{ to } 1{,}500) \times 10^{-7}$	B
Gneiss	$(2.36 \text{ to } 236) \times 10^{-10}$	A
Gneiss	$(4.7 \text{ to } 260{,}000) \times 10^{-10}$	G
Granite	$(4.25 \text{ to } 23.6) \times 10^{-13}$	A
Granite, weathered	1.62×10^{-5}	B
Granite, weathered	$(3.3 \text{ to } 52) \times 10^{-6}$	D
Granite, fractured	$(3 \text{ to } 9) \times 10^{-4}$	O
Granite, fractured	$(1 \text{ to } 100) \times 10^{-7}$	O
Gravel	$(4.75 \text{ to } 9{,}950) \times 10^{-4}$	B
Gravel	$(0.3 \text{ to } 31.2) \times 10^{-3}$	D
Gravel (very well sorted)	4.16×10^{-1}	C
Gravel, very fine	$(3.77 \text{ to } 5.19) \times 10^{-3}$	A
Gravel, fine	$(5.19 \text{ to } 7.55) \times 10^{-3}$	A
Gravel, fine	5.21×10^{-3}	B
Gravel, medium	$(7.55 \text{ to } 10.4) \times 10^{-3}$	A
Gravel, medium	3.13×10^{-3}	B
Gravel, coarse	$(1.04 \text{ to } 1.42) \times 10^{-2}$	A
Gravel, coarse	1.74×10^{-3}	B
Gravel, very coarse	$(1.42 \text{ to } 1.89) \times 10^{-2}$	A
Greenstone	$(5.7 \text{ to } 10{,}000) \times 10^{-8}$	G
Hematite	$(9.43 \text{ to } 42{,}500) \times 10^{-13}$	A
Igneous and metamorphic rocks	$(9.4 \text{ to } 19{,}000) \times 10^{-10}$	G
Igneous, coarse-grained rock (granite, diorite, gabbro)	$(4.2 \text{ to } 42{,}000) \times 10^{-9}$	G
Igneous, tight, fine-grained rock (rhyolite, trachyte, basalt)	$(6.1 \text{ to } 13{,}000) \times 10^{-7}$	G
Igneous, fine-grained, cavernous rock	$(1.16 \text{ to } 81.2) \times 10^{-3}$	F
Limestone, karst and reef	$(1.0 \text{ to } 10{,}000) \times 10^{-6}$	C
Limestone	$(4.72 \text{ to } 943) \times 10^{-12}$	A
Limestone	1.09×10^{-5}	B
Limestone (0.16 porosity)	1.36×10^{-6}	C
Limestone, karst	$(2.31 \text{ to } 47{,}500) \times 10^{-6}$	B
Limestone, argillaceous	9.68×10^{-10}	C
Limestone and dolomite	$(5.79 \text{ to } 47{,}500) \times 10^{-10}$	B
Limestone	$(1.3 \text{ to } 1.7) \times 10^{-8}$	K
Limestone	4.7×10^{-6}	L
Limestone	1.1×10^{-6}	I
Limestone, fractured	$(1.8 \text{ to } 47.5) \times 10^{-4}$	M
Limestone, fractured and calcareous sandstone	1.47×10^{-4}	N
Loess	$(4.72 \text{ to } 142) \times 10^{-7}$	A

TABLE B.3 ***(Continued)***

Material	Hydraulic Conductivity in [m/s]	Reference
Peat	6.60×10^{-5}	B
Quartzite	$(1.9 \text{ to } 26{,}000) \times 10^{-9}$	G
Sand	$(4.75 \text{ to } 9{,}950) \times 10^{-6}$	B
Very fine sand	$(4.72 \text{ to } 47.2) \times 10^{-6}$	A
Very fine sand	$(4.75 \text{ to } 13.9) \times 10^{-6}$	B
Very fine sand	9.58×10^{-5}	C
Sand, very fine and fine	$(9.38 \text{ to } 27.8) \times 10^{-6}$	B
Sand, fine	$(4.72 \text{ to } 47.2) \times 10^{-5}$	A
Sand, fine	$(2.31 \text{ to } 6.60) \times 10^{-5}$	B
Sand, fine	$(0.2 \text{ to } 189) \times 10^{-6}$	C
Sand, fine and medium	$(4.75 \text{ to } 11.6) \times 10^{-5}$	B
Sand, medium	$(4.72 \text{ to } 21.2) \times 10^{-4}$	A
Sand, medium	$(9.38 \text{ to } 23.1) \times 10^{-5}$	B
Sand, medium	$(0.9 \text{ to } 567) \times 10^{-6}$	D
Sand, medium (well sorted)	2.52×10^{-3}	C
Sand, medium and coarse	$(1.85 \text{ to } 4.75) \times 10^{-4}$	B
Sand, coarse	$(2.12 \text{ to } 3.07) \times 10^{-3}$	A
Sand, coarse	$(3.82 \text{ to } 9.38) \times 10^{-4}$	B
Sand, coarse	$(0.9 \text{ to } 6{,}610) \times 10^{-6}$	C
Sand, coarse (well sorted)	3.00×10^{-2}	C
Sand, coarse and very coarse	$(7.06 \text{ to } 18.5) \times 10^{-4}$	B
Sand, very coarse	$(3.07 \text{ to } 3.77) \times 10^{-3}$	A
Sand, very coarse	$(1.39 \text{ to } 4.28) \times 10^{-3}$	B
Sand and gravel	$(9.38 \text{ to } 231) \times 10^{-5}$	B
Sand, very coarse, and gravel, very fine	$(2.78 \text{ to } 7.06) \times 10^{-3}$	B
Sand, Scituate	$(4.05 \text{ to } 9.49) \times 10^{-5}$	B
Sand, Plum Island	$(1.85 \text{ to } 2.66) \times 10^{-4}$	B
Sand, Fort Peck	$(1.74 \text{ to } 2.89) \times 10^{-5}$	B
Sand, Ottowa sand	$(5.56 \text{ to } 8.45) \times 10^{-5}$	B
Sand, Union Falls	$(4.17 \text{ to } 9.95) \times 10^{-4}$	B
Sand, Franklin Falls	$(9.03 \text{ to } 15.0) \times 10^{-6}$	B
Sand, dam filters	$(1.50 \text{ to } 99.5) \times 10^{-5}$	B
Sandstone	$(3.3 \text{ to } 54{,}000) \times 10^{-9}$	G
Sandstone	$(1.42 \text{ to } 14{,}200) \times 10^{-9}$	A
Sandstone	$(4.75 \text{ to } ?) \times 10^{-8}$	B
Sandstone (0.29 porosity)	2.32×10^{-5}	C
Sandstone	3.4×10^{-7}	E
Sandstone, fine-grained	2.31×10^{-6}	B
Sandstone, fine	$(0.5 \text{ to } 2{,}270) \times 10^{-8}$	D
Sandstone, medium-grained	3.59×10^{-5}	B
Sandstone, silty	2.52×10^{-8}	C
Sandstone, coarse	1.07×10^{-5}	C

TABLE B.3 (*Continued*)

Material	Hydraulic Conductivity in [m/s]	Reference
Sandstone arkosic, siltstone and shale	$(4.7 \text{ to } 710{,}000) \times 10^{-10}$	G
Sandstone	$(2.4 \text{ to } 140) \times 10^{-6}$	R
Schist	2.31×10^{-6}	B
Schist	$(0.002 \text{ to } 1{,}130) \times 10^{-8}$	D
Schist	$(4.7 \text{ to } 120{,}000) \times 10^{-9}$	G
Schist and Gneiss, fractured, crystalline	3.6×10^{-7}	S
Shale	$(2.4 \text{ to } 26) \times 10^{-6}$	G
Shale	$(1.16 \text{ to } 475{,}000) \times 10^{-13}$	B
Shale	2.0×10^{-8}	E
Silt	$(4.72 \text{ to } 47.2) \times 10^{-7}$	A
Silt	9.26×10^{-7}	B
Silt	$(0.09 \text{ to } 7{,}090) \times 10^{-9}$	D
Silt, loess	$(9.95 \text{ to } 174{,}000) \times 10^{-10}$	B
Sandy silt	$(6.94 \text{ to } 30.1) \times 10^{-11}$	B
Silt, Boston	$(9.95 \text{ to } 1{,}970) \times 10^{-11}$	B
Silt, North Carolina	$(5.56 \text{ to } 1{,}270) \times 10^{-9}$	B
Silty sand	$(6.94 \text{ to } 69{,}400) \times 10^{-8}$	B
Siltstone	$(0.1 \text{ to } 142) \text{ to } 10^{-10}$	D
Siltstone-shale	2.0×10^{-8}	E
Siltstone-shale	2.8×10^{-7}	E
Slate	$(4.72 \text{ to } 14.2) \times 10^{-7}$	A
Slate	9.26×10^{-10}	B
Sodium montmorillonite	1.85×10^{-10}	B
Till	$(1.42 \text{ to } 236) \times 10^{-9}$	A
Tuff	$(1.42 \text{ to } 47{,}200) \times 10^{-10}$	A
Tuff	2.31×10^{-6}	B

Notes: ? indicates that upper or lower limits were not available.

References:

(A) California Dept. of Water Resources, 1968.

(B) Daly, C., 1982. Evaluation of procedures for determining selected aquifer parameters. Prepared for U.S. Army Toxic Hazardous Materials Agency, CR REL Report 82-41.

(C) Davis, S., and J. M. DeWiest, 1966. *Hydrogeology*. New York: Wiley, p. 463.

(D) Morris, D. A., and A. I. Johnson, 1967. Summary of hydrological and physical properties of rock and soil materials as analyzed by the hydrologic laboratory of the U.S. Geological Survey. USGS Water Supply Paper 1839-D.

(E) Golder Associates, 1977. Development of site suitability criteria for a high level waste repository. Lawrence Livermore Laboratory Report, UCRL-13793.

(F) Daly, C., 1982. Evaluation of procedures for determining selected aquifer parameters. Prepared for U.S. Army Toxic Hazardous Materials Agency, CR REL Report 82-41.

(G) Rasmussen, W. C., 1964. Permeability and storage of heterogeneous aquifers in the U.S. *International Association of Scientific Hydrologic Publication 64*, pp. 317–325.

TABLE B.3 (*Continued*)

(H) Claasen, H. C., and E. H. Cordes, 1975. Two-well recirculating tracer test in fractured carbonate rock. *Nevada, Hydrol. Sci. Bull.* 20(3), 367–382.

(I) Kreft, A., A. Lenda, B. Turek, A. Zuber, and K. Czauderna, 1974. Determination of effective porosities by the two-well pulse method. *Isot. Tech. Groundwater Hydrol., Proc. Symp.*, vol. 2, pp. 295–312.

(J) Walter, G. B., 1983. Convergent flow tracer test at H-6: Waste isolation pilot plant (WIPP), southeast New Mexico (draft), Hydro Geochem, Inc., Tucson, AZ.

(K) Bredehoeft, J. D., and G. F. Pinder, 1973. Mass transport in flowing groundwater, *Water Resources Res.*, 9(1), 144–210.

(L) Fenske, P. R., 1973. Hydrology and radionuclide transport, monitoring well HT-2m, Tatum Dome, Mississippi. Proj. Rep. 25, Tech. Rep. NVD-1253-6. Cent. for *Water Resources Res.*, Desert Res. Inst., Univ. of Nev. Syst., Reno.

(M) Rabinowitz, D. D., and G. W. Gross, 1972. Environmental tritiuim as a hydrometeorologic tool in the Roswell Basin, New Mexico. Tech. Completion Rep. OWRR:A-037-NMEX, New Mexico, *Water Resources Res. Inst.*, Las Cruces.

(N) Segol, G., and G. F. Pinder, 1976. Transient simulation of saltwater intrusion in southeastern Florida. *Water Resources Res.* 12(1), 65–70.

(O) Dieulin, A., 1981. Lixiviation in situ d'un gisement d'uranium en milieu granitique, Draft Rep. LHM/RD/81/63, Ecole Natl. Super. des Mines de Paris, Fontainebleau, France.

(P) Ivanovitch, M., and D. B. Smith. Determination of aquifer parameters by a two-well pulsed method using radioactive tracers. *J. Hydrol.* 36(1/2), 35–45.

(Q) Grove, D. B., 1977. The use of Galerkin finite-element methods to solve mass transport equations, Rep. USGS/WRD/WRI-78/011. USGS, Denver, CO. (Available as NTIS PB 277-532 from Natl. Tech. Inf. Serv., Springfield, VA.)

(R) Oakes, D. B., and D. J. Edworthy, 1977. Field measurement of dispersion coefficients in the United Kingdom. *Ground Water Quality, Measurement, Prediction, and Prediction*, Reading, England: Water Research Centre, pp. 327–340.

(S) Webster, D. S., J. F. Procter, and J. W. Marine, 1970. Two-well tracer test in fractured crystalline rock. USGS, Water Supply Paper, 1544-I.

TABLE B.4 Vertical Hydraulic Conductivity

Material	Horizontal Conductivity K_h in [m/s]	Vertical Conductivity K_v in [m/s]	K_v/K_h	Reference
Anhydrite	$(1.0 \text{ to } 100) \times 10^{-14}$	$(1.0 \text{ to } 100) \times 10^{-15}$	0.1	K
Chalk	$(1.0 \text{ to } 100) \times 10^{-10}$	$(5.0 \text{ to } 100) \times 10^{-11}$	0.5	K
Clay, Boston	—	—	0.3 to 1.4	B
Clay, marine	—	—	0.8	C*
Clay, marine	—	—	0.95	D*
Clay, soft	—	—	0.66	E*
Clay, varved	—	—	0.3 to 0.66	F*
Clay, varved	—	—	0.66	G*
Clay, varved	—	—	0.025 to 0.25	H*
Clay, varved	—	—	0.066 to 0.3	I*
Dolomite	$(1.0 \text{ to } 100) \times 10^{-9}$	$(5.0 \text{ to } 500) \times 10^{-10}$	0.5	K
Limestone	$(1.0 \text{ to } 100) \times 10^{-9}$	$(5.0 \text{ to } 500) \times 10^{-10}$	0.5	K
Limestone, fractured and calcareous sandstone	4.5×10^{-3}	9×10^{-6}	0.002	L
Salt	1.0×10^{-14}	1.0×10^{-14}	1.0	K
Sand, silt and clay	5×10^{-4}	5.1×10^{-5}	0.1	M
Sandstone	$(5 \text{ to } 1{,}000) \times 10^{-13}$	$(2.5 \text{ to } 500) \times 10^{-13}$	0.5	K
Shale	$(1 \text{ to } 100) \times 10^{-14}$	$(1 \text{ to } 100) \times 10^{-15}$	0.1	K
Sandstone	3.4×10^{-7}	3.4×10^{-7}	1.0	J*
Siltstone-shale	2.1×10^{-6}	2.1×10^{-7}	0.1	A
Silt, organic	—	—	0.6 to 0.8	A
Siltstone-shale	2.8×10^{-7}	3.0×10^{-8}	0.107	A
Shale	2.0×10^{-8}	1.0×10^{-8}	0.5	A

*Summarized in R. E. Olsen and D. E. Daniel, 1981, Measurement of the hydraulic conductivity of fine-grained soils, in *Permeability and Groundwater Transport*, ASTM STP 746, edited by T. F. Zimmie and C. O. Riggs, Washington, D.C.: American Society for Testing and Materials, pp. 18–64.

TABLE B.4 ***(Continued)***

References:

(A) Golder Associates, 1977. Development of site suitability criteria for a high level waste reposistory. Lawrence Livermore Laboratory Report, UCRL-13793.

(B) Haley and Aldrich, 1969. Engineering properties of foundation soils at Long Creek–Fore River areas and back cover. Report no. 1, Maine State Highway Commission.

(C) Lumb, P., and J. K. Holt, 1968. *Geotechnique* 18, 25–36.

(D) Subbaraju, B. H., T. K. Natarajan, and R. K. Bhandari, *Proc., 8th International Conference of Soil Mechanics and Foundations Engineers*, Moscow, vol. 2.2, pp. 217–220.

(E) Bazett, D. J., and A. F. Brodie, 1961. *Ontario Hydro Research News* 13 (4), 1–6.

(F) Chan, H. T., and T. C. Kenney, 1973. *Canadian Geotechnical J.* 10 (3), 453–472.

(G) Kenney, T. C., and H. T. Chan, 1973. *Canadian Geotechnical J.* 10 (3), 473–488.

(H) Casagrande, L., and Poulos, S. J., 1969. *Canadian Geotechnical J.* 6 (3), 287–326.

(I) Wu, T. H., N. Y. Chang, and E. M. Ali, 1978. *J. Geotechnical Engineers Division*. American Society of Chemical Engineers, vol. 104, no. GT7, pp. 899–905.

(J) Tsien, S. I., 1955. Stabilization of marsh deposit. *Highway Res. Board, Bull.* 115, 15–43.

(K) Domenico, P. A., and F. W. Schwartz, 1990. *Physical and Chemical Hydrogeology*. New York: Wiley, p. 824.

(L) Segol, G., and G. F. Pinder, 1976. Transient simulation of saltwater intrusion in southeastern Florida. *Water Resources Res.* 12(1), 65–70.

(M) Papadopulos, S. S., and S. P. Larson, 1978. Aquifer storage of heated water: II. Numerical simulation of field results. *Groundwater* 16(4), 242–248.

TABLE B.5 Specific Storage

Material	Specific Storage in [l/m]
Clay, plastic	$(2.6 \text{ to } 20) \times 10^{-3}$
Clay, stiff	$(1.3 \text{ to } 2.6) \times 10^{-3}$
Clay, medium hard	$(9.2 \text{ to } 13) \times 10^{-4}$
Gravel, dense sandy	$(4.9 \text{ to } 10) \times 10^{-5}$
Rock, fissured, jointed	$(3.3 \text{ to } 69) \times 10^{-6}$
Rock, sound	$<3.3 \times 10^{-6}$
Sand, loose	$(4.9 \text{ to } 10) \times 10^{-4}$
Sand, dense	$(1.3 \text{ to } 2.0) \times 10^{-4}$

Reference:
Domenico, P. A., 1972. *Concepts and Models in Groundwater Hydrology*. New York: McGraw-Hill.

TABLE B.6 Unsaturated Hydraulic Conductivity Relationships

Relationship	Reference
$K(\theta) = K_s[(\theta - \theta_S)/(\theta_S - \theta_r)]^{3.5}$	A
$K(h) = a\lvert h\rvert^{-b}$	B
$K(h) = K_s[\exp(-a\lvert h\rvert)]$	C
$K(h) = [(\lvert h\rvert/a)^b + 1]^{-1}$	C
$K(h) = a\{\cosh[(h/h_{cr})^b] - 1\}/\{\cosh[(h/h_{cr})^b] + 1\}$	D
$K(h) = \exp[a(\lvert h\rvert - \lvert h_{cr}\rvert)]$ for $h \geq h_1$	E
$K(h) = K_s(\lvert h\rvert/\lvert h_1\rvert)^{-b}$ for $h < h_1$	E
$K(h) = (\lvert h\rvert/\lvert h_{cr}\rvert)^{-b}$	F
$K(h) = K_s(a)/(a + \lvert h\rvert^b)$	G
$K(h) = a\theta_s^{\,a} S_r^{\,b}$	H
$K(\theta) = a[\exp(b)\theta(h)]$	I
$K(h) = K_s[\exp(bh)]$	I
$K(h) = a[(\theta(h))^b]$	I

Note:

K = hydraulic conductivity in [L/T]
θ = moisture content in [l]
h = pressure head (pore pressure) in [L]
K_s = saturated hydraulic conductivity in [L/T]
θ_s = saturated moisture content in [l]
θ_r = residual moisture content in [l]
S_r = degree of saturation in [l]
a, b, h_{cr}, h_1 = empirical constants

TABLE B.6 (*Continued*)

References:

(A) Averjanov, S. F., 1950. About permeability of subsurface sites in case of incomplete saturation. *Engineering Collection*, vol. 7, quoted in Polubarinova Kochina, 1962, *The Theory of Groundwater Movement*. Princeton, NJ: Princeton Univ. Press.

(B) Wind, G. P., 1955. Flow of water through plant roots, *Netherlands J. Agricultural Sci.*, 3, 259–264.

(C) Gardner, W. R., 1958. Some steady-state solutions to the unsaturated flow equation with application to evaporation from the water table. *Soil Sci.* 85, 228–232.

(D) King, L. G., 1964. Imbibition of fluids by porous solids. Ph.D. thesis, Colorado State Univ., Fort Collins, CO.

(E) Rijtema, P. E., 1965. An analysis of actual evapotranspiration. Agricultural Research Report 659. Center for Agricultural Publication and Documentation, Wageningen, The Netherlands.

(F) Brooks, R. H., and A. T. Corey, 1966. Properties of porous media affecting fluid flow, *J. Irrigation Drainage Division, Am. Soc. Civil Engineers* 92(IR2), 61–68.

(G) Haverkamp, R., M. Vauclin, J. Touma, P. J. Wierenga, and G. Vachaud, 1977. A comparison of numerical simulation models for one-dimensional infiltration. *Soil Sci. Soc. of America J.* 41, 285–294.

(H) Pickens, J. F., R. W. Gillham, and D. R. Cameron, 1979. Finite-element analysis of the transport of water and solutes in tile-drained soils. *J. Hydrol.* 40, 243–264.

(I) van Genuchten, M. T., 1980. A closed form equation for predicting the hydraulic conductivity of unsaturated soils. *Soil Sci. Soc. of America J.* 44, 892–898.

TABLE B.7 Moisture Content Relationships

Relationship	Reference
$\theta(h) = \theta_r + a(\theta_s - \theta_r)/[a + \lvert h\rvert^b]$	A
$\theta(h) = \theta_r + a(\theta_s - \theta_r)/(a + (\ln\lvert h\rvert)^b)^m$	B
$S_r(h) = (\theta_0/\theta_s)\{\cosh[(h/h_{cr})^m + a] - b\}/\{(\cosh[h/h_{cr})^m + a] + b\}$	C
where $b = (\theta_0 - \theta_r)/(\theta_0 + \theta_r)$	
$\theta(h) = \theta_r + (\theta_s - \theta_r)/[1 + a(\lvert h\rvert)^b]^m$ where $m = 1 - 1/b$	D

Note:

θ = moisture content in [l]

h = pressure head (pore pressure) in [L]

θ_r = residual moisture content in [l]

θ_s = saturated moisture content in [l]

a, b, m, h_{cr}, h_1 = empirical constants

References:

(A) Brutsaert, W., 1966. Probability laws for pore-size distributions. *Soil Sci.* 101, 85–92.

(B) Haverkamp, R., M. Vauclin, J. Touma, P. J. Wierenga, and G. Vachaud, 1977. A comparison of numerical simulation models for one-dimensional infiltration. *Soil Sci. Soc. of America J.* 41, 285–294.

(C) Pickens, J. F., R. W. Gillham, and D. R. Cameron, 1979. Finite-element analysis of the transport of water and solutes in tile-drained soils. *J. Hydrol.* 40, 243–264.

(D) van Genuchten, M. T., 1980. A closed form equation for predicting the hydraulic conductivity of unsaturated soils. *Soil Sci. Soc. of America J.* 44, 892–898.

APPENDIX C
EPA Priority Pollutants

TABLE C.1 Characteristics of Priority Pollutants

Priority Pollutant	Chemical Abstract No. (CAS)	Type of Chemical Substance
Acenaphthene	83-32-9	Aromatic
Acenaphthylene	208-96-8	Aromatic
Acrolein	107-02-8	Pesticide
Acrylonitrile	107-13-1	Organic
Aldrin (HHDN)	309-00-2	Pesticide
Anthracene	120-12-7	Aromatic
Antimony	7440-360-0	Metal
Arsenic	7440-38-2	Metal
Asbestos	1332-21-4	Mineral
Beryllium	7440-41-7	Metal
Benzene (Benzol)	71-43-2	Aromatic
Benzidine	92-87-5	Substitute aromatic
Benzo [*a*] anthracene	56-55-3	Aromatic
Benzo [*b*] fluoranthene (B[*b*]F)	205-99-2	Aromatic
Benzo [*k*] fluoranthene (B[*k*]F)	207-08-9	Aromatic
Benzo [*ghi*] perylene	191-24-2	Aromatic
Benzo[*a*]pyrene (B[*a*]P)	50-32-8	Aromatic
Bis (2-chloroethoxy) methane	111-91-1	Chlorinated ether
Bis (2-chloroethyl) ether (Chlorex)	111-44-4	Chlorinated ether
Bis (chloromethyl) ether (BCME)	542-88-1	Chlorinated ether
Bis (2-chloroisopropyl) ether	108-60-1	Chlorinated ether
Bis (2-ethylhexyl) phthalate (DEHP)	117-81-7	Phthalate ester
Bromoform	75-25-2	Chlorinated alkane
4-Bromophenyl phenyl ether	101-55-3	Chlorinated ether
Butyl benzyl phthalate (BBP)	85-68-7	Phthalate ester
Cadmium	7440-43-9	Metal
Carbon tetrachloride (CTET)	56-23-5	Chlorinated alkane
Chlordane	57-74-9	Pesticide
Chlorobenzene	108-90-7	Chlorinated aromatic
Chlorodibromomethane	124-48-8	Chlorinated alkane
Chloroethane	75-00-3	Chlorinated alkane
2-Chloroethyl vinyl ether	110-75-8	Chlorinated ether
Chloroform	67-66-3	Chlorinated alkane
p-Chloro-m-cresol	59-50-7	Phenol
2-Chloronaphthalene	91-58-7	Chlorinated biphenol
2-Chlorophenol	95-57-8	Phenol
4-Chlorophenyl phenyl ether	7005-72-3	Chlorinated esther
Chromium	7440-47-3	Metal
Chrysene	218-01-9	Aromatic
Copper	7440-50-8	Metal
Cyanide	57-12-5	Miscellaneous
4,4-DDD (TDE)	72-54-8	Pesticide
4,4-DDE	72-55-9	Pesticide
4,4-DDT	50-29-3	Pesticide
Dibenzo[*a,h*]anthracene (DBA)	53-70-3	Chlorinated aromatic
1,2-Dichlorobenzene (DCB, 1,2)	95-50-1	Chlorinated aromatic

Density in [g/cm^3]	Kinematic Viscosity in [cp]	Solubility in [ppm]	Log K_{ow} in [l]	Vapor Pressure in [mmHg]	Henry Coefficient in [atm.m^3/mol]
1.189		3.42	4.33	1 $E^{-2,-3}$	3.93–9.1 E^{-5}
		3.93	4.07	1 $E^{-2,-3}$	1.9 E^{-4}, 1.45 E^{-3}
0.8427		2.1 E^{5}	−0.09, 1.02	215–220	9.7 E^{-5}, 5.7 E^{-5}
0.8004		7.9 E^{4}	−0.92, −0.14	100	6.3 E^{-5}, 8.8 E^{-5}
		0.01, 27–180	5.3	6 E^{-6}	2.1 E^{-3}, 1.4 E^{-5}
1.25		1.29, 0.075	4.54, 4.45	1.95 E^{-4}	8.6 E^{-5}
6.691		6000		0.0	
5.727		1000		0.0	
				0.0	
1.848			0.0		
0.879	0.6028	820, 1750	1.95–2.13	60.0	2.67–5.3 E^{-3}
1.250		400	1.81, 1.34	5 E^{-4}	3 E^{-7}
		0.01–0.057	5.61	5 E^{-9}	1 E^{-6}
					1.22 E^{-5}
		4.3 E^{-3}	6.84	9.59 E^{-11}	3.87 E^{-5}
		2.6 E^{-4}	7.23	1.3 E^{-10}	1.44 E^{-7}
		0.38, 3 E^{-3}	6.0–6.5	5 E^{-9}	5 E^{-2}
		8.1 E^{4}	1.26	<0.1	2.86 E^{-7}
1.222	1 E^{4}	1.58	0.71	1.3 E^{-5}	
1.271		2.2 E^{4}	−0.38	30	2.1 E^{-4}
1.1122		1.7 E^{3}	2.58	0.85	1.1 E^{-4}
		0.4–1.3	5.3, 8.7	2.0 E^{-7}	3 E^{-7}
2.89		3190; 3010	2.3	5.6	5.8–6.3 E^{-4}
1.423			4.28	1.5 E^{-3}	1 E^{-4}
		2.9	4.8, 5.8	8.6 E^{-6}	8.3 E^{-6}
8.65		1700		0.0	
1.59	0.965	1160, 800	2.62–2.95	56	2.5 E^{-2}
1.59	69 poise	1.85	2.78–6.00	1 E^{-7}	9.4 E^{-5}
1.106	0.799	∞500	2.18–3.79	8.8	3.58–4.0 E^{-3}
2.445		4 E^{3}	2.09	15, 76	8.4–9.9 E^{-4}
0.917	0.279	5740	1.54, 1.39	1000	0.148
		15,000	1.28	26.75	2.5 E^{-7}
1.489	0.596	8200	1.97	150	2.88–3.4 E^{-3}
1.215		3850	2.95, 3.10	0.05	2.5 E^{-6}
		6.74	4.12	0.017	5.4 E^{-4}
1.241	2.250	28,500	2.15, 2.19	2.2	2.1 E^{-5}, 7.56 E^{-6}
		3.3	4.08	0.0027	2.1 E^{-4}
7.2					
1.274		0.0015	5.61	1 E^{-11}, E^{-6}	1.05 E^{-6}
8.96				0.0	
1.52–1.86					
		0.02–0.1	5.99, 6.1, 5.1	1.89 E^{-6}	2.2 E^{-8}
		0.0013	4.3, 4.8, 5.7	6.5 E^{-6}	6.8 E^{-5}
1.54		0.0034	3.98, 5.13, 6.0	1.9 E^{-7}	1.6–3.89 E^{-5}
		5 E^{-4}	5.97, 6.5	1 E^{-10}	7.3 E^{-8}
1.305	1.324	100, 133	3.38, 3.56	1	1.77–1.9 E^{-3}

TABLE C.1 **(*Continued*)**

Priority Pollutant	Chemical Abstract No. (CAS)	Type of Chemical Substance
1,3-Dichlorobenzene (DCB, 1,3)	541-73-19	Chlorinated aromatic
1,4-Dichlorobenzene (DCB, 1,4)	106-46-7	Chlorinated aromatic
3,3-Dichlorobenzidene	91-94-1	Substituted aromatic
Dichlorobromomethane	75-27-4	Chlorinated alkane
Dichlorodifluoromethane	75-71-8	Chlorinated alkane
1,1-Dichloroethane (DCA, 1,1)	75-34-3	Chlorinated alkane
1,2-Dichloroethane (DCA, 1,2)	107-06-2	Chlorinated alkane
1,2-*trans*-Dichloroethylene	540-59-0	Chlorinated alkane
1,1-Dichloroethylene (DCE, 1,1)	75-35-4	Chlorinated alkane
2,4-Dichlorophenol (DCP, 2,4)	120-83-2	Phenol
1,2-Dichloroprane	78-87-5	Chlorinated alkane
1,3-Dichloropropene	542-75-6	Chlorinated alkane
Dieldrin (HEOD)	60-57-1	Pesticide
Diethyl phthalate (DEP)	84-66-2	Phthalate ester
2,4-Dimethylphenol (2,4-Xylenol)	105-67-9	Phenol
Dimethyl phthalate (DMP)	131-11-3	Phthalate ester
Di-*n*-Butyl phthalate (DBP)	84-74-2	Phthalate ester
4,6-Dinitro-o-cresol (DNOC)	534-52-1	Phenol
2,4-Dinitrophenol (DNP, 2,4)	51-28-5	Phenol
2,4-Dinitrotoluene (DNT)	121-14-2	Substituted aromatic
2,6-Dinitrotoluene	606-20-2	Substituted aromatic
Di-n-octyl phthalate (DOP)	117-84-0	Phthalate ester
1,2-Diphenylhydrazine	122-66-7	Substituted aromatic
A-Endosulfan-α	115-29-7	Pesticide
B-Endosulfan-β	115-29-7	Pesticide
Endosulfan sulfate	1031-07-8	Pesticide
Endrin	72-20-8	Pesticide
Endrin aldehyde	7421-93-4	Pesticide
Ethylbenzene	100-41-4	Aromatic
Fluoranthene	206-44-0	Aromatic
Fluorene	86-73-7	Aromatic
Heptachlor (E 3314)	76-44-8	Pesticide
Heptachlor epoxide	1024-57-3	Pesticide
Hexachlorobenzene (HCB)	118-4-1	Chlorinated aromatic
Hexachlorobutadiene (HCBD)	87-68-3	Chlorinated alkane
Hexachlorocyclohexane-α (BHC)	319-84-6	Pesticide
Hexachlorocyclohexane-β (BHC)	319-86-8	Pesticide
Hexachlorocyclohexane-γ (lindane)	58-89-9	Pesticide
Hexachlorocyclopentadiene (HCCPD)	77-47-4	Pesticide
Hexachloroethane (HCE)	67-72-1	Chlorinated alkane
Indeno [1,2,3-cd] pyrene (IP)	193-39-5	Chlorinated alkane
Isophorone	78-59-1	Aromatic
Lead	7439-92-1	Pesticide
Mercury	7439-97-6	Metal
Methyl bromide	74-83-9	Chlorinated alkane
Methyl chloride	74-87-3	Chlorinated alkane

Density in [g/cm³]	Kinematic Viscosity in [cp]	Solubility in [ppm]	Log K_{ow} in [l]	Vapor Pressure in [mmHg]	Henry Coefficient in [atm.m³/mol]
1.288	1.04	69	3.38, 3.56	2.28	2.7–3.61 E^{-3}
1.458	0.72	49	3.39, 3.53	0.6	2.1–3.0 E^{-3}
		4	3.02		8 E^{-7}
1.98		4.5 E^{3}	1.88, 2.1	50	2.1 E^{-4}, 2.41 E^{-3}
1.329		280	2.16	4250	2.1, 2.98
1.74	0.505	5500	1.79	180	5.1–5.4 E^{-3}
1.25	0.887	9200	1.48	60	9.14 E^{-4}
1.25	0.4	600	1.48	200	
1.218	0.44	400	1.48	591	1.7 E^{-1}
1.383		4600	2.75, 3.23	0.12, 0.059	4.2 E^{-5}, 2.8 E^{-6}
1.16		2700	2.28, 2.02	42	2 E^{-3}
1.217		2700	1.98	25	
1.75		0.1, 186–200	3.69–5.48	1.8 E^{-7}	1.7–4.57 E^{-7}
1.120		210–896	2.47, 3.22	3.5 E^{-3}	1.2 E^{-6}
1.036		590	2.3, 2.5, 2.42	0.0621	1.7 E^{-5}
1.19	9.18	5000	1.56, 212	4.2 E^{-3}	2.15 E^{-6}, 4.2 E^{-7}
1.046		13	5.2, 5, 56	1 E^{-5}	6.3 E^{-5}, 2.8 E^{-7}
			2.85	5 E^{-2}	4 E^{-5}
1.683		5600	1.51, 1.54	1.49 E^{-5}	6.45 E^{-10}
1.521		270	2.01, 1.97	1.3 E^{-3}	1 E^{-5}, 4.5 E^{-6}
			2.05	0.018	1 E^{-5}, 1 E^{-6}
		3	9.2, 9.87	1.4 E^{-4}	1.7 E^{-5}
			3.03		3.4 E^{-9}
		0.26–0.6	−1.69, 3.55	1 E^{-5}	1 E^{-5}
		0.06–0.10	−1.69, 3.62	1 E^{-5}	1.91 E^{-5}
		0.117–0.22	−1.3, 3.66	1 E^{-5}	2.6 E^{-5}
		0.25	3.21–5.6	1 E^{-7}	4 E^{-7}
				2 E^{-7}	2 E^{-9}
0.866	0.678	140	3.15–3.24	7	5.7–6.6 E^{-3}
1.252		0.265	5.33, 4.90	1 $E^{-4, -6}$	6.5 E^{-6}
1.203		1.69–1.98	4.18–4.38	1 $E^{-4, -2}$	2.1 E^{-4}, 6.4 E^{-5}
1.579		0.056	3.87–5.44	3 E^{-4}	2.3–4.1 E^{-3}
		0.11–0.35		3 E^{-4}	3.9 E^{-4}
2.044		0.006–0.11	5.23–6.18	1.09 E^{-5}	6.8 E^{-4}
1.675		2	3.74–4.78	0.15	0.0256
		1.21–2.0	3.81		
		0.13–0.7	3.78, 3.89	2.8 E^{-7}	4.5 E^{-7}, 1.1 E^{-2}
				1.7 E^{-5}	2 E^{-7}
		2.15–12	3.72	1.6 E^{-4}	7.8 E^{-6}
1.702		0.8, 1.8	3.99	0.081	0.016
2.09		50	3.34–4.62	0.4	
			7.66	1 E^{-10}	6.95 E^{-8}
0.92		12,000	1.7	0.38	4.2–5.75 E^{-6}
11.35		600		0.0	
13.55		300		1.6 E^{-3}	
1.73		900	1.19	1420	9.3 E^{-2}

TABLE C.1 (*Continued*)

Priority Pollutant	Chemical Abstract No. (CAS)	Type of Chemical Substance
Methylene chloride	75-09-2	Chlorinated alkane
Naphthalene	91-20-3	Aromatic
Nickel	7440-02-0	Metal
Nitrobenzene	98-95-3	Substituted aromatic
2-Nitrophenol	88-75-5	Phenol
4-Nitrophenol	100-02-7	Phenol
n-Nitrosodimethylamine	62-75-9	Organic
n-Nitrosodi-n-propylamine	621-64-7	Organic
n-Nitrosodiphenylamine	86-30-6	Organic
PCB-1016		Chlorinated biphenol
PCB-1221	111-042-82	Chlorinated biphenol
PCB-1232	111-411-65	Chlorinated biphenol
PCB-1242	534-692-19	Chlorinated biphenol
PCB-1248	126-722-96	Chlorinated biphenol
PCB-1254	110-976-91	Chlorinated biphenol
PCB-1260	110-968-25	Chlorinated biphenol
Pentachlorophenol (PCP)	87-86-5	Phenol
Phenanthrene	85-01-8	Aromatic
Phenol	108-95-2	Phenol
Pyrene	129-00-0	Aromatic
Selenium	7782-49-2	Metal
Silver	7440-22-4	Metal
2,3,7,8-Tetrachloridibenzo-*p*-dioxin (TCDD, Dioxin)	1746-01-6	Chlorinated organic
1,1,2,2-Tetrachloroethane	630-20-6	Chlorinated alkane
Tetrachloroethylene (PCE, PER)	127-18-4	Chlorinated alkane
Thallium	7440-28-0	Metal
Toluene	108-88-3	Aromatic
Toxaphene (PCC)	8001-35-2	Pesticide
1,2,4-Trichlorobenzene (TCB)	120-82-1	Chlorinated aromatic
1,1,1-Trichloroethane (TCA)	71-55-6	Chlorinated alkane
1,1,2-Trichloroethane	79-00-5	Chlorinated alkane
Trichloroethylene (TCE,TRI)	79-01-6	Chlorinated alkane
Trichlorofluoromethane (Freon 11)	75-69-4	Chlorinated alkane
2,4,6-Trichlorophenol	88-06-2	Phenol
Vinyl chloride (MVC)	75-01-4	Chlorinated alkane
Zinc	7440-66-6	Metal

Notes:

- Most properties are at approximately 20°C.
- The table is generated using various sources and includes both measured and calculated values.
- While this table was prepared with care, the accuracy of these data is not guaranteed.
- Most of these parameters are measured in pure water and may not be appropriate for the investigated environmental conditions.
- For much data, information differs in the literature. Consult chemical reference books for detailed information, such as the following

Density in [g/cm^3]	Kinematic Viscosity in [cp]	Solubility in [ppm]	Log K_{ow} in [l]	Vapor Pressure in [mmHg]	Henry Coefficient in [atm.m^3/mol]
0.991		5300–7250	0.91	3765	
1.326	0.45	16,700; 20,200	1.25	362	2–2.5 E^{-3}
1.145	0.78	31, 34	4.33	0.0492, 0.23	3.6–4.6 E^{-4}
8.902		2500		0.0	
1.2	1.634	1900	1.85, 2.93	0.15	1.1–1.31 E^{-5}
1.657		2100	1.76, 2.00	1.0	7.6 E^{-5}
1.479		1.6 E^4	2.91, 1.91	2.2	2.5 E^{-5}
			0.06	8.1	2.3 E^{-5}
				0.4	6.3 E^{-6}
			2.57		
1.33			>5.58	4 E^{-4}	3.3 E^{-4}
1.15		15–40	2.8, 4.09	6.7 E^{-3}	1.7 E^{-4}
1.24				4.06 E^{-3}	1.13 E^{-5}
1.35		0.13–0.34	4.11, 5.58	4.06 E^{-3}	1.98 E^{-3}, 5.73 E^{-4}
1.41				4.94 E^{-4}	3–3.51 E^{-3}
1.50		0.012–0.31	6.03, 6.47	7.71 E^{-5}	2.3–2.8 E^{-3}
1.58		0.0027	6.11, 7.14	4.05 E^{-5}	6.1–7.1 E^{-3}
1.978		5.0	5.01–5.24	1 E^{-4}	2.1–2.8 E^{-6}
1.025		1.18	4.46	6.8 E^{-4}	1.3–2.3 E^{-4}
1.07	4.076	8.2 E^4	−0.22, 1.46	0.2–0.53	2.7–4.54 E^{-7}
1.277		0.032–0.13	4.88–5.96	6.85 E^{-7}	5.1 E^{-6}
4.79				0.0	
		0.2, 2 E^{-4}	6.84	1 E^{-6}	2 E^{-3}
1.6	1.844	2900, 3000	2.56, 2.39	5, 6.5	3.8–4.0 E^{-4}
1.6	1.93	150–200		14	
11.85				0.0	
0.867	0.552	470	2.69, 2.89	10	5.7 E^{-3}
		0.3–0.74	3.3	1 E^{-6}	6.3 E^{-2}, 0.21
1.446		19, 30	4.26, 4.05	0.42, 0.29	2.3 E^{-3}
1.35	0.903	4400, 5497	2.17–2.51	100	3.6–4.92 E^{-3}
1.44	0.12	4500, 1100	2.17, 2.07	19	1.2 E^{-2}, 7.43 E^{-4}
1.456	0.566	1100	2.29, 2.42	20	1 E^{-2}, 9.1 E^{-3}
1.494		1100	2.53	667.4	1.1 E^{-1}, 5.8 E^{-2}
1.49		800	3.72, 3.61	0.012	4 E^{-6}
0.9121		60	0.6, 1.23	2660	6.4, 8.14 E^{-2}
7.13				0.0	

References:

Callahan, M. M. et al., 1979. Water-related environmental fate of 129 priority pollutants. NTIS PB80-204381.

Lyman, W. J., W. F. Ruhl, and D. H. Rosenblatt, 1982. *Handbook of Chemical Property Estimation Methods.* New York: McGraw-Hill.

Verschueren, K., 1983. *Handbook of Environmental Data on Organic Chemicals*, 2d ed. New York: Van Nostrand Reinhold.

Weast, R. C., (ed). 1985–1986. *Handbook of Chemistry and Physics.* Cleveland, OH: CRC Press.

TABLE C.2 Transport and Fate of Priority Pollutants

Priority Pollutant	Advection	Sorption	Decay/ Degradation	Hydrolysis	Volatilization
Acenaphthene	?	Y	Y	N	N
Acenaphthylene	?	Y	Y	N	N
Acrolein	?	N	Y	N	?
Acrylonitrile	?	N	?	N	Y
Aldrine (HHDN)	?	Y	Y	N	?
Anthracene	?	Y	Y	N	?
Antimony	Y	Y	Y	Y	?
Arsenic	Y	Y	Y	Y	Y
Asbestos	Y	N	N	N	N
Beryllium	Y	Y	N	Y	N
Benzene (Benzol)	?	?	?	N	Y
Benzidine	?	Y	?	N	N
Benzo [*a*] anthracene	?	Y	Y	N	N
Benzo [*b*] fluoranthene (B[*b*]F)	?	Y	Y	N	?
Benzo [*k*] fluoranthene (B[*k*]F)	?	Y	Y	N	?
Benzo [*ghi*] perylene	?	Y	?	N	?
Benzo[*a*]pyrene (B[*a*]P)	?	Y	T	N	?
Bis (2-chlorethyoxy) methane	Y	N	?	?	N
Bis (2-chloroethyl) ether (Chlorex)	?	N	N	N	Y
B (chloromethyl) ether (BCME)	N	N	N	Y	N
Bis (2-chloroisopropyl) ether	?	N	?	?	Y
Bis (2-ethylhexyl) phthalate (DEHP)	Y	Y	Y	N	N
Bromoform	?	?	?	N	Y
4-Bromophenyl phenyl ether	?	?	?	N	?
Butyl benzyl phthalate (BBP)	?	?	Y	N	?
Cadmium	Y	Y	N	Y	N
Carbon tetrachloride (CTET)	?	N	N	N	Y

Chlordane	?	?	?	N	?
Chlorobenzene	?	?	?	N	Y
Chlorodibromomethane	?	?	?	N	?
Chloroethane	?	N	N	N	Y
2-Chloroethyl vinyl ether	?	N	?	?	Y
Chloroform	?	N	N	N	Y
p-Chloro-m-cresol	?	N	?	N	N
2-Chloronaphthalene	?	?	Y	N	?
2-Chlorophenol	?	N	?	N	N
4-Chlorophenyl phenyl ether	?	?	?	N	?
Chromium	Y	Y	N	Y	N
Chrysene	?	Y	Y	N	?
Copper	Y	Y	N	Y	N
Cyanide	N	N	Y	?	Y
4,4-DDD (TDE)	?	Y	?	?	Y
4,4-DDE	?	Y	?	N	Y
4,4-DDT	?	Y	?	?	Y
Dibenzo[*a,h*]anthracene (DBA)	?	Y	?	N	?
1,2-Dichlorobenzene (DCB, 1,2)	?	?	?	N	Y
1,3-Dichlorobenzene (DCB, 1,3)	?	?	?	N	?
1,4-Dichlorobenzene (DCB, 1,4)	?	?	?	N	?
3,3-Dichlorobenzidene	?	Y	N	N	N
Dichlorobromomethane	?	?	?	N	?
Dichlorodifluoramethane	?	?	?	N	Y
1,1-Dichloroethane (DCA,1,1)	?	N	N	N	Y
1,2-Dichloroethane (DCA,1,2)	?	N	?	N	Y
1,2-*trans*-Dichloroethylene	?	N	N	N	Y
1,1-Dichloroethylene (DCE-1,1)	?	N	N	N	Y
2,4-Dichlorophenol (DCP-2,4)	?	N	Y	N	N
1,2-Dichloropropane	?	?	?	?	Y
1,3-Dichloropropene	?	?	?	?	Y

TABLE C.2 (*Continued*)

Priority Pollutant	Advection	Sorption	Decay/ Degradation	Hydrolysis	Volatilization
Dieldrin (HEOD)	?	Y	?	N	?
Diethyl phthalate (DEP)	Y	?	?	N	N
2,4-Dimethylphenol (2,4-Xylenol)	?	?	?	N	N
Dimethyl phthalate (DMP)	Y	?	?	?	N
Di-*n*-Butyl phthalate (DBP)	Y	Y	?	N	N
4,6-Dinitro-o-cresol (DNOC)	?	Y	?	N	N
2,4-Dinitrophenol (DNP,2,4)	?	Y	?	?	N
2,4-Dinitrotolluene (DNT)	?	Y	?	N	N
2,6-Dinitrotoluene	?	Y	?	N	N
Di-n-octyl phthalate (DOP)	?	?	?	N	N
1,2-Diphenylhydrazine	?	Y	?	N	N
A-Endosulfan-α	?	Y	Y	Y	?
B-Endosulfan-β	?	Y	Y	Y	?
Endosulfan sulfate	?	Y	Y	Y	?
Endrin	?	?	?	N	?
Endrin aldehyde	?	?	?	N	?
Ethylbenzene	?	?	?	N	Y
Fluoranthene	?	Y	Y	N	N
Fluorene	?	Y	Y	N	N
Heptachlor (E 3314)	?	N	N	Y	?
Heptachlor epoxide	?	Y	Y	N	?
Hexachlorobenzene (HCB)	?	Y	N	N	?
Hexachlorobutadiene (HCBD)	N	Y	?	?	?
Hexachlorocyclohexane-α (BHC)	?	?	Y	N	?
Hexachlorocyclohexane-β (BHC)	?	?	Y	N	?
Hexachlorocyclohexane-δ (BHC)	?	?	Y	N	?
Hexachlorocyclohexane-γ (lindane)	?	?	Y	N	?

Hexachlorocyclopentadiene (HCCPD)	N	Y	N	Y	Y
Hexachloroethane (HCE)	N	?	?	?	?
Indeno [1,2,3-cd] pyrene (IP)	?	Y	?	N	?
Isophorone	Y	N	?	N	?
Lead	Y	Y	Y	Y	?
Mercury	Y	Y	Y	Y	Y
Methyl bromide	?	N	N	Y	Y
Methyl chloride	?	N	N	N	Y
Methylene chloride	?	N	N	N	Y
Naphthalene	?	?	Y	N	?
Nickel	Y	Y	N	Y	N
Nitrobenzene	?	?	?	N	?
2-Nitrophenol	?	Y	N	?	N
4-Nitrophenol	?	Y	N	?	N
n-Nitrosodimethylamine	?	N	N	N	N
n-Nitrosodi-n-propylamine	?	N	N	N	N
n-Nitrosodiphenylamine	?	?	?	N	N
PCB-1016	?	Y	N	N	Y
PCB-1221	?	Y	N	N	Y
PCB-1232	?	Y	N	N	Y
PCB-1242	?	Y	N	N	Y
PCB-1248	?	Y	N	N	Y
PCB-1254	?	Y	N	N	Y
PCB-1260	?	Y	N	N	Y
Pentachlorophenol (PCP)	Y	Y	Y	N	N
Phenanthrene	?	Y	Y	N	N
Phenol	?	N	Y	N	?
Pyrene	?	Y	Y	N	?
Selenium	Y	Y	Y	Y	?
Silver	Y	Y	N	Y	N

TABLE C.2 (*Continued*)

Priority Pollutant	Advection	Sorption	Decay/ Degradation	Hydrolysis	Volatilization
2,3,7,8-Tetrachlorodibenzo-*p*-dioxin (TCDD, Dioxin) 1,1,2,2-Tetrachloroethane	Y	Y	?	N	?
Tetrachloroethylene (PCE, PER)	?	N	N	N	?
Thallium	Y	Y	N	Y	N
Toluene	?	?	?	N	Y
Toxaphene (PCC)	Y	Y	?	N	?
1,2,4-Trichlorobenzene (TCB)	?	?	?	N	?
1,1,1-Trichloroethane (TCA)	?	N	N	N	Y
1,1,2-Trichloroethane	?	N	N	N	Y
Trichloroethylene (TCE, TRI)	?	N	?	N	Y
Trichlorofluoromethane (Freon 11)	?	?	?	N	Y
2,4,6-Trichlorophenol	?	?	?	N	N
Vinyl chloride (MVC)	N	N	N	?	Y
Zinc	Y	Y	N	Y	N

Note: These data continue to be evaluated and some of the information in this table may, at the time of reading, be out of date.

Reference:

Callahan, M. A., et al., 1979. Water-related environmental fate of 129 priority pollutants. EPA Contract No. 68-01-3852, 68-01-3867.

APPENDIX D
Compilation of Typical Values of Transport Parameters

TABLE D.1 Molecular Diffusion Coefficients

Diffusion Coefficients in Water at 25°C

Cation	D_0 in $[10^{-10}\ m^2 s]$	Anion	D_0 in $[10^{-10}\ m^2 s]$	Reference
H^+	93.1	OH^-	52.7	A
Na^+	13.3	F^-	14.6	A
K^+	19.6	Cl^-	20.3	A
Rb^+	20.6	Br^-	20.1	A
Cs^+	20.7	HS^-	17.3	A
		HCO_3^-	11.8	A
Mg^{2+}	7.05			A
Ca^{2+}	7.93	CO_3^{2-}	9.55	A
Sr^{2+}	7.94	SO_4^{2-}	10.7	A
Ba^{2+}	8.48			A
Ra^{2+}	8.89			A
Mn^{2+}	6.88			A
Fe^{2+}	7.19			A
Cr^{2+}	5.94			A
Fe^{3+}	6.07			A

Diffusion Coefficients of I^- in Granite

Material	Effective Diffusion Coefficient (Porosity Excluded) D_0 in $[m^2/s]$	Intrinsic Diffusion Coefficient (Porosity Included) D_0 in $[m^2/s]$	Reference
Ossian	$(2.9\text{ to }51) \times 10^{-11}$	$(6.4\text{ to }16) \times 10^{-13}$	B
Scottish Lowland	$(1.0\text{ to }1.8) \times 10^{-10}$	$(2.7\text{ to }4.0) \times 10^{-13}$	B
Skene Complex	$(6.1\text{ to }62) \times 10^{-11}$	$(2.8\text{ to }7.4) \times 10^{-13}$	B
Cornish Carnmenellis	$(26\text{ to }5.3) \times 10^{-11}$	$(23\text{ to }3.2) \times 10^{-14}$	B
Granite in Sweden	$(1.0\text{ to }70) \times 10^{-14}$		C

Diffusion Coefficients of Cl^- and ^{36}Cl in Clay

Material	Effective Diffusion Coefficient (Porosity Excluded) D_0 in $[m^2/s]$	Reference
Clay glacial till	50×10^{-11}	D
Silty clay	10.0×10^{-10}	E
	6.0×10^{-10}	
Silty clay	$(5.5\text{ to }8.0) \times 10^{-10}$	F
	$(7.4\text{ to }7.9) \times 10^{-10}$	
Silty clay	11.0×10^{-10}	F
Silty clay	9.2×10^{-10}	F
Glaciolacustrine clay	5.8×10^{-10}	F
Varved glaciolacustrine clay and silt	5.8×10^{-10}	F
Glaciomarine clay	2.0×10^{-10}	F

TABLE D.1 ***(Continued)***

References:

(A) Li, Y. H., and S. Gregory, 1974. Diffusion of ions in seawater and in deep-sea sediments. Oxford: Pergamon Press.

(B) Lever, D. A., and M. H. Bradbury, 1985. Rock-matrix diffusion and its implications for radionuclide migration. *Mineralogical Magazine* 49, 245–254.

(C) Skagius, K., and I. Neretnieks, 1986. Porosities and diffusivities of some nonsorbing species in crystalline rocks. *Water Resources Res.* 22(3), 389–398.

(D) Grisak, G. E., and J. F. Pickens, 1980. Solute transport through fractured media 1 & 2. *Water Resources Res.* 16(4), 719–739.

(E) Crooks, V. E., and R. M. Quicgley, 1984. Saline leachate migration through clay: A comparative laboratory and field investigation. *Can. Geotech. J.* 21, 349–362.

(F) Desaulniers, D. E., 1986. Groundwater origin, geochemistry and solute transport in three major glacial clay plains of east-central North America. Ph.D. thesis, Dept. of Earth Sciences, Univ. Waterloo, Ontario, p. 445.

TABLE D.2 Mechanical Dispersion Coefficients

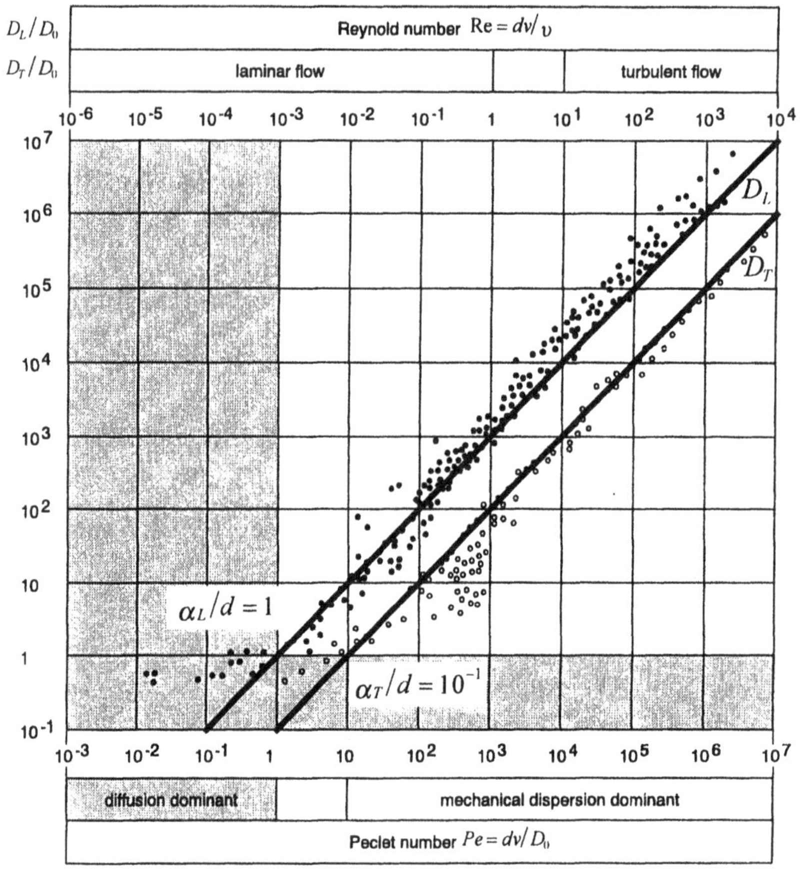

• = longitudinal dispersion experminent		∘ = transverse dispersion experminent	
D_0	= molecular diffusion coefficient	υ	= kinematic viscosity
D_L	= longitudinal dispersion coefficient	α_L	= longitudinal dispersivity
D_T	= transverse dispersion coefficient	α_T	= transverse dispersivity
d	= average grain diameter	v	= transport velocity

Adopted from
Bear, J., 1972, Dynamics of fluids in porous media. New York, American Elsevier, 764 p., and Spitz, K., 1985. Dispersion in porösen Medien: Einfluß von Inhomogenitäten und Dichteunterschieden. Mitteilungen des Instituts für Wasserbau, Universität Stuttgart, Heft 60.

TABLE D.2 (*Continued*)

References:

Barovic, G., 1979. Einfluß der Sorption auf Transportvorgänge im Grundwasser. Report of the Instituts für Wasserwirtschaft, Hydrologie und Landwirtschaftlichen Wasserbaus, Universität Hannover, no. 6.

Bear, J., 1972. *Dynamics of Fluids in Porous Media.* New York: American Elsevier.

Bernard, R. A., and R. H. Wilhelm, 1950. Turbulent diffusion in fixed beds of packed solids. *Chem. Eng. Progress* 46(5), 233–244.

Blackwell, R. J., J. R. Rayne, and W. M. Terry, 1959. Factors influencing the efficiency of miscible displacement. *Trans. AIME* 216, 1.

Brigham, W. E., J. N. Dew, and P. W. Reed, 1961. Experiments on mixing during miscible displacement in porous media. *Soc. Petrol. Eng. J.*, 1–8.

Carberry, J. J., and R. H. Bretton, 1958. Axial dispersion of mass in flow through fixed beds. *Am. Inst. of Chem. Eng. J.* 4(3), 367–375.

Day, P. R., 1956. Dispersion of a moving salt-water boundary advancing through saturated sand. *Trans. AGU* 37(5), 595–601.

Ebach, E. A., and R. R. White, 1958. Mixing of liquids flowing through beds of packed solids. *Am. Inst. of Chem. Eng. J.* 4(2), 161–169.

Grane, F. E., and G. H. F. Gardner, 1961. Measurements of transverse dispersion in granular media. *J. Chem. Eng. Data* 6(2), 283–287.

Harleman, D. R. F., P. F. Mehlhorn, and R. R. Rumer, Jr., 1963. Dispersion permeability correlation in porous media. *J. Hydraulic Div., ASCE* 89, 67–85.

Hoopes, J. A., and D. R. F. Harleman, 1965. Waste water recharge and dispersion in porous media. MIT Hydrodynamics Lab., no. 75, p. 166.

Klotz, D., 1973. Untersuchungen zur dispersion in porösen medien, *Z. Deutsch. Geol. Ges.* 124, 523–533.

Li, W. H., and F. H. Lai, 1966. Experiment on lateral dispersion in porous media. *J. Hydraulic Div., ASCE* 92, 141–149.

Liles, A. W., and E. J. Geankoplis, 1957. Axial diffusion of liquids in packed beds and end effects. *Am. Inst. of Chem. Eng. J.* 3(1), 591–595.

List, E. J., and N. K. Brooks, 1967. Lateral dispersion in saturated porous media. *J. Geophysical Res.* 72(10), 2531–2541.

Raimondi, P., G. H. G. Gardner, and C. B. Petrick. Effect of pore structure and molecular diffusion on the mixing of miscible liquids flowing in porous media. A.I. Ch. E.—S.P.E. Joint Symp. on Fundamental Concepts of Miscible Fluid Displacement, Part II, San Francisco, Pre-print 43.

Rifai, M. N. E., W. J. Kaufman, and D. K. Todd, 1956. Dispersion phenomenon in laminar flow through porous media. Canal Seepage Research, Progress Report, 2, Univ. of California, Berkeley, p. 157.

Rumer, R. R., Jr., 1972. On the derivation of a convective-dispersion equation by spatial averaging. *Developments in Soil Science: 2. Fundamentals of Transport Phenomena in Porous Media.* Amsterdam: Elsevier.

Simpson, E. S., 1962. Transverse dispersions in liquid flow through porous media. Prof. Paper 411-C, USGS, Washington, DC.

Sitz, K., 1985. Dispersion in porösen Medien: Einfluß von Inhomogenitäten und Dichteunterschieden. Report of the Instituts für Wasserbau, Universität Stuttgart, no. 60.

van der Poel, C., 1962. Effect of lateral diffusivity on miscible displacement in horizontal reservoirs. *Soc. of Petrol. Eng. J.*, 316–326.

TABLE D.3 Field-Scale Dispersivities

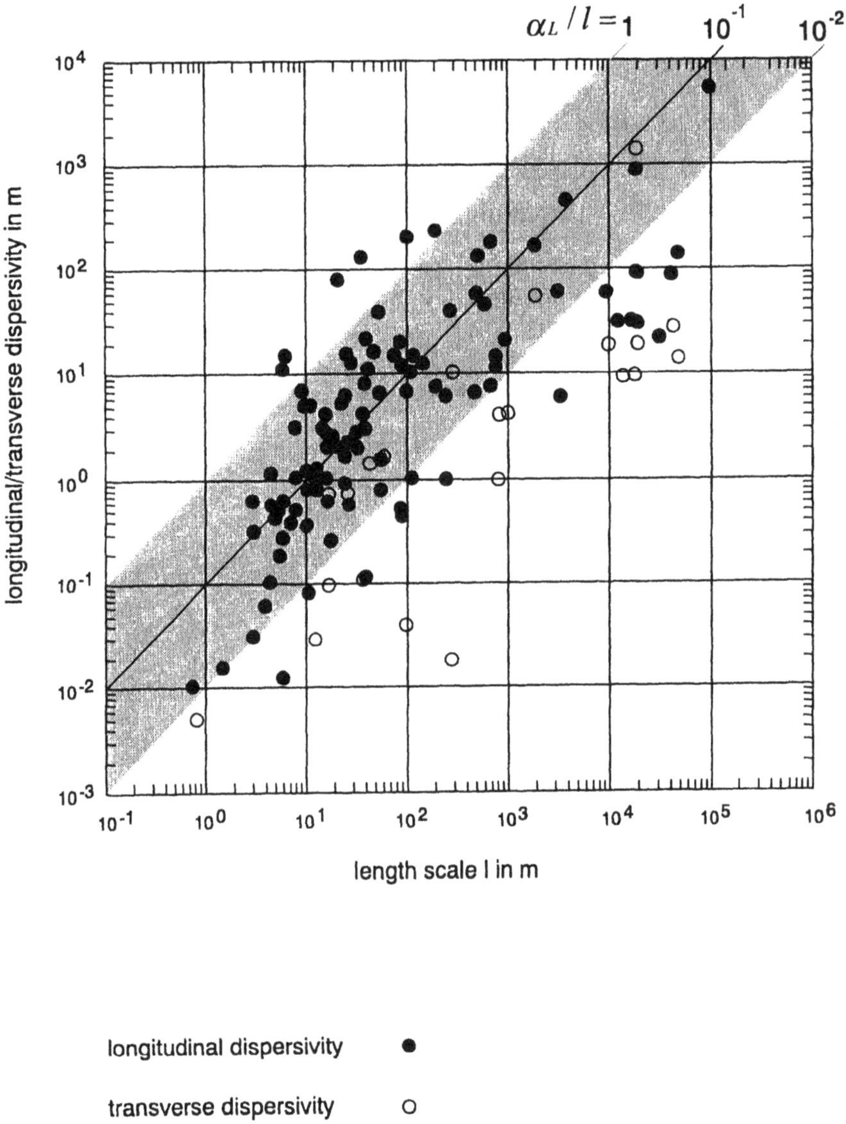

Adopted from
Gelhar et al., 1992. A critical review of data on field-scale dispersion in aquifers. Water Resources Research, Vol. 28(7), p.1955-1974.

TABLE D.3 (*Continued*)

Material	Migration Distance in [m]	Dispersivity in [m]	References
Alluvium	40	3	Iris 1980
Alluvium	15	3	Dieulin 1980
Alluvium	—	0.03 to 0.5	Pickens et al. 1977
Alluvium, derived from tuff	91	10 to 30	Daniels 1981, 1982
Alluvium	18,000	30.5	Konikow and Bredehoeft 1974
Alluvium	—	0.1 to 12	Fried 1975
Alluvium	13,000	30.5	Konikow 1976
Alluvium	—	30.5	Konikow 1977
Alluvium (gravels)	290	41	New Zealand Ministry of
Alluvium (gravels)	25	0.3 to 1.5	Work and Development 1977
Alluvium	6.4	15.2	Robson 1974
Alluvium	10,000	61	Robson 1974
Alluvium	3200	61	Robson 1978
Basalt, brecciated	17.1	0.60	Gelhar 1982
Basalt, lava, and sediments	20,000	91	Grove 1977
Basalt, lava, and sediments	20,000	910	Robertson 1974
Chalk	8	1.0	Ivanovitch and Smith 1978
Chalk, fractured	8	3.1	Ivanovitch and Smith 1978
Crystalline rock, fractured	538	134	Webster et al. 1970
Dolomite, fractured	23	5.2	Bentley and Walter 1983
Dolomite, fractured limestone	122	15	Claasen and Cordes 1975
Dolomite, fractured	55	38.1	Grove and Beetem 1971
Dolomite	250	7	Lenda and Zuber 1970
Dolomite, fractured	21.3	2.1	Kreft et al. 1974
Granite, fractured	5	0.5	Dieulin 1981
Granite, fractured	17	2	Goblet 1982
Gravel, fluvioglacial	10	5	Klotz et al. 1980
Gravel with cobbles	54 to 9	1.4 to 11.5	New Zealand Ministry of Work and Development 1977
Gravel	700	130 to 234	Werner et al. 1983
Limestone	2,000	170	Bredehoeft and Pinder 1973

TABLE D.3 (*Continued*)

Material	Migration Distance in [m]	Dispersivity in [m]	References
Limestone	91	11.6	Fenske 1973
Limestone	41.5	20.8	Kreft et al. 1974
Limestone, fractured	32,000	23	Rabinowitz and Gross 1972
Limestone, fractured	490	6.7	Segol and Pinder 1976
Sandstone	3 to 6	0.16 to 0.6	Oates and Edworthy 1977
Sandstone and alluvial Sediments	50,000	200	Gupta et al. 1975
Sandstone with silt and clay layers	28	1.0	Harpaz, 1965
Sand, glaciofluvial	90	0.5	Rajarim and Gelhar, 1991
Sand	13	1.0	Sauty et al. 1978
Sand, glaciofluvial	11	0.08	Sudicky et al. 1983
Sand, glaciofluvial	700	7.6	Seykes et al. 1982, 1983
Sand	100,000	5600 to 40,000	Wood 1981
Sand, glaciofluviatile	600	30 to 60	Egboka et al. 1981
Sand, glaciofluviatile	90	0.43	Freyberg 1986
Sand, medium to coarse	250	0.96	Garabedian et al. 1988
Sand, medium, layered	38.3	4.0	Huyakan et al. 1986
Sand, fluvial	25	1.6	Kies 1981
Sand	6	0.18	Kreft et al. 1974
Sand	6	0.01	Lee et al. 1980
Sand, fine with glacial till	4	0.06	Leland and Hittel 1981
Sand	2 to 8	0.01 to 0.42	Meyer et al. 1981
Sand	13 to 32.5	0.8 to 2.7	Molinari and Peaudecerf 1977
Sand, fluvial	40	0.06 to 0.16	Moltyaner and Killey 1988
Sand, medium, to fine	57.3	1.5	Papadopulos and Larson 1978
Sand	3	0.03	Pickens and Grisak 1981
Sand	8	0.5	Pickens and Grisak 1981
Sand, gravel and silt	11 to 43	2 to 11	Roberts et al. 1981
Sand and gravel	25 to 150	11 to 25	Sauty 1977
Sand gravel	—	2 to 3	Lau et al. 1957
Sand, silt and clay	57.3	0.76	Seykes et al. 1983
Sand and gravel	43,400	91.4	Vaccaro and Bolke 1983
Sand, silt and gravel	16	1	Valocchi et al. 1981
Sand and gravel	18.3	0.26	Wiebenga et al. 1967
Sand, silt and gravel	79.2	15.2	Wilson 1971
Sand and gravel	1.52	0.015	Wood and Ehrlich 1978
Sand and gravel, very heterogeneous	200	7.5	Adams and Gelhar 1991
Sand and gravel, glaciofluvial	20,000	30.5	Ahlstran et al. 1977
Sand and gravel, glaciofluvial	3500	6	Bierschenk 1959

TABLE D.3 (*Continued*)

Material	Migration Distance in [m]	Dispersivity in [m]	References
Sand and gravel, glaciofluviatile	4000	460	Cole 1972
Sand, glaciofluvial	600	30 to 60	Egboka et al. 1981
Sand, glaciofluvial	90	0.43	Freyberg 1986
Sand and gravel with cobbles	6	11	Fried and Ungemach 1971
Sand and gravel with clay lenses, alluvial	800	15	Fried 1975
Sand and gravel with clay lenses, alluvial	1000	12	Fried 1975
Sand and gravel, layered and silty	10.4	0.7	Hoehn 1983
Sand and gravel, layered and silty	100	6.7	Hoehn and Santschi 1987
Sand and gravel, layered and silty	100	10.0	Hoehn and Santschi 1987
Sand and gravel, layered and silty	500	58.0	Hoehn and Santschi 1987
Sand and gravel with clay lenses	19	2 to 3	Lan et al. 1957
Sand and gravel	16.4	2.13 to 3.35	Naymik and Barcelona 1981
Schistgneiss, fractured	—	134.1	Webster et al. 1970
Till, glacial	—	3.0 to 6.1	Schwartz 1977

References:

Ahlstrom, S. W., H. P. Foote, R. C. Arnett, C. P. Cole, and R. J. Serne, 1977. Multicomponent mass transport model: Theory and numerical implementations (discrete-particle-random-walk-version). Report. BNWL-2127. Battelle Pac. Northwest Lab., Richland, WA.

Bentley, H. W., and G. R. Walter, 1983. Two-well recirculating tracer tests at H-2: Waste Isolation Pilot Plant (WIPP), southwest New Mexico. Draft paper. Hydro Geochem., Inc., Tucson, AZ.

Bierschenk, W. H., 1959. Aquifer characteristics and ground-water movement at Hanford. Report HW-60601, Hanford At. Products Oper., Richland, WA.

Bredehoeft, J. D., and G. F. Pinder, 1973. Mass transport in flowing groundwater. *Water Resources Res.*, 9(1), 144–210.

Claasen, H. C., and E. H. Cordes, 1975. Two-well recirculating tracer test in fractured carbonate rock. *Nevada, Hydrol. Sci. Bull.* 20(3), 367–382.

Daniels, W. R. (ed.), 1981. Laboratory field studies related to the radionuclide migration project. Progress Report LA-8670-PR Los Alamos Sci. Lab., Los Alamos, NM.

Daniels, W. R. (ed.), 1982. Laboratory field studies related to the radionuclide migration project. Draft. Progress Report LA-9192-PR. Los Alamos Sci. Lab., Los Alamos, NM.

Dieulin, A., 1980. Propagation de pollution dans un aquifere alluvial: L'effet de parcours. Doctoral dissertation. Univ. Pierre et Marie Curie-Paris VI and l'Ecole Natl. Super. des Mines de Paris, Fontainebleau, France.

Dieulin, A., 1981. Lixiviation in situ d'un gisement d'uranium en milieu granitique. Draft Report LHM/RD/81/63. Ecole Natl. Super. des Mines de Paris, Fontainebleau, France.

TABLE D.3 ***(Continued)***

Egboka, B. C. E., J. A. Cherry, R. N. Farvolden, and E. O. Frind, 1983. Migration of contaminants in groundwater at a landfill: A case study. 3. Tritium as an indicator of dispersion and recharge. *Hydrol.*, 63, 51–80.

Fenske, P. R., 1973. Hydrology and radionuclide transport. monitoring well HT-2m. Tatum Dome, Mississippi, Proj. Report 25, Tech. Report NVD-1253-6. Cent. for Water Resour. Res., Desert Res. Inst., Univ. of Nevada Syst., Reno.

Freyberg, D. L., 1986. A natural gradient experiment on solute transport in a sand aquifer: 2. Spatial movements and the advection and dispersion of nonreactive racers. *Water Resources Res.* 22(13), 2031–2046.

Fried, J. J., 1975. *Groundwater Pollution*, New York: Elsevier.

Fried, J. J., and P. Ungemach, 1971. Determination in situ du coefficient de dispersion longitudinale d'un milieu poreux naturel. *C.R. Acad. Sci.*, Ser. 2, 272, 1327–1329.

Garabedian, S. P., L. W. Gelhar, and M. A. Celia, 1988. Large-scale dispersive transport in aquifers: Field experiments and reactive transport theory. Report 315, Ralph M. Parsons Lab. for Water Resour. and Hydrodyn., MIT, Cambridge.

Gelhar, L. W., 1982. Analysis of two-well tracer tests with a pulse input. Report RHO-BW-CR-131 P, Rockwell Intl., Richland, WA.

Goblet, P., 1982. Interpretation d'experiences de tracage en milieu granitique (site B). Report LHM/RD/82/11. Cent. d'Inf Geol., Ecole Natl. Super. des Mines de Paris. Fontainebleau, France.

Grove, D. B., 1977. The use of Galerkin finite-element methods to solve mass transport equations. Report USGS/WRD/WRI-78/011. USGS, Denver, CO. (Available as NTIS PB 277-532 from Natl. Tech. Inf. Serv., Springfield, VA.)

Grove, D. B., and W. A. Beetem, 1971. Porosity and dispersion constant calculations for a fractured carbonate aquifer using the two-well tracer method. *Water Resources Res.* 7(1), 128–134.

Gupta, S. K., K. K. Tanji, and J. N. Luthin, 1975. A three-dimensional finite element groundwater model. Report UCAL-WRC-C-152, California Water Resour. Cent., Univ. of California, Davis. (Available as NTIS PB 248-925 from Natl. Tech. Inf. Serv., Springfield, VA.)

Halevy, E., and A. Nir, 1962. Determination of aquifer parameters with the aid of radioactive tracers. *J. Geophys. Res.* 67(5), 2403–2409.

Harpaz, Y., 1965. Field experiments in recharge and mixing through wells. Underground Water Storage Study Tech. Report 17. Publ. 483, Tahal-Water Plann. for Isr., Tel Aviv.

Helweg, O. J., and J. W. Labadie, 1977. Linked models for managing river basin salt balance. *Water Resources Res.* 13(2), 329–336.

Hoehn, E., 1983. Geological interpretation of local-scale tracer observations in a river-ground water infiltration system. Draft report. Swiss Fed. Inst. Reactor Res. (EIR), Würenlingen, Switzerland.

Hoehn, E., and P. H. Santschi, 1987. Interpretation of tracer displacement during infiltration of river water to groundwater. *Water Resources Res.* 23(4), 633–640.

Huyakorn, P. S., P. F. Anderson, F. J. Motz, O. Guven, and J. G. Melville, 1986. Simulations of two-well tracer tests in stratified aquifers at the Chalk River and the Mobile sites. *Water Resources Res.* 22(7), 1016–1030.

Iris, P., 1980. Contribution a l'étude de la valorisation energetique des aquifères peu profonds, thèse de docteur-ingenieur. Ecole des Mines de Paris, Fontainebleau, France.

Ivanovitch, M., and D. B. Smith, 1978. Determination of aquifer parameters by a two-well pulsed method using radioactive tracers. *J. Hydrol.* 36(1/2), 35–45.

Kies, B., 1981. Solute transport in unsaturated field soil and in groundwater. Ph.D. dissertation. Dept. of Agron, New Mexico, State Univ., Las Cruces.

TABLE D.3 (*Continued*)

Klotz, D., K. P. Seiler, H. Moser, and F. Neumaier, 1980. Dispersivity and velocity relationship fiom laboratory and field experiments. *J. Hydrol.* 45(3/4), 169–184.

Konikow, L. F., 1977. Modeling chloride movement in the alluvial aquifer at the Rocky Mountain Arsenal, Colorado. USGS Water-Supply Paper 2044.

Konikow, L. F., and J. D. Bredehoeft, 1974. Modeling flow and chemical quality changes in an irrigated stream-aquifer system. *Water Resources Res.*, 10(3), 546–562.

Kreft, A., A. Lenda, B. Turek, A. Zuber, and K. Czauderna, 1974. Determination of effective porosities by the two-well pulse method. Isot. Tech. Groundwater Hydrol., Proc. Symp., 2, pp. 295–312.

Lau, L. K., W. J. Kaufman, and D. K. Todd, 1957. Studies of dispersion in a radial flow system. Canal Seepage Research: Dispersion phenomena in flow through porous media. Progress Report 3. I.E.R. Ser. 93. Issue 3, Sanit. Eng. Res. Lab., Dept. of Eng. and School of Public Health, Univ. of California, Berkeley.

Lau, L. K., W. J. Kaufman, and D. K. Todd, 1957. Dispersion phenomena in flow through porous media. Sanit. Eng. Res. Lab. Progress Report 3, Univ. of California, Berkeley.

Lee, D. R., J. A. Cherry, and J. F. Pickens, 1980. Groundwater transport of a salt tracer through a sandy lakebed. *Limnol. Oceanogr.* 25(1), 46–61.

Leland, D. F., and D. Hillel, 1981. Scale effects on measurement of dispersivity in a shallow, unconfined aquifer, paper presented at Chapman Conference on Spatial Variability in Hydrologic Modeling. AGU, Fort Collins, CO., July 21–23.

Lenda, A., and Zuber, A., 1970. Tracer dispersion in groundwater experiments. Isot. Hydrol. Proc. Symp. 1970. pp. 619–641.

Mercado, A., 1966. Recharge and mining tests at Yavne 20 well field. Underground Water Storage Study Tech. Report 12. Publ. 611, Tahal-Water Plann. for Isr., Tel Aviv.

Meyer, B. R., C. A. R. Bain, A. S. M. DeJesus, and D. Stephenson, 1981. Radiotracer evaluation of groundwater dispersion in a multilayered aquifer. *J. Hydrol.* 50(1/3), 259–271.

Molinari, J., and P. Peaudecerf, 1977. Essais conjoints en laboratoire et sur le terrain en vue d'une approach simplefiee de la prévision des propagations de substances miscibles dans les aquifères réels. Paper presented at Symposium on Hydrodynamic Diffusion and Dispersion in Porous Media. Int. Assoc. for Hydraul. Res., Pavis, Italy.

Moltyaner, G. L., and R. W. D. Killey, 1988a. Twin Lake tracer tests: Longitudinal dispersion. *Water Resources Res.* 24(10), 1613–1627.

Moltyaner, G. L., and R. W. D. Killey, 1988b. Twin Lake tracer tests: Transverse dispersion. *Water Resources Res.* 24(10), 1628–1637.

Naymik, T. G., and M. J. Barcelona, 1981. Characterization of a contaminant plume in ground water, Meredosia, Illinois. *Groundwater* 19(5), 517–526.

New Zealand Ministry of Works and Development, Water and Soil Division, 1977. Movement of contaminants into and through the Heretaunga Plains aquifer. Report. Wellington.

Oakes, D. B., and D. J. Edworthy, 1977. Field measurement of dispersion coefficients in the United Kingdom. *Ground Water Quality, Measurement, Prediction, and Protection.* Reading, England: Water Research Centre, pp. 327–340.

Papadopulos, S. S., and S. P. Larson, 1978. Aquifer storage of heated water: II. Numerical simulation of field results. *Groundwater* 16(4), 242–248.

Pickens, J. F., and G. E. Grisak, 1981. Scale dependent dispersion in a stratified granular aquifer. *Water Resource Res.* 17(4), 1191–1211.

Pickens, J. F., W. F. Merritt, and J. A. Cherry, 1977. Field determination of the physical contaminant-transport parameters in a sandy aquifer. Proc. IAEA Advisory Group Meeting on the Use of Nuclear Techniques in Water Pollution Studies, Vienna, Austria.

TABLE D.3 (*Continued*)

Pinder, G. F., 1973. A Galerkin-finite element simulation of groundwater contamination on Long Island. *Water Resources Res.* 9(6), 1657–1669.

Rabinowitz, D. D., and G. W. Gross, 1972. Environmental tritium as a hydrometeorologic tool in the Roswell Basin, New Mexico. Tech. Completion Report OWRR: A-037-NMEX, New Mexico Water Resources Res. Inst., Las Cruces.

Rajaram, H., and L. W. Gelhar, 1991. Three-dimensional spatial moments analysis of the Borden tracer test. *Water Resources Res.* 27(6), 1239–1251.

Roberts, P. V., M. Reinhard, G. D. Hopkins, and R. S. Summers, 1981. Advection-dispersion-sorption models for simulating the transport of organic contaminants. Paper presented at International Conference on Ground Water Quality Research, Rice Univ., Houston, TX.

Robertson, J. B., 1974. Digital modeling of radioactive and chemical waste transport in the Snake River Plain aquifer of the National Reactor Testing Station, Idaho. USGS Open File Report, IDO-22054.

Robertson, J. B., and J. T. Barraclough, 1973. Radioactive and chemical waste transport in groundwater of National Reactor Testing Station: 20-year case history and digital model. Underground Waste Manage. Artif. Recharge Prepr. Pap. Int. Symp. 2nd, 1, pp. 291–322.

Robson, S. G., 1974. Feasibility of digital waste quality modeling illustrated by application at Barstow, California. USGS Water Resources Investigations, pp. 46–73.

Robson, S. G., 1978. Application of digital profile modeling techniques to ground water solute transport at Barstow, California. USGS Water Supply Pap., 2050.

Rousselot, D., J. P. Sauty, and B. Gaillard, 1977. Etude hydrogéologique de la zone indurtrielle de Blyes-Saint-Vulbas, rapport préliminaire no. 5: Caracteristiques hydrodynamiques du systeme aquifère. Rep. Jal 77/33. Bur. de Rech. Geol. et Mtn., Orleans, France.

Sauty, J. P., 1977. Contribution à l'identification des paramèters de dispersion dans les aquifères par interprétation des experiences de tracage, dissertation. Univ. Sci. et Med. et Inst. Natl. Polytech. de Grenoble, Grenoble, France.

Schwartz, F. W., 1977. Macroscopic dispersion in porous media: The controlling factors. *Water Resources Res.* 13(4). p. 743–752.

Segol, G., and G. F. Pinder, 1976. Transient simulation of saltwater intrusion in southeastem Florida. *Water Resources Res.* 12(1), 65–70.

Sudicky, E. A., J. A. Cherry, and E. O. Frind, 1983. Migration of contaminants in groundwater at a landfill: A case study, 4, a natural-gradient dispersion test. *J. Hydrol.* 63, 81–108.

Sykes, J. F., S. B. Pahwa, R. B. Lantz, and D. S. Ward, 1982. Numerical simulation of flow and contaminant migration at an expensively monitored landfill. *Water Resources Res.* 18(6), 1687–1704.

Sykes, J. F., S. B. Pahwa, D. S. Ward, and Lantz, D.S., 1983. The validation of SWENT, a geosphere transport model. *Scientific Computing*, ed. R. Stapleman et al., Amsterdam: IMAES/North-Holland, pp. 351–361.

Vaccaro, J. J., E. L. and Bolke, 1983. Evaluation of water quality characteristics of part of the Spokane aquifer, Washington and Idaho, using a solute transport digital model, USGS Open File Report, pp. 82–769.

Walter, G. B., 1983. Convergent flow tracer test at H-6: Waste isolation pilot plant (WIPP), southeast New Mexico. Draft. Hydro Geochem, Inc., Tucson, AZ.

Webster, D. S., J. F. Procter, and I. W. Marine, 1970. Two-well tracer test in fractured crystalline rock. USGS Water Supply, Paper, 1544-I.

Werner, A, et al., 1983. Nutzung von Grundwasser für Wärmepumpen, Versickerimgstest Aefligen, Versuch 2, 1982/83. Water and Energy Management Agency of the State of Bern, Switzerland.

TABLE D.3 (*Continued*)

Wiebenga, W. A., et al., 1967. Radioisotopes as groundwater tracers. *J. Geophys. Res.* 72(16), 4081–4091.

Wilson, L. G., 1971. Investigations on the subsurface disposal of waste effluents at inland sites. Res. Develop. Progress Report 650. U.S. Dept. of Interior, Washington DC.

Wood, W., 1981. A geochemical method of determining dispersivity in regional groundwater systems. *J. Hydrol.* 54(1/3), 209–224.

Wood, W. W., and G. G. Ehrlich, 1978. Use of baker's yeast to trace microbial movement in ground water. *Groundwater* 16(6), 398–403.

TABLE D.4 K_d Values for Metals

Element	Material	K_d in [ml/g]	Comments	Reference
Americium	Clay	50,000	pH = 5 to 8	Van Dalen et al. 1975
	Sand	400	pH = 5 to 8	Van Dalen et al. 1975
	Silt loam	971 to 4830		Sheppard et al. 1976
	Loamy sand	714		Sheppard et al. 1976
	Sand	249 to 476		Sheppard et al. 1976
	Basalt	50	Batch exp.	Moody 1982
	Granite	200	Batch exp.	Moody 1982
	Salt	300	Batch exp.	Moody 1982
	Tuff	50	Batch exp.	Moody 1982
Antimony	Sand	0		Batelle Pacific NW Labs. 1978
Arsenic	Basalt	1.4		Essington and Nork 1969
	Loam	0.14		Batelle Pacific NW Labs. 1978
	Sand	>39		Radian Corp. 1975
	Loam	>78		Radian Corp. 1975
	Loam	15	Calcareous	Dames & Moore, 1980
	Clay	>39		Radian Corp. 1975
	Clay	0.1 to 1.6	pH = 5.0	Griffin et al. 1977
Cadmium	Sand	0.68	Noncalcareous	Dames & Moore, 1980
	Loam	0.8	Calcareous	Dames & Moore, 1980
	Loam	4.5	Calcareous	Dames & Moore, 1980
	Clay	0.17	pH = 5.0	Griffin et al. 1977
Caesium	Alluvium	121 to 3165		Nork et al. 1962
	Tuff	12,100 to 17,800		Goldberg et al. 1962
	Carbonate	13.5		Nork and Fenske 1970
	Granodiorite	8 to 1810		Beetem et al. 1962
	Granite	34.3		Nork and Fenske 1970
	Basalt	792 to 9520		Angelo et al. 19962
	Basalt	6.5 to 280		Essington and Nork 1969

Siltstone	309		Nork and Fenske 1970
Sandstone	102		Nork and Fenske 1970
Rhyolite	4 to 100,000		Berak 1963
Volcanic glass	6 to 84		Berak 1963
Rhyodacite	41 to 96		Berak 1963
Quartz porphyry	32 to 730		Berak 1963
Trachyte	16 to 89		Berak 1963
Andesite	52 to 54		Berak 1963
Dolerite	14 to 800		Berak 1963
Picrite	30 to 1300		Berak 1963
Periodotite	96		Berak 1963
Pyroxenite	<2.0		Berak 1963
Olivine basalt	75 to 1685		Berak 1963
Phonolite	34 to 1600		Berak 1963
Leucitite	398 to 3746		Berak 1963
Yermiculite	52,000	Distilled water	Jackson and Inch 1980
Illite	26,000	Distilled water	Jackson and Inch 1980
Kaolinite	2500	Distilled water	Jackson and Inch 1980
Silty clay	3000		Wilding and Rhodes 1963
Snadstone	102 to 389	Fine	Sokol 1970
Siltstone	309 to 541	Black, shaley	Sokol 1970
Sandstone	346 to 630	Very fine, silty	Sokol 1970
Sandstone	613 to 10,300	Calcite cement	Janzer et al. 1962
Siltstone	1034 to 13,400	Sandy	Janzer et al. 1962
Siltstone	27 to 170	Gypsum bearing	Janzer et al. 1962
Dolomite	51 to 6657		Janzer et al. 1962
Gypsum rock	419 to 3600		Janzer et al. 1962
Siltstone	661 to 8250	Dolomite cement	Janzer et al. 1962
Basalt	300	Batch exp.	Moody 1982
Granite	300	Batch exp.	Moody 1982
Dome salt	1	Batch exp.	Moody 1982
Bedded salt	800	Batch exp.	Moody 1982
Tuff	100	Batch exp.	Moody 1982

TABLE D.4 (*Continued*)

Element	Material	K_d in [ml/g]	Comments	Reference
Chromium	Sand	1.3		Radian Corp. 1975
	Loam	2.6		Radian Corp. 1975
	Clay	2.6		Radian Corp. 1975
	Clay	30.2	pH = 5.0	Griffin et al. 1977
	Clay	6.2	pH = 5.0	Griffin et al. 1977
Carbon	Basalt	0	Batch exp.	Moody 1982
	Granite	0	Batch exp.	Moody 1982
	Salt	0	Batch exp.	Moody 1982
	Tuff	0	Batch exp.	Moody 1982
Chromium 3	Clay		Very high	Versar 1979
Chromium 6	Clay		Very low	Versar 1979
Curium	Silt loam	704 to 1,330		Sheppard et al. 1976
	Loamy sand	106		Sheppard et al. 1976
	Sand	1240 to 1850		Sheppard et al. 1976
	Clay loam	73.7 to 97.7	pH = 1.2	Nishita et al. 1976
	Clay loam	90,000 to 110,000	pH = 5.5	Nishita et al. 1976
	Clay loam	61,225 to 81,633	pH = 7.08	Nishita et al. 1976
	Clay loam	1634 to 1918	pH = 9.43	Nishita et al. 1976
	Clay loam	355 to 361	pH = 11.25	Nishita et al. 1976
	Basalt	50	Batch exp.	Moody 1982
	Granite	200	Batch exp.	Moody 1982
	Salt	300	Batch exp.	Moody 1982
	Tuff	50	Batch exp.	Moody 1982
Europium	Quartz sand	228 to 705	pH = 7.7	Baetsle and Dejonghe 1962
	Basalt	50	Batch exp.	Moody 1982
	Granite	100	Batch exp.	Moody 1982
	Salt	50	Batch exp.	Moody 1982
	Tuff	50	Batch exp.	Moody 1982

Iodine	Tuff	1.1		Goldberg et al. 1962
	Basalt	0	Batch exp.	Moody 1982
	Granite	0	Batch exp.	Moody 1982
	Salt	0	Batch exp.	Moody 1982
	Tuff	0	Batch exp.	Moody 1982
Lead	Sand	190	Noncalcareous	Dames & Moore, 1980
	Loam	314	Calcareous	Dames & Moore, 1980
	Loam	323	Calcareous	Dames & Moore, 1980
	Clay	15.2	pH = 5.0	Griffin et al. 1977
	Basalt	25	Batch exp.	Moody 1982
	Granite	5	Batch exp.	Moody 1982
	Salt	2	Batch exp.	Moody 1982
	Tuff	25	Batch exp.	Moody 1982
Mercury	Sand	5.2		Radian Corp. 1975
	Loam	130		Radian Corp. 1975
	Clay	>65		Radian Corp. 1975
Molybdenum	Basalt	4 to 100	Batch exp.	Moody 1982
	Granite	1 to 25	Batch exp.	Moody 1982
	Salt	1, 25, 0	Batch exp.	Moody 1982
	Tuff	4 to 100	Batch exp.	Moody 1982
Neptunium	Silt loam	20.2 to 127		Sheppard et al. 1976
	Loamy sand	15.4		Sheppard et al. 1976
	Sand	32.4 to 37.2		Sheppard et al. 1976
	Basalt	3 to 500	Batch exp.	Moody 1982
	Granite	1 to 500	Batch exp.	Moody 1982
	Salt	7 to 300	Batch exp.	Moody 1982
	Tuff	3 to 500	Batch exp.	Moody 1982
Nickel	—		Very low	Versar 1979
	Basalt	50	Batch exp.	Moody 1982
	Granite	10	Batch exp.	Moody 1982
	Salt	6	Batch exp.	Moody 1982
	Tuff	50	Batch exp.	Moody 1982

TABLE D.4 ***(Continued)***

Element	Material	K_d in [ml/g]	Comments	Reference
				Moody 1982
Niobium	Basalt	100	Batch exp.	Moody 1982
	Granite	100	Batch exp.	Moody 1982
	Salt	50	Batch exp.	Moody 1982
	Tuff	100	Batch exp.	Moody 1982
Palladium	Basalt	50	Batch exp.	Moody 1982
	Granite	10	Batch exp.	Moody 1982
	Salt	3	Batch exp.	Moody 1982
	Tuff	50	Batch exp.	Moody 1982
Protactinium	Basalt	100	Batch exp.	Moody 1982
	Granite	100	Batch exp.	Moody 1982
	Salt	50	Batch exp.	Moody 1982
	Tuff	100	Batch exp.	Moody 1982
Plutonium	Sand	2 to 3	pH < 3	Knoll 1969
	Coarse sand	5.6		Evans 1956
	Fine sand	10		Evans 1956
	Coarse silt	11.3		Evans 1956
	Medium silt	33.5		Evans 1956
	Fine silt	48.8		Evans 1956
	Coarse clay	80.9		Evans 1956
	Clay loam	386 to 474	pH = 1.21	Nishita et al. 1976
	Clay loam	3067 to 3105	pH = 7.08	Nishita et al. 1976
	Clay loam	1.4 to 4.6	pH = 10.31	Nishita et al. 1976
	Basalt	100 to 20,000	Batch exp.	Moody 1982
	Granite	100 to 50,000	Batch exp.	Moody 1982
	Salt	50 to 50,000	Batch exp.	Moody 1982
	Tuff	0 to 100	Batch exp.	Moody 1982
Radium	Tuff	6700	Batch exp.	Stead 1964

	Basalt	50	Batch exp.	Moody 1982
	Granite	50	Batch exp.	Moody 1982
	Salt	5	Batch exp.	Moody 1982
	Tuff	200		Moody 1982
Samarium	Basalt	50	Batch exp.	Moody 1982
	Granite	100	Batch exp.	Moody 1982
	Salt	50	Batch exp.	Moody 1982
	Tuff	50	Batch exp.	Moody 1982
Selenium	Sand	3.2		Versar 1979
	Sand	15.6	Noncalcareous	Dames & Moore, 1980
	Loam	16		Radian Corp. 1975
	Loam	1088	Calcareous	Dames & Moore, 1980
	Clay	29		Radian Corp. 1975
	Clay	1.4	pH = 7.0	Griffin et al. 1977
	Clay	0.56	pH = 7.0	Griffin et al. 1977
	Basalt	5 to 20	Batch exp.	Moody 1982
	Granite	2	Batch exp.	Moody 1982
	Salt	20, 100, 20	Batch exp.	Moody 1982
	Tuff	2	Batch exp.	Moody 1982
Strontium	Basalt	16 to 135		Angelo et al. 1962
	Quartz sand	1.7 to 38	pH = 7.7	Baetsle et al. 1963
	Granodiorite	11 to 23		Beetem et al. 1962
	Dolomite	5 to 14	TDS = 4250	Stead 1963
	Alluvium	48 to 2454		Nork et al. 1971
	Silty clay	220 to 240		Wilding and Rhodes 1963
	Sandstone	1.13 to 2.08	Fine	Sokol 1970
	Siltstone	8.32 to 9.56	Black, shaley	Sokol 1970
	Sandstone	7.76 to 9.79	Very fine, silty	Sokol 1970
	Limestone	8.32	Shaley	Nork and Fenske 1970
	Sandstone	1.37		Nork and Fenske 1970
	Tuff	260		Nork and Fenske 1970

TABLE D.4 (*Continued*)

Element	Material	K_d in [ml/g]	Comments	Reference
	Tuff	1700 to 4300		Kaufman 1963
	Limestone	0.19		Nork and Fenske 1970
	Granodiorite	4 to 23		Beetem et al. 1962
	Granite	1.7		Nork and Fenske 1970
	Basalt	16 to 135		Angelo et al. 1962
	Basalt	220		Essington and Nork 1969
	Rhyolite	<56.0		Berak 1963
	Volcanic glass	3 to 32		Berak 1963
	Rhyodacite quartz	11 to 27		Berak 1963
	Porophy	6 to 47		Berak 1963
	Trachyte	12 to 39		Berak 1963
	Andesite	11 to 50		Berak 1963
	Spillite	8		Berak 1963
	Dolerite	<45		Berak 1963
	Basalt	6 to 187		Berak 1963
	Picrite	0 to 108		Berak 1963
	Teschenite	22 to 74		Berak 1963
	Peridotite	35		Berak 1963
	Pyroxenite	0		Berak 1963
	Olivine basalt	321 to 334		Berak 1963
	Phonolite	7 to 140		Berak 1963
	Tephrite	26 to 143		Berak 1963
	Basanite	<59		Berak 1963
	Nephelinite	5 to 60		Berak 1963
	Leucitite	16 to 82		Berak 1963
	Melilite	29		Berak 1963
	Montmorillonite	506	pH = 7.5	Jackson and Inch 1980
	Illite	117 (760)	pH = 7.5 (10)	Jackson and Inch 1980

	Kaolinite	55 (257)	pH = 7.5 (10)	Jackson and Inch 1980
	Alumina	2100 (34,000)	pH = 7.5 (10)	Jackson and Inch 1980
	Muscovite	82	pH = 7.5	Jackson and Inch 1980
	Biotite	48	pH = 7.5	Jackson and Inch 1980
	Quartz	0	pH = 7.5	Jackson and Inch 1980
	Basalt	100	Batch exp.	Moody 1982
	Granite	12	Batch exp.	Moody 1982
	Salt	5	Batch exp.	Moody 1982
	Tuff	100	Batch exp.	Moody 1982
Technetium	Basalt	0 to 20	Batch exp.	Moody 1982
	Granite	4	Batch exp.	Moody 1982
	Salt	2	Batch exp.	Moody 1982
	Tuff	0 to 10	Batch exp.	Moody 1982
Thorium	Basalt	500	Batch exp.	Moody 1982
	Granite	500	Batch exp.	Moody 1982
	Dome salt	50	Batch exp.	Moody 1982
	Bedded salt	100	Batch exp.	Moody 1982
	Tuff	500	Batch exp.	Moody 1982
Tin	Basalt	100 to 10	Batch exp.	Moody 1982
	Granite	500 to 10	Batch exp.	Moody 1982
	Salt	1 to 50	Batch exp.	Moody 1982
	Tuff	500 to 50	Batch exp.	Moody 1982
Uranium	Clay	39	River sediment	Racon 1973
	River peat	33		Racon 1973
	Clay	16		Racon 1973
	Altered schist	270		Racon 1973
	Quartz	0		Racon 1973
	Calcite	7		Racon 1973
	Illite	139		Racon 1973
	Basalt	6	Batch exp.	Moody 1982
	Granite	4	Batch exp.	Moody 1982

TABLE D.4 *(Continued)*

Element	Material	K_d in [ml/g]	Comments	Reference
	Salt	1	Batch exp.	Moody 1982
	Tuff	4	Batch exp.	Moody 1982
Zirconium	Tuff	260 to 350		Dlouhy 1967
	Basalt	500	Batch exp.	Moody 1982
	Granite	500	Batch exp.	Moody 1982
	Salt	500	Batch exp.	Moody 1982
	Tuff	500	Batch exp.	Moody 1982

References:

Angelo, C. G., et al., 1962. Summary of K_d's for fission products between groundwater and basaltic rocks, in Geology of the U18a site, Buckboard Mesa. NTS Part VI, USGS Report project Danny Boy, WT 1828, pp. 66–81.

Baetsle, L., and P. Pejonghe, 1962. Investigations on the movement of radioactive substances in the ground: III. Practical aspects of the program and physiochemical considerations. *Ground Disposal of Radioactive Wastes.* TID-7128, pp. 198–210.

Baetsle, L., et al., 1963. Present status of the study program on the movement of radioelements in the soil at Mol. EURAEC 416.

Batelle Pacific NW Labs., 1978. Radionuclide interactions with soil and rock media. Report EPA 520/6-78-007.

Beetem, W. A., et al., 1962. Summary of distribution coefficients for fission products between groundwater and granitic rocks. Climax Stock. NTS, USGS Tech. Letter NTS-13.

Berak, L., 1963. The sorption of microstrontium and microcaesium on the silicate minerals and rocks. UJV-528-63.

Chiou, C. T., L. J. Peters, and V. H. Freed, 1979. A physical concept of soil-water equilibria for nonionic organic compounds. *Science* 206, 831–832.

Dames & Moore, 1980. Unpublished collection of data from attenuation study conducted for Exxon Corporation. Job number 08837-090-007.

Dlouhy, Z., 1967. Movement of radionuclides in the aerated zone. *Disposal of Radioactive Wastes in the Ground.* IAEA-SM 93/18.

DOE, 1993. Track 2 sites: Guidance for assessing low probability hazard site at the INEL. Draft. DOE/ID-10389.

Essington, E. H., and W. E. Nork, 1969. Radionuclide contamination evaluation—Milrow event. NVO-1229-117.

Evans, E. J., 1956. Plutonium retention in Chalk River soil. CRHP 660.

Goldberg, M.C., et al., 1962. The effect of sodium ion concentration on distribution coefficients for tuffs from NTS. USGS Tech. letter NTS-16.

Griffin, R. A., et al., 1977. Attenuation of pollutants in municipal landfill leachate by clay minerals: 2. Heavy metal adsorption. Enviroinmental Geology Notes series.

Jackson, R. E., and K. F. Inch, 1980. Hydrogeochemical processes affecting the migration of radionuclides in a fluvial sand aquifer at the Chalk River Nuclear Laboratories. NHRI Paper no. 7.

Janzer, V. J., et al., 1962. Summary of distribution coefficients for fission products between groundwater and rocks from Project Gnome. Hydrologic and Geologic Studies for Project Gnome, part IV., USGS.

Kaufman, W. J., 1963. An appraisal of the distribution coefficient for estimating underground movement of radioisotopes. HNS 1229-21.

Karickhoff, S. W., et al., 1979. Sorption of hydrophobic pollutants on natural sediments. *Water Resources Res.* 13, 241–248.

Kenega, E. E., 1980. Predicted bioconcentration factors and soil sorption coefficients of pesticides and other chemicals. *Ecotoxicol. Environ. Saf.* 4, 26–38.

Knoll, K. C., 1969. Reactions for organic wastes in soils. BNWL-860.

Mehran, M., et al., 1987. Distribution coefficient of trichloroethylene in soil-water systems. *Groundwater* 25, 275–282.

Moody, J. B., 1982. Radionuclide migration/retardation: Research and development technology status report. Office of Nuclear Waste Isolation, Battelle Memorial Inst., ONWI-321.

Nishita, H., et al., 1976. Extractability of 238Pu and 242Cm from a contaminated soil as a function of pH and certain soil components, HNO3-NaOH system. Presented at annual meeting of Soil Science Society of America, Houston, TX.

Nork, W. E., and P. R. Fenske, 1970. Radioactivity in water—Project Rulison. NVO-1229-131.

Nork, W. E., et al., 1971. Radioactivity in water. Central Nevada test area. NVO-1229-175.

Racon, D., 1973. The behaviour in underground environments of uranium and thorium discharged by the nuclear industry. *Environmental Behaviour of Radionuclides Released by the Nuclear Industry.* Vienna, Austria: IAEA-SM-172/55, pp. 333–346.

Radian Corp., 1975. The environmental effects of trace elements in the pond disposal of ash and flue gas desulphurization sludge. Report prepared for the Electric Power Research Institute, Research Project 202.

Sheppard, J. C., et al., 1976. Determination of distribution ratios and diffusion coefficients of neptunium, americium and curium in soil-aquatic environments. RLO-2221-T-12-2.

Sokol, D., 1970. Groundwater safety evaluation—Project Gasbuggy. PNE-1009.

Stead, F. W., 1963. Tritium distribution in groundwater around large underground fusion explosions. *Science* 142, 1163–1165.

Stead, F. W., 1964. Distribution in groundwater of radionuclides from underground nuclear explosions. Proc. Third Plowshare Symp. Engineering with Nuclear Explosives, April 21–23. TID-7965, pp. 127–138.

Van Dalen, A., et al., 1975. Distribution coefficients for some radionuclides between saline water and clays, sandstones and other samples from the Dutch subsoil. Reactor Centrum Nederland, pp. 75–109.

Versar, Inc., 1979. Water related environmental fate of 129 priority pollutants. EPA-440/4-79-029, Versar Inc., Springfield, VA.

Wilding, M. W., and D. W. Rhodes, 1963. Removal of radioisotopes from solution by earth materials from eastern Idaho. IDO-14624.

TABLE D.5 K_d Values for Organics

Contaminant	Material/Comments	K_d in [ml/g]	Reference
Benzene	Medium sand, glacio-fluvia (f_{oc} = 0.00017)	0.031	D
	Fine to medium-grained sand Glacio-fluvial (f_{oc} = 0.00023)	0.079	D
	Fine sand, glacio-fluvial (f_{oc} = 0.00026)	0.051	D
	Fine to medium sand, glacio-fluvial (f_{oc} = 0.00028)	0.07	D
	Fine to medium sand, glacio glacial (f_{oc} = 0.0006)	0.885	D
	Medium sand, glacio-fluvial (f_{oc} = 0.00065)	0.15	D
	Fine-grained sand, glacio lacustrine (f_{oc} = 0.00102)	1.8	D
	Silt, lacustrine (f_{oc} = 0.00108)	0.092	D
Bromoform	Fine to medium sand, glacio fluvial	1.9 to 2.8	B
	Sand, gravel, silt	6.0	C
Carbon tetrachloride	Fine to medium sand, glacio fluvial	1.8 to 2.5	B
Chlorobenzene	Sand, gravel, silt	33	C
Chloroform	Sand, gravel, silt	2.5 to 3.8	C
Dibromochloromethane	Sand, gravel, silt	6.0	C
Dichlorobenzene, 1, 2	Sand, gravel, silt	100	C
	Fine to medium sand, glacio fluvial	3.9 to 9.0	B
Hexachloroethane	Fine to medium sand, glacio fluvial	5.1	B
Naphtalene	Medium sand, glacio-fluvial (f_{oc} = 0.00017)	0.12	D
	Fine- to medium-grained sand glacio-fluvial (f_{oc} = 0.00023)	0.648	D
	Fine sand, glacio-fluvial (f_{oc} = 0.00026)	0.28	D
	Fine to medium sand, glacio fluvial (f_{oc} = 0.00028)	0.353	D
	Fine to medium sand, glacio glacial (f_{oc} = 0.0006)	9.03	D
	Medium sand, glacio-fluvial (f_{oc} = 0.00065)	2.01	D

TABLE D.5 (*Continued*)

Contaminant	Material/Comments	K_d in [ml/g]	Reference
	Fine-grained sand, glacio lacustrine (f_{oc} = 0.00102)	21.8	D
	Silt, lacustrine (f_{oc} = 0.00108)	1.57	D
Tetrachloroethylene	Fine to medium sand, glacio fluvial	2.7 to 5.9	B
Toluene	Medium sand, glacio-fluvia (f_{oc} = 0.00017)	0.057	D
	Fine- to medium-grained sand glacio-fluvial (f_{oc} = 0.00023)	0.157	D
	Fine sand, glacio-fluvial (f_{oc} = 0.00026)	0.072	D
	Fine to medium sand, glacio fluvial (f_{oc} = 0.00026)	0.072	D
	Fine to medium sand, glacio glacial (f_{oc} = 0.0006)	4.152	D
	Medium sand, glacio-fluvial (f_{oc} = 0.00065)	0.289	D
	Fine-grained sand, glacio lacustrine (f_{oc} = 0.00102)	8.296	D
	Silt, lacustrine (f_{oc} = 0.00108)	0.199	D
Trichloroethane	Clay	0.01 to 0.18	A
	Silt	0.03	A
	Sandy silt	0.01 to 3.2	A
	Clay	0.02	A
	Silty sand	0.12 to 0.51	A
	Sandy clay	0.09 to 0.17	A
	Fine sand	0.07 to 0.7	A
	Sandy gravel	0.15	A
	Gravelly sand	0.57	A
	Average	0.18	A
Trichloroethane,1,1,1	Sand, gravel, silt	12	C
Xylene, -P	Medium sand, glacio-fluvial (f_{oc} = 0.00017)	0.119	D
	Fine- to medium-grained sand glacio-fluvial (f_{oc} = 0.00023)	0.383	D
	Fine sand, glacio-fluvial (f_{oc} = 0.00026)	0.190	D
	Fine to medium sand, glacio fluvial (f_{oc} = 0.00028)	0.261	D
	Fine to medium sand, glacio glacial (f_{oc} = 0.0006)	7.536	D

TABLE D.5 (*Continued*)

Contaminant	Material/Comments	K_d in [ml/g]	Reference
⋮	Medium sand, glacio-fluvial (f_{oc} = 0.00065)	0.805	D
	Fined grained sand, glacio lacustrine (f_{oc} = 0.00102)	11.976	D
	Silt, lacustrine (f_{oc} = 0.00108)	0.569	D

Note: K_d values can also be estimated from the equations given in Appendix D.7.

References:

(A) Mehran, M., et al., 1987. Distribution coefficient of trichloroethylene in soil water systems. *Groundwater* 25, 275–282.

(B) Roberts, P. V., et al., 1986. *Water Resources Research*, vol. 22. Washington, DC: AGU, p. 2047–2058.

(C) Roberts, P. V., M. Reinhard, G. D. Hopkins, and R. Scott Summers, 1985. Advection-dispersion-sorption models for simulating the transport of organic contaminants. *Groundwater Quality* eds. C. H. Ward, W. Giger, and P. L. McCarty. New York: Wiley.

(D) Gillham, R. W., S. F. O'Hannesin, C. J. Ptacek, and J. F. Barker, 1987. Evaluation of small scale retardation tests for BTX in groundwater. Pace Report No. 87-2. Inst. for Groundwater Res. University of Waterloo, Ontario (unpublished).

TABLE D.6 Organic Carbon Content of Sediments f_{oc}

Material	Organic Carbon Content in [g/g]	Reference
Medium sand, glacio-fluvial (North Bay)	0.00017	A
Fine- to medium-grained sand, glacio-fluvial (Woolwich)	0.00023	A
Fine sand, glacio-fluvial (Chalk River)	0.00026	A
Fine to medium sand, glacio-fluvial (Borden)	0.00028	A
Fine to medium sand, fluvio-glacial (Ottawa–Gloucester)	0.00060	A
Medium sand, glacio-fluvial (Cambridge)	0.00065	A
Fine-grained sand, glacio-lacustrine	0.00102	A
Silt, lacustrine (Wildwood)	0.00108	A
Sand, gravel (Glatt Valley)	0.0004 to 0.0073	B
Sand, gravel (Aare Valley)	0.0023	B
Sediment from a eutrophic lake	0.019 to 0.058	B

References:

(A) Gillham, R. W., S. F. O'Hannesin, C. J. Ptacek, and J. F. Barker, 1987. Evaluation of small scale retardation tests for BTX in groundwater. Pace Report 87-2. Inst. for Groundwater Res., University of Waterloo, Ontario (unpublished).

(B) Schwarzenbach, R. P., and J. Westall, 1981. Transport of nonpolar organic compounds from surface water to groundwater-laboratory studies. *Environ. Sci. Technol.* 15, 1360–1367.

TABLE D.7 Empirical K_{oc} Relationships

Empirical Equations by Which K_{oc} Can Be Estimated from K_{ow}	Reference
$\log(K_{oc}) = 1.000 \log K_{ow} - 0.21$	Karickhoff, et al. 1979
$\log(K_{oc}) = 0.544 \log K_{ow} + 1.377$	Kenaga and Goring 1980
$\log(K_{oc}) = 1.029 \log K_{ow} - 0.18$	Roa and Davidson 1980
$\log(K_{oc}) = 0.940 \log K_{ow} + 0.22$	Roa and Davidson 1980
$\log(K_{oc}) = 0.989 \log K_{ow} - 0.346$	Karickhoff 1981
$\log(K_{oc}) = 0.937 \log K_{ow} - 0.006$	Lyman 1982
$\log(K_{oc}) = 0.720 \log K_{ow} + 0.49$	Schwarzenbach and Westall 1981
$\log(K_{oc}) = 1.000 \log K_{ow} - 0.317$	Hassett et al. 1980
$\log(K_{oc}) = 0.524 \log K_{ow} + 0.618$	Briggs 1973
$\log(K_{oc}) = 0.830 \log K_{ow} + 0.3$	Matthes et al. 1985
$\ln(K_{oc}) = \ln K_{ow} - 0.7301$	McCall et al. 1983
$K_{oc} = 0.63 K_{ow}$	Karickhoff et al. 1979
Empirical Equations by Which K_{oc} Can Be Estimated from S	**Reference**
$\log(K_{oc}) = 3.803 - 0.557 \log(S_1)$	Chiou et al. 1979
$\log(K_{oc}) = 0.440 - 0.540 \log(S_2)$	Karickhoff et al. 1979
$\log(K_{oc}) = 3.640 - 0.550 \log(S_3)$	Kenega 1980
$\log(K_{oc}) = 4.273 - 0.686 \log(S_3)$	Means et al. 1980
$\log(K_{oc}) = 3.950 - 0.620 \log(S_3)$	Hassett et al. 1983
Empirical Equation by Which K_d Can Be Estimated from K_{oc}	
$K_d = K_{oc} f_{oc}$	

Note:

- K_{oc} = organic carbon partition coefficient between soil organic carbon and water
- K_{ow} = octanol-water partition coefficient
- S_1 = solubility of compound in water in [micromoles/l]
- S_2 = solubility of compound in water in [mole fraction]
- S_3 = solubility of compound in water in [mg/l]
- K_d = adsorption distribution coefficient in [ml/g]
- f_{oc} = fraction of organic carbon in soil in [g/g]

References:

Briggs, G. G., 1973. Molecular structure of herbicides and their sorption by soils. *Nature*, 223, 1288.

Chiou, C. T., L. J. Peters, and V. H. Freed, 1979. A physical concept of soil-water equilibria for non-ionic organic compounds. *Science* 206 (16), 831–832.

Hassett, J. J., J. C. Means, W. L. Banwart, and S. G. Wood, 1980. Sorption properties of sdediments and energy-related pollutants. U.S. Environmental Protection Agency, EPA-600/3-80-041.

Hassett, J. J., W. L. Banwart, and R. A. Griffin, 1983. Correlation of compound properties with sorption characteristics of nonpolar compounds by soils and sediments: Concepts and limitations. *Environment and Solid Wastes: Characterization, Treatment and Disposal*, eds. C. W. Francis and S. I. Auerback. Boston: Butterworth, pp. 161–178.

TABLE D.7 (*Continued*)

Karickhoff, S. W., 1981. Semi-empirical estimation of sorption of hydrophobic pollutants on natural sediments and soils. *Chemosphere* 10, 833–846.

Karickhoff, S. W., D. S. Brown, and T. A. Scott, 1979. Sorption of hydrophobic pollutants on natural sediments. *Water Res.* 13, 241–248.

Kenaga, E. E., 1980. Predicted bioconcentration factors and soil sorption coefficients of pesticides and other chemicals. *Ecotoxicology and Environ. Saf.* 4, 26–38.

Kenaga, E. E., and C. A. I. Goring, 1980. Relationship between water solubility, soil sorption, octanol-water partitioning and bioconcentration of chemicals in biota. Third Aquatic Toxicology Symposium, Proc. American Society of Testing and Materials, No. STP 707, pp. 78–115.

Lyman, W. J., 1982. Adsorption coefficient for soils and sediment. *Handbook of Chemical Property Estimation Methods*. eds. W. J. Lyman et al., New York: McGraw-Hill, pp. 4.1–4.33.

Matthes, G., M. Isenbeck, A. Pekdeger, D. Schenk, and J. Schröter, 1985. Der Stofftransport im Grundwasser und die Wasserschutzgebietsrichtlinie W101. Statusbericht und Problemanalyse. Umweltbundesamt Ber., 7/85, p. 181.

McCall, P. J., R. L. Swann, and D. A. Laskowski, 1983. Partition models for equilibrium distribution of chemicals in environmental compartments. In: Fate of chemicals in the environment, eds., R. L. Swann and A. Eschenroder. Washington, DC: American Chemical Society, pp. 105–123.

Means, J. C., S. G. Wood, J. J. Hassett, and W. L. Banwart, 1980. Sorption of polynuclear aromatic hydrocarbons by sediments and soils. *Environ. Sci. Technol.* 14(12), 1524–1528.

Rao, P. S. C., and J. M. Davidson, 1980. Estimation of pesticide retention and transformation parameters required in nonpoint source pollution models. *Environmental Impact of Nonpoint Source Pollution*, eds. M. R. Overcash and J. M. Davidson, Ann. Arbor, MI: Ann Arbor Science Publishers, pp. 23–26.

Schwarzenbach, R. P., and J. Westall, 1981. Transport of nonpolar organic compounds from surface water to groundwater: Laboratory sorption stdies. *Environ. Sci. Technol.* 15(11), 1360–1367.

TABLE D.8 Radioactive Half-Lifes

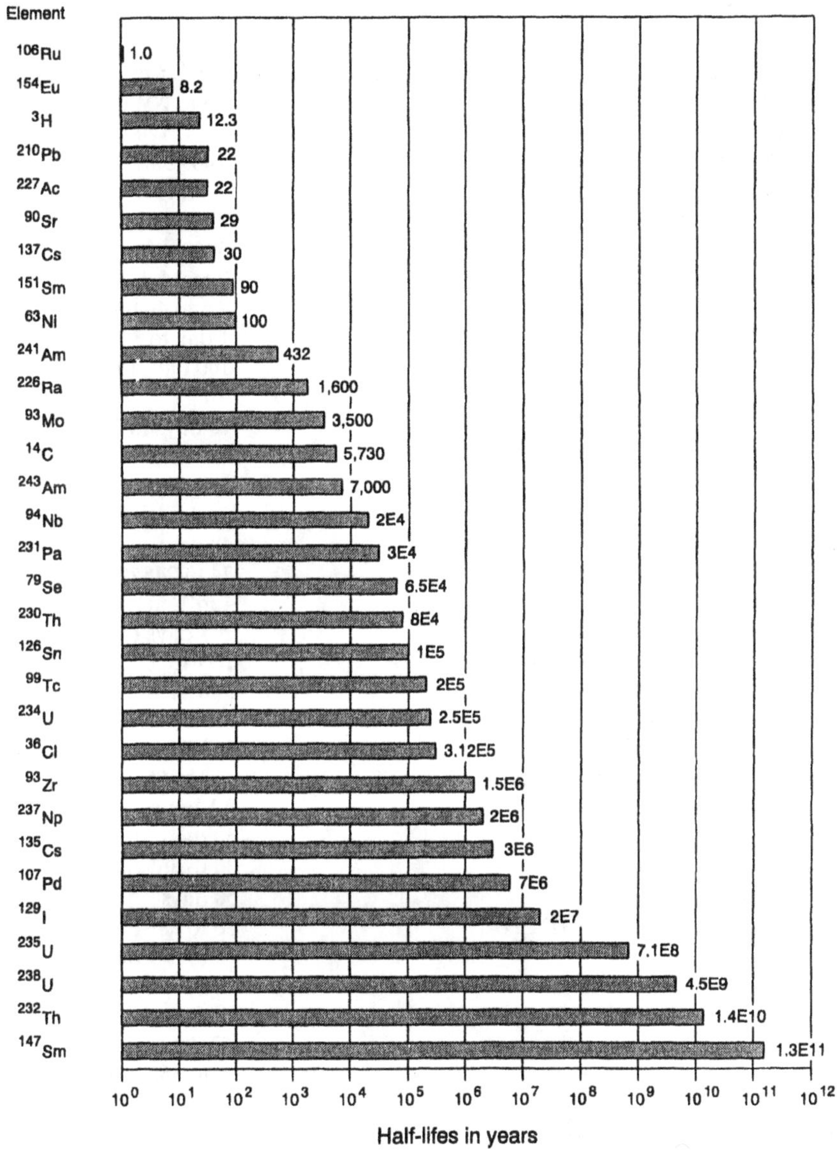

Modified from Moody, J.B., 1982. Radionuclide migration retardation: Research and development technology status report. Office of Nuclear Waste Isolation, Battele Memorial Institute, ONWI-321, 61p.

TABLE D.9 Half-Lifes for Organics

Compound	Half-Lifes in [Days] or [Years]
Acenaphthene	25 to 204 days
Acenaphthylene	85 to 120 days
Acidic acid t-butyl ester	140 years
Acidic acid ethyl ester	2 years
Acidic acid phenyl ester	38 days
Acetamide	3950 years
Acrolein	7 to 24 days
Acrylonitrile	2.5 to 46 days
Aldrin (HHDN)	1 day to 3.2 years
Anthracene	100 days to 2.5 years
Benzene (Benzol)	long
	10 days to 2 years
	8.6 days
	120 days
	68 days
	48 days
	24 to 248 days
Benzidine	4 to 16 days
Benzo [*a*] anthracene	204 days to 3.73 years
Benzo [*b*] fluoranthene (B[*b*]F)	1.97 to 3.34 years
Benzo [*k*] fluoranthene (B[*k*]F)	5 to 11.7 years
Benzo [*ghi*] perylene	3.2 to 3.6 years
Benzo[*a*]pyrene (B[*a*]P)	114 days to 2.9 years
Bis (2-chloroethyl) ether (Chlorex)	56 to 365 days
Bis (chloromethyl) ether (BCME)	0.1 hours
Bis (2-chloroisopropyl) ether	50 days
	36 to 365 days
Bis (2-chloropropyl) ether	50 days
Bis (2-ethylhexyl) phthalate (DEHP)	10 to 389 days
	10 to 389 days
Bromoform	686 years
	58 days
	0.3 days
	56 to 365 days
Butyl benzyl phthalate (BBP)	2 to 180 days
Carbon tetrachloride (CTET)	3 to 360 days
	2.6 days
	0.3 days
	7000 years
	14 days
Chlordane	1.5 to 7.6 years
Chlorobenzene	136 to 300 days
	5.3 days
	37 days
	118 days

Comment	Reference
Estimated	Howard et al. 1991
Estimated	Howard et al. 1991
	Mabey and Mill 1978
	Mabey and Mill 1978
	Mabey and Mill 1978
	Mabey and Mill 1978
Estimated	Howard et al. 1991
Estimated	Howard et al. 1991
Estimated	Howard et al. 1991
Estimated	Howard et al. 1991
Field observations	Piet and Smeenk 1985
Estimated	Howard et al. 1991
Soil incubation study	Tabak et al. 1981
Static-culture flask biodegradation test	Zoeteman et al. 1981
Field observation	Baker et al. 1985
Field observation	Baker et al. 1985
Soil incubation study	Baker et al. 1985
Estimated	Howard et al. 1991
Estimated	Howard et al. 1991
Estimated	Howard et al. 1991
Estimated	Howard et al. 1991
Soil die-away test	Coover and Sims 1987
Estimated	Howard et al. 1991
Estimated	Howard et al. 1991
Estimated	Howard et al. 1991
Field observations	Piet and Smeenk 1985
	Howard et al. 1991
Field observations	Piet and Smeenk 1985
Estimated	Howard et al. 1991
	Howard et al. 1991
	Mabey and Mill 1978
Static-culture flask diodegradation test	Tabak et al. 1981
Test using glass bead columns	Bower et al. 1983
	Howard et al. 1991
Estimated	Howard et al. 1991
Estimated	Howard et al. 1991
Static-culture flask diodegradation test	Tabak et al. 1981
Test using glass bead columns	Bouwer et al. 1983
	Vogel et al. 1987
Laboratory tests	Wood et al. 1985
Estimated	Howard et al. 1991
Estimated	Howard et al. 1991
Static-culture flask biodegradation test	Taba et al. 1981
Field observation	Zoeteman et al. 1981
Soil incubation study	Wilson et al. 1983

TABLE D.9 ***(Continued)***

Compound	Half-Lifes in [Days] or [Years]
	240 days
	37 days
Chlorodibromomethane	14 to 180 days
	274 years
Chlorocresol	1 day
Chloroethane	14 to 56 days
	10 days
Chloroform	56 days
	5 years
	7.4 days
	16.4 days
	0.4 days
	113.7 days
	3500 years
	1.3 years
	236 days
p-Chloro-m-cresol	
2-Chloronaphthalene	
2-Chlorophenol	2.5 days
4-Chlorophenyl phenyl ether	
Chrysene	2.04 to 5.48 years
Cyanide	
4,4-DDD (TDE)	70 days to 31.3 years
4,4-DDE	16 days to 31.3 years
4,4-DDT	16 days to 31.3 years
Dibenzo[*a,h*[anthracene (DBA)	1.98 to 5.15 years
Dichlorobenzene	7.8 days
	110 days
1,2-Dichlorobenzene (DCB)	
1,3-Dichlorobenzene (DCB)	56 to 365 days
1,4-Dichlorobenzene (DCB)	56 to 365 days
3,3-Dichlorobenzidene	5 to 365 days
Dichlorobromomethane	11.3 days
	137 years
Dichlorodifluormethane	56 to 365 days
Dichloromethane	
1,1-Dichloroethane (DCA,1.1)	64 to 144 days
	9.5 days
	159 days
1,2-Dichloroethane (DCA,1.2)	100 to 365 days

Comment	Reference
Soil incubation study	Wilson et al. 1982
Aquatic	Zoetman et al. 1981
Estimated	Howard et al. 1991
	Mabey and Mill, 1978
	Peit and Smeenk 1985
Estimated	Howard et al. 1991
Laboratory tests	Wood et al. 1985
Estimated	Howard et al. 1991
Grab sample data	Wilson, J. T. et al. 1983
Static-culture flask biodegradation test	Tabak et al. 1981
Soil/sediment or soil/water incubation study	Parson et al. 1984
Testing using glass bead columns	Bouwer et al. 1983
Soil incubation study	Bouwer et al. 1986
	Mabey and Mill 1978
	Vogel et al. 1987
Laboratory tests	Wood et al. 1985
Static-culture flask biodegration test	Tabak et al. 1981
Estimated	Howard et al. 1991
Soil test under field conditions	Lichtenstein and Schultz 1959
	Stewart and Chisholm 1971
Soil test under field conditions	Lichtenstein and Schultz 1959
	Stewart and Chisholm 1971
Field observation	Lichtenstein and Schultz 1959
	Stewart and Chisholm 1971
Estimated	Howard et al. 1991
Static-culture flask biodegradation test	Tabak et al. 1981
Field observation	Zoeteman et al. 1981
Estimated	Howard et al. 1991
Estimated	Howard et al. 1991
Estimated	Howard et al. 1991
Static-culture flask biodegradation test	Tabak et al. 1981
	Mabey and Mill 1978
Estimated	Howard et al. 1991
Soil grab sample data	Henson et al. 1989
	Wilson et al. 1983
Static-culture flask biodegradation test	Tabak et al. 1981
Soil incubation study	Wilson et al. 1982
Soil grab sample data	Henson et al. 1989

TABLE D.9 (*Continued*)

Compound	Half-Lifes in [Days] or [Years]
	18.5 days
	0.3 days
	<60 days
1,2-*trans*-Dichloroethylene	long
	55 to 266 days
	55 to 266 days
	6.2 days
	76.8 days
	139 days
1,1-Dichloroethylene (DCE-1,1)	53 days
	56 to 132 days
1,2-Dichloroethylene	5 days
	2.4 days
	76.8 days
	110 days
1,2-*cis*-Dichloroethylene	7.2 days
	140 days
2,4-Dichlorophenol (DCP-2,4)	5.5 to 43 days
1,2-Dichloropropane	0.9 to 7.1 years
1,3-Dichloropropene	
1,2-Dichloropropylene	5.5 to 11.3 days
Dieldrin (HEOD)	1 day to 6 years
Diethyl phthalate (DEP)	6 to 112 days
	long
2,4-Dimethylphenol (2,4-Xylenol)	2 to 14 days
Dimethyl phthalate (DMP)	2 to 14 days
4,6-Dinitro-o-cresol (DNOC)	2.8 to 42 days
2,4-Dinitrophenol (DNP-2,4)	2.8 to 17.5 days
2,4-Dinitrotoluene (DNT)	2 days to 1 year
2,6-Dinitrotoluent	2 days to 1 year
Di-*n*-butyl phthalate	long
Di-*n*-octyl phthalate (DOP)	14 to 365 days
A-Endosulfan-α	<9.1 days
B-Endosulfan-β	<9.1 days
Ethylbenzene	6 to 228 days
	2.6 days
	37 days
	126 days
	37 days
	long
Ethylbromide	26 days
Ethylmethyl	1 day
Fluoranthene	0.8 to 2.4 years
Fluorene	64 to 120 days

Comment	Reference
	Wilson et al. 1983
Static-culture flask biodegradation test	Tabak et al. 1981
Test using glass bead columns	Bouwer et al. 1983
	Wood et al. 1985
Laboratory tests	Wood et al. 1985
Laboratory tests	Wood et al. 1985
	Howard et al. 1991
Static-culture flask biodegradation test	Tabak et al. 1981
Soil incubation study	Wilson et al. 1982
Soil/sediment or soil/water incubation study	Barrio-Lage et al. 1986
Soil grab sample data	Wilson et al. 1986
Static-culture flask biodegradation test	Tabak et al. 1981
Soil/sediment or soil/water incubation study	Fogel et al. 1986
Soil incubation study	Wilson et al. 1986
Soil/sediment or soil/water incubation study	Barrio-Lage et al. 1986
Static-culture flask biodegradation test	Tabak et al. 1981
Soil/sediment or soil/water incubation study	Barrio-Lage et al. 1986
Estimated	Howard et al. 1991
Estimated	Howard et al. 1991
Estimated	Howard et al. 1991
Observed	Howard et al. 1991
Estimated	Howard et al. 1991
Estimated	Howard et al. 1991
Field observations	Piet and Smeents 1985
Estimated	Howard et al. 1991
Estimated	Howard et al. 1991
Estimated	Howard et al. 1991
Estimated	Howard et al. 1991
Estimated	Howard et al. 1991
Estimated	Howard et al. 1991
	Piet and Smeents 1985
Estimated	Howard et al. 1991
Estimated	Howard et al. 1991
Estimated	Howard et al. 1991
Estimated	Howard et al. 1991
Static-culture flask biodegradation test	Tabak et al. 1981
Natural soil groundwater system	Zoeteman et al. 1981
Soil incubation study	Wilson et al. 1986
Aquatic	Zoeteman et al. 1981
Field observations	Piet and Smeents 1985
	Mabey and Mill 1978
Field observations	Piet and Smeents 1985
Estimated	Howard et al. 1991
Estimated	Howard et al. 1991

TABLE D.9 (*Continued*)

Compound	Half-Lifes in [Days] or [Years]
Heptachlor (E 3314)	1 to 5.4 days
Heptchlor epoxide	1 day to 3 years
Hexachlorobenzene (HCB)	5.3 to 11.4 years
	9.8 days
Hexachlorobutadiene (HCBD)	56 to 365 days
e-BHC-α	13.8 to 240 days
b-BHC-β	13.8 to 248 days
r-BHC(lindane)-γ	5.9 to 240 days
g-BHC-δ	13.8 to 248 days
Hexachloro-cyclopentadiene (HCCPD)	5.9 to 240 days
Hexachloroethane (HCE)	13.8 to 200 days
Indeno [1,2,3-cd] pyrene	7.2 to 56 days
Isophorone	56 to 365 days
Isopropylbromide	3.29 to 4.0 years
Methyl bromide	14 to 56 days
	2 days
	14 to 38 days
	20 days
	36 days
Methyl chloride	14 to 56 days
	1 day
	<11 days
	334 days
Methylene chloride	704 years
	11 days
Methylthiobenzothiazole, 2	3 days
Naphthalene	1 to 258 days
	1 day
	0.9 days
	110 days
	1 day
	1 day
	1 to 14 days
	long
N-ethylacetamide	70,000 years
Nitrobenzene	2 days to 1.08 years
2-Nitrophenol	14 to 28 days
4-Nitrophenol	1.5 to 9.8 days
n-Nitrosodimethylamine	42 to 365 days
n-Nitrosodi-n-propylamine	42 to 365 days
n-Nitrosodiphenylamine	20 to 68 days
Parathion	130 days
Pentachlorophenol (PCP)	46 days to 4.2 years

Comment	Reference
Estimated	Howard et al. 1991
Estimated	Howard et al. 1991
Estimated	Howard et al. 1991
Static-culture flask biodegradation test	Tabak et al. 1981
Estimated	Howard et al. 1991
Soil die-away test	Macrae et al. 1984
Soil die-away test	Macrae et al. 1984
Soil die-away test	Macrae et al. 1984
Soil die-away test	Macrae et al. 1984
Estimated	Howard et al. 1991
Estimated	Howard et al. 1991
Estimated	Howard et al. 1991
Estimated	Howard et al. 1991
Estimated	Howard et al. 1991
	Howard et al. 1991
Estimated	Howard et al. 1991
	Mabey and Mill 1978
Estimated	Howard et al. 1991
Static-culture flask biodegradation test	Tabak et al. 1981
	Vogel et al. 1987
Laboratory tests	Wood et al. 1985
	Mabey and Mill 1978
	Mabey and Mill 1978
Laboratory tests	Wood et al. 1985
Field observations	Piet and Smeents 1985
Estimated	Howard et al. 1991
Static-culture flask biodegradation test	Tabak et al. 1981
Batch test using groundwater	Kappeler et al. 1978
Natural soil groundwater system	Zoeteman et al. 1981
Soil incubation study	Wilson et al. 1985
Soil percolation study	Kappeler et al. 1978
Soil microcosm study	U.S.EPA 1988
Field observations	Piet and Smeents 1985
	Mabey and Mill 1978
Estimated	Howard et al. 1991
Estimated	Howard et al. 1991
Soil die-away test	Paris et al. 1983
Estimated	Howard et al. 1991
Estimated	Howard et al. 1991
Estimated	Howard et al. 1991
	Tinsley 1979
Water/sediment grab sample data	Baker and Mayfield 1980
	Delaune et al. 1983

TABLE D.9 (*Continued*)

Compound	Half-Lifes in [Days] or [Years]
	21 to 1087 days
	14 to 26 days
	11 days
	13.7 days
	6.5 to 8 days
Phenanthrene	32 days to 1.1 years
Phenol	0.5 to 7 days
	1.4 days
Propylbromide	26 days
Pyrene	1.15 to 10.4 years
2,3,7,8-Tetrachlorodibenzo-*p*-dioxin (TCDD)	2.29 to 3.23 years
1,1,2,3-Tetrachloroethane	0.6 to 66.8 days
Tetrachloroethylene (PCE, PER)	1 to 2 years
	10.1 days
	300 days
	0.2 to 1.8 days
	0.7 days
	12.8 days
	34 days
	0.7 years
	long
Toluene	7 to 28 days
	37 days
	1 day
	39 days
	37 days
	8 days
	126 days
Toxaphene (PCC)	
Trichloroacetamide	83 days
1,2,4-Trichlorobenzene (TCB)	56 to 365 days
1,1,1-Trichloroethane (TCA)	140 to 545 days
	17 years
	230 days
	27.8 days
	300 days
	0.35 days
	12.1 days
	24 days
	0.5, 1.7, 2.5 years
	16 days
1,1,2-Trichloroethane	0.37 to 2 years
	170 years

Comment	Reference
Soil microcosm study	U.S.EPA 1988
Soil microcosm study	Lamar et al. 1990
Soil microcosm study	Lamar et al. 1990
Field observation	Lamar et al. 1990
Anaerobic sewage sludge used as inoculum	Mikesell et al. 1988
Estimated	Howard et al. 1991
Static-culture flask biodegradation test	Tabak et al. 1981
	Mabey and Mill 1978
Estimated	Howard et al. 1991
Estimated	Howard et al. 1991
Estimated	Howard et al. 1991
Estimated	Howard et al. 1991
Static-culture flask biodegradation test	
Natural soil groundwater system	Roberts et al. 1982
Test using glass bead columns	Vogel et al. 1985
Test using glass bead columns	Bouwer et al. 1983
Soil/sediment or soil/water incubation study	Parson et al. 1984
Laboratory tests	Wood et al. 1985
	Vogel et al. 1987
	Piet and Smeents, 1985
Soil grab sample data	Wilson et al. 1983
	Swindoll et al. 1987
Field observation	Zoeteman et al. 1981
Batch test using groundwater	Kappeler et al. 1978
Field observation	Baker et al. 1985
Groundwater system	Baker et al. 1985
Field observation	Baker et al. 1987
Soil incubation study	Wilson et al. 1986
	Mabey and Mill 1978
Estimated	Howard et al 1991
Soil grab sample data	Wilson et al. 1983
	Vogel et al. 1987
	Roberts et al. 1982
Static-culture flask biodegradation test	Tabak et al. 1981
Field observation	Roberts et al. 1982
Test using glass bead columns	Bouwer et al. 1983
Static-culture flask biodegradation test	Tabak et al. 1981
	Wood et al. 1985
	Vogel et al. 1987
Laboratory tests	Wood et al. 1985
Groundwater die-away study	Wilson et al. 1984
	Vogel et al. 1987

TABLE D.9 (*Continued*)

Compound	Half-Lifes in [Days] or [Years]
	24 days
Trichloroethylene (TCE)	230 days
	33 days
	43 days
	6.8 days
	2.4 days
	300 days
	88 days
	0.9 years
	43 days
	100 to 1640 days
Trichlorofluoromethane (Freon 11)	1 to 2 years
2,4,6-Trichlorophenol	14 days to 5 years
Trimethylphosphate	1.2 years
Vinyl chloride (MVC)	56 days to 7.92 years
	2.3 days
	long
Xylene-m/p	long
Xylene-o	1.2 days
	11 days
	32 days
	31 days
	30 days
	126 days

Notes:

- Many of these citations were compiled from Howard et al. (1991) and from Dragun (1988).
- Estimates of half-lifes were tabulated for organics in which disappearance rates were reported in the literature.
- Actual half-lifes for organics in soil and groundwater account for a great number of environmental factors, and therefore the given figures are not necessarily representative of an organic compound's actual persistence within an aquifer.
- When the half-life of a compound is denominated "long," no evident decrease of concentration has been observed during the period of observation.

References:

Baker, J. F., and G. C. Patrick, 1985. Natural attenuation of aromatic hydrocarbons in shallow sand aquifer. Proc. NWWA/API Conference on Petroleum Hydrocarbons and Organic Chemicals in Groundwater—Prevention, Detection, and Restoration., November 13–15, 1985, Houston, TX., Dublin, OH: National Water Well Association.

Baker, J. F., G. C. Patrick, and D. Major, 1987. Natural attentuation of aromatic hydrocarbons in a shallow sand aquifer. *Ground Water Monitor. Rev.* 7, 64–71.

Comment	Reference
Laboratory tests	Wood et al. 1985
	Roberts et al. 1982
	Barrio-Lage et al. 1986
	Wood et al. 1985
Static-culture flask biodegradation test	Tabak et al. 1981
Soil/sediment or soil/water incubation study	Fogel et al. 1986
Field observation study	Roberts et al. 1982
Soil incubation study	Wilson et al. 1986
	Vogel et al. 1987
	Wood et al. 1985
	Vogel et al. 1987
	Dilling et al. 1975
	Barrio-Lage et al. 1986
Estimated	Howard et al. 1991
Estimated	Howard et al. 1991
	Mabey and Mill 1978
Estimated	Howard et al. 1991
Soil/sediment or soil/water incubation study	Fogel et al. 1986
Field observations	Wood et al. 1985
Field observations	Piet and Smeents 1985
Batch test using groundwater	Kappeler et al. 1978
Field observation	Zoeteman et al. 1981
Field observation	Baker et al. 1985
Groundwater system	Baker et al. 1985
Field observation	Baker et al. 1987
Soil incubation study	Wilson et al. 1982

Baker, M. D., and C. I. Mayfield, 1980. Microbial and non-biological decomposition of chlorophenols and phenols in soil. *Water Air Soil Poll.* 13, 411.

Barrio-Lage, G., F. Z. Parson, R. S. Nassar, and P. A. Lorenzo, 1986. Sequential dehalogenation of chlorinated ethenes. *Environ. Sci. Technol.* 20, 96–99.

Bouwer, E. J., and P. L. McCarty, 1983. Transformation of 1- and 2-carbon halogenated aliphatic organic compounds under methanogenic conditions. *Appl. Environ. Microb.* 45, 1286–1294.

Coover, M. P., and R. C. C. Sims, 1987. The effects of temperature on polycyclic aromatic hydrocarbon persistence in a unacclimated agricultural soil. *Haz. Waste Haz Mat.* 4, 69–82.

DeLaune, R. D., R. P. Gambrell, and K. S. Reddy, 1983. Fate of pentachlorophenol in estuarine sediment. *Environ. Poll.* Series B. 6, 297–308.

Dilling, W. L., N. B. Tefertiller, and G. J. Kallos, 1975. Evaporation rates and reactivities of methylene chloride, chloroform, 1,1,l-trichloroethane, trichloroethylene, tetrachloroethylene and other chlorinated compounds in dilute aqueous solutions. *Environ. Sci. Technol.* 9, 833–838.

Dragun, J., 1988. Microbial degradation of petroleum products in soil. Proc. Conference on Environmental and Public Health Effects of Soils Contaminated with Petroleum Products, October 30–31, 1985, Amherst, MA. New York: Wiley.

TABLE D.9 (*Continued*)

Fogel, M. M., A. R. Taddeo, and S. Fogel, S., 1986. Biodegradation of chlorinated ethenes by a methane-utilizing mixed culture. *Appl. Environ. Microb.* 52, 720–724.

Henson, J. M., M. V. Yates, and J. W. Cochran, 1989. Metabolism of chlorinated methanes, ethanes and ethylenes by a mixed bacterial culture growing of methane. *J. Industrl. Microb.* 4, 29–35.

Howard, P. H., R. S. Boethling, W. F. Jarvis, W. M. Meylan, and E. M. Michalenko, 1991. *Handbook of Environmental Degradation Rates*. Chelsea, MI: Lewis Publishers, p. 725.

Kappeler, T., and L. Wuhrmann, 1978. Microbial degradation of the water soluble fraction of gas-oil-II. Bioassays with pure strains. *Water Resources Res.* 12, 335–342.

Kappeler, T., and Wuhrmann, L., 1978. Microbial degradation of the water soluble fraction of gas-oil-I. *Water Resources Res.* 12, 327–333.

Lamar, R. T., M. J. Larsen, and T. K. Kirk, 1990. Sensitivity to and degradation of pentachlorophenol by phanerochaete spp. *Appl. Environ. Microb.* 56, 3519–3526.

Lamar, R. T., J. A. Glaser, and T. K. Kirk, 1990. Fate of pentachlorophenol (PCP) in sterile soils inoculated with the white-rot basidiomycete phanerochaete chrysosporium: Mineralization, volatilization and depletion of PCP. *Soil Biol. Biochem.* 4, 433–440.

Lamar, R. T., and D. M. Dietrich, 1990. In situ depletion of pentachlorophenol from contaminated soil by phanerochaete spp. *Appl. Environ. Microb.* 56, 3093–3100.

Lichtenstein, E. P., and K. R. Schultz, 1959. Persistence of some chlorinated hydrocarbon insecticides influenced by soil types, rates of application and temperature. *J. Econ. Entomol.* 52, 124–131.

Mabey, W., and T. Mill, 1978. Critical review of hydrolysis of organic compounds in water under environmental conditions. *J. Phys. Chem. Ref. Data*, 7, 383–415.

Mikesell, M. D., and S. A. Boyd, 1988. Enhancement of pentachlorophenol degradation in soil through induced anaerobiosis and bioaugmentation with anaerobic sewage sludge. *Environ. Sci. Technol.* 22, 1411–1414.

Parson, F., P. R. Wood, and J. DeMarco, 1984. Transformations of tetrachloroethene and trichloroethene in microcosoms and groundwater. *J. AWWA* 76 56–59.

Piet, G. J., and J. G. M. M. Smeenk, 1985. Behavior of organic pollutants in pretreated Rhine water during dune infiltration. *Ground Water Quality*, eds. by C. H. Ward et al. New York: Wiley, pp. 122–144.

Roberts, P. V., J. E. Schreiner, and G. D. Hopkins, 1982. Field study of organic water quality changes during groundwater recharge in the Palo Alto baylands. *Water Res.* 16, 1025–1035.

Stewart, D. K. R., and D. Chrisholm, 1971. Long-term persistence of BHC, DDT and chlordane in a sandy loam clay. *Can. J. Soil Sci.* 61, 379–383.

Strand, S. E., and L. Shippert, 1986. Oxidation of chloroform in an aerobic soil exposed to natural gas. *Appl. Environmental Microb.* 52, 203–205.

Swindoll, C. M., C. M. Aelion, and F. K. Pfaender, 1987. Inorganic and organic amendment effects of the biodegradation of organic pollutants by groundwater microorganisms. *Amer. Soc. Microb. Abstr.*, 87th Annl. Mtg., Atlanta, GA., p. 298.

Tabak, H. H., S. A. Quave, C. I. Mashni, and E. F. Barth, 1981. Biodegradability studies with organic priority pollutant compounds. *J. Water Poll. Control Fed.* 53, 1503–1518.

Tinsley, I. J., 1979. *Chemical Concepts in Pollutant Behaviour.* New York: Wiley-Interscience.

U.S. EPA, 1988. Characterization and laboratory soil treatability studies for creosote and pentachlorophenol sludges and contaminated soil. EPA/600/2-88/055, Washington, DC.

Vogel, T. M., and P. L. McCarty, 1985. Biotransformation of tetrachloroethylene to trichloroethylene, dichloroethylene, vinyl chloride, and carbon dioxide under mathanogenic conditions. *Appl. and Environ. Microb.* 49, 1080–1083.

Vogel, T. M., and P. L. McCarty, 1987. Abiotic and biotic transformations of 1,1,1-trichloroethane under methanogenic conditions. *Environ. Sci. Technol.* 12, 1208–1213.

TABLE D.9 (*Continued*)

Vogel, T. M., C. S. Criddle, and P. L. McCarty, 1987. Transformations of halogenated aliphatic compounds. *Environ. Sci. and Technol.* 21, 722–736.

Wilson, B. H., G. B. Smith, and J. F. Rees, 1986. Biotransformations of selected alkylbenzenes and halogenated aliphatic hydrocarbons in methanogenic aquifer materials: A microcosom study. *Environ. Sci. Technol.* 20, 997–1002.

Wilson J. T., J. F. McNabb, D. L. Balkwill, and W. C. Ghiorse, 1983. Enumeration and characterization of bacteria indigenous to a shallow water-table aquifer. *Ground Water* 21, 134–142.

Wilson, J. T., J. F. McNabb, B. H. Wilson, and M. J. Noonan, 1982. Biotransformation of selected organic pollutants in groundwater. *Dev. Indust. Microb.* 24, 225–233.

Wilson, J. T., J. F. McNabb, J. W. Cochran, T. H. Wang, M. B. Tomson, and P. B. Bedient, 1985. Influence of microbial adaptation on the fate of organic pollutants in groundwater. *Environ. Toxi. Chem.* 4, 721–726.

Wood, P. R., R. F. Land, and I. L. Payan, 1985. Anaerobic transformation, transport, and removal of volatile chlorinated organics in ground water. *Ground Water Quality*, eds. C. H. Ward et al. New York: Wiley, pp. 493–511.

Zoeteman, B. C. J., E. De Greef, and F. J. J. Brinkmann, 1981. Persistency of organic contaminants in groundwater, lessons from soil pollution incidents in the Netherlands. *Sci. Total Environ.* 21, 187–202.

APPENDIX E
EPA Drinking Water Standards

TABLE E.1 U.S. EPA Drinking Water Standards, May 1995

Chemical	Maximum Contaminant Level (MCL) in [mg/l]
Alachlor	0.002
Aldicarb	0.007
Aldicarb sulfone	0.007
Aldicarb sulfoxide	0.007
Atrazine	0.003
Antimony	0.005
Arsenic	0.05
Asbestos (fibers per liter)	7×10^{-3}
Barium	2
Benzene	0.005
Benzo(a)pyrene	0.002
Beryllium	0.004
Bromate	0.01
Bromodichloromethane	0.1
Bromoform	0.1
Cadmium	0.005
Carbofuran	0.04
Carbon tetrachloride	0.005
Chloral hydrate	0.06
Chloramine	4
Chlordane	0.002
Chlorine	4
Chlorine dioxide	0.8
Chlorite	1
Chlorodibromomethane	0.1
Chloroform	0.1
Chromium (total)	0.1
Copper	1.3
Cyanide	0.2
2,4-Dichlorophenoxyacetic acid (2,4-D)	0.07
Dalapon	0.2
Di(2-ethylhexyl)adipate	0.4

TABLE E.1 (*Continued*)

Chemical	Maximum Contaminant Level (MCL) in [mg/l]
Dibromochloropropane	0.0002
Dichloroacetic acid	0.06
Dichlorobenzene *p*-	0.075
Dichlorobenzene *m*-	0.6
Dichlorobenzene *o*-	0.6
Dichloroethane(1,2-)	0.005
Dichloroethylene(1,1-)	0.007
Dichloroethylene(*cis*-1,2)	0.07
Dichloroethylene(*trans*-1,2-)	0.1
Dichloromethane	0.005
Dichloroprane(1,2-)	0.005
Diethylhexyl phthalate	0.006
Dinoseb	0.007
Diquat	0.02
Endothall	0.1
Endrin	0.002
Ethylbenzene	0.7
Ethylene dibromide	0.00005
Fluoride	4
Glyphosate	0.7
Heptachlor	0.0002
Heptachlor epoxide	0.001
Hexachlorobenzene	0.05
Hexachlorocyclopentadiene	0.0002
Lead	0.015
Lindane	0.0002
Mercury	0.04
Methoxychlor	0.1
Monochlorobenzene	0.1
Nickel	10
Nitrate (as N)	1
Nitrite (as N)	10
Nitrate + nitrite (as N)	0.2
Oxamyl	0.001
Pentachlorophenol	0.5
Picloram	0.0005
Polychlorinated biphenyls	0.05
Selenium	0.004
Simazine	0.1
Styrene	500
Sulfate	3×10^{-8}
2,3,7,8-TCDD (dioxin)	0.005
Tetrachloroethylene	0.002
Thallium	1
Toluene	0.003

TABLE E.1 (*Continued*)

Chemical	Maximum Contaminant Level (MCL) in [mg/l]
Toxaphene	0.05
2,4,5-TP	0.05
Trichloroacetic acid	0.06
Trichlorobenzene(1,2,4-)	0.07
Trichloroethane(1,1,1-)	0.2
Trichloroethane(1,1,2-)	0.005
Trichloroethylene	0.005
Vinyl chloride	0.002
Xylenes	10
Radionuclides	
Beta particle and photon activity (excluding radium 228)	4 mrem ede/year
Gross alpha particle (excluding radium 226, uranium and radon 222)	15 pCi/L
Radium 226/228	20 pCi/L
Radon 222	300 pCi/L
Uranium	20 μg/l

Note: The U.S. EPA safe drinking water hotline for phone inquiries: 1-800-426-4791.

GLOSSARY

Actual transport velocity. True but unknown transport velocity of a solute in the pore space.

Adsorption. Adhesion of molecules or ions on the grain surface.

Adsorption isotherm. Relationship at equilibrium between the concentration of a solute in the adsorbed phase and in the adjacent water phase.

Advection. Mass transport caused by the bulk movement of flowing groundwater.

Advection-dispersion model. Transport model which accounts for advective and dispersive transport.

Analog model. Groundwater model that originates from the similarity of its physical process to groundwater flow (e.g., electrical analog model).

Analytical model. Analytical solutions for flow or transport problems.

Aquiclude. Very low permeability unit that will not transmit water freely.

Aquifer. Permeable unit that can yield water in usable quantities when tapped by a well.

Aquifer approach. *See* depth-integration.

Aquifer performance test. Test to identify unknown aquifer parameters, usually transmissivity and storativity.

Aquitard. Low-permeability unit that is less permeable than aquifers and more permeable than aquicludes. Water cannot be produced economically through wells but flow is significant enough to feed adjacent aquifers.

Areal source. Any source including farmland, industrial areas, and mining sites from which water and/or contaminants are generated over a large area.

Artificial recharge. Addition of water to the groundwater system by human activities.

Aspect ratio. Ratio of maximum to minimum element dimensions.

Automated calibration. Model calibration using inverse models.

Average transport velocity. Average transport velocity calculated as the Darcy velocity divided by the effective porosity.

Bandwidth. Maximum number of columns between the first and last nonzero entry along any row of the coefficient matrix.

Backward differences. *See* implicit model.

Basis function. Interpolation functions used in FE models to define primary variable within the element.

Barrier boundary. *See* no-flux boundary.

Bench-scale model. *See* porous media model.

Biodegradation. Reduction of solutes due to enzyme-catalyzed transformation.

Black box model. *See* mass balance model.

Block-centered model. Type of FD models in which solutions are calculated at nodes in the center of the model blocks.

Boundary. Interface between the model area and the surrounding environment.

Boundary condition. Flow or transport condition on the model boundary.

Calculation-mesh cell size. Space discretization in FD models on which the numerical solution is based.

Calibration. Repeatedly altering the model input data until the simulated and observed values (hydraulic heads or concentrations) are fitted within a chosen tolerance.

Catchment area. Land area from which precipitation drains into a well, reservoir, pond, lake, or stream.

Cauchy's condition. *See* semipermeable boundary or mixed condition.

Cell. Subdomain in FD models.

Cell aspect ratio. Ratio of cell length to width.

Code. Set of instructions used to solve an equation on a computer.

Coefficient matrix. Matrix in the numerical matrix equation that comprises all information on the aquifer characteristics and dicretization.

Compiler. Program that transforms a code into an executable file.

Computer program. *See* code.

Conceptual model. Simplified representation of the investigated groundwater system for which an approximate solution is sought.

Confined aquifer. Aquifer confined at the top and base by less permeable units; ground water in a confined aquifer is under pressure greater than atmospheric.

Conservative solute. Solute that remains stable and does not change due to physical, biological, or chemical processes.

Contaminant. Solute that causes groundwater contamination.

Contamination. Undesirable changes in groundwater quality that do not render the groundwater unfit for domestic consumption.

Convergence. Behavior of numerical solution during iterative solution or over time.

Convergence criterion. Criterion used to judge convergence of solution.

Courant criterion. Limitation of Courant number to control stability of numerical solution.

Courant number. Ratio of convective to time dependent terms in the transport equation. Can be expressed as ratio of transport velocity times time step to discretization length.

Crank-Nicholson scheme. Numerical scheme that makes use of the arithmetic mean of the known hydraulic heads or concentrations at the beginning of the time step and the unknown ones at the end of the numerical solution procedure.

Cross-sectional model. *See* vertical-plane model.

Darcy's law. Empirical law that describes groundwater movement at a macroscopic scale.

Decay. Reaction of the first order (e.g., radioactive decay) that results in a steady decrease of concentration.

Density-dependent flow. Groundwater flow caused or influenced by density differences.

Density-stratified flow. Groundwater flow with stable density difference between groundwater layers (e.g., example freshwater above seawater).

Depth-integration. Averaging of the three-dimensional equation for flow or transport in the vertical dimension reducing the governing equations to the two horizontal dimensions.

Desorption. Release of molecules or ions from the rock or soil grain surface.

Diffusion. Net flux of solutes from a zone of higher concentration to a zone of lower concentration.

Dirichlet's condition. *See* prescribed head or prescribed concentration boundary.

Discharge node. Node in an FD model that represents a discharge well or other discharge conditions.

Discrete fracture model. Model to simulate flow or transport in a fracture network that accounts explicitly for flow and/or transport in fractures.

Discretization. *See* space and time discretization.

Dispersion. Irregular spreading of solutes due to aquifer heterogeneities at pore-grain scale (mechanical dispersion) or at field scale (macroscopic dispersion).

Dispersion coefficient. Empirical coefficient used in Fick's law to describe dispersion.

Dispersivity. Empirical aquifer property controlling the magnitude of dispersion.

DNAPL. Dense, nonaqueous phase liquid.

Dual porosity. Aquifer in which total porosity is composed of two parts: fracture pore volume and grain or rock pore space.

Dual porosity model. Model to simulate transport in an aquifer with pronounced bimodal distribution of hydraulic conductivities and porosities.

Dupuit assumption. Assumption of essentially horizontal flow.

Effective porosity. Part of total porosity that transmits water freely.

Electrical analog model. Model that is based on the similarity between electrical flow and groundwater flow.

Element. Subdomain in FE model.

Empirical model. Representation of physical or chemical processes by generalities or simplifications.

Equivalent porous medium model. Model for heterogeneous aquifers such as fractured rock where the aquifer is approximated by a porous medium with equivalent macroscopic characteristics.

Error criterion. Criterion used to judge convergence of the numerical solution.

Execution time. Time required by the numerical model to perform one model run.

Explicit model. Model that makes use of the known hydraulic heads or concentrations at the beginning of the time step in the numerical solution procedure.

Fick's law. Empirical law to express the diffusive and/or dispersive movement of solutes.

Field-scale dispersion. *See* macrodispersion.

Finite-difference (FD) model. Numerical approximation of the governing equation by expanding the equation term by term into a Taylor series.

Finite-element (FE) model. Numerical approximation of the governing equation by integration rather than differentiation.

First-kind (or type) boundary condition. Prescribed head or prescribed concentration boundary.

Flow line. Line shown in a flow net representing the path of a solute under conditions of steady-state flow.

Flow net. Net of equipotential and flow lines intersecting at right angles.

Flow path. Direction of movement of groundwater and any contaminants that may be contained therein.

Forward differences. *See* explicit model.

Fourier's condition. *See* semipermeable boundary or mixed boundary conditions.

Fourier number. Ratio of dispersive to time dependent terms in the transport equation. Also, ratio of Courant number to Peclet number. Can be expressed as ratio of dispersion coefficient times time step to the square of the discretization length.

Galerkin method. Weighting method in FE models in which the basic functions are chosen as weighting functions.

Geometrical model. Groundwater model that approximates flow and/or transport at pore-grain scale.

Gigo rule. Garbage in, garbage out. Reminder that numerical results cannot be more accurate than the model input data allow.

Grid. Array of blocks (FD model) or elements (FE model) used to subdivide the model area.

Groundwater divide. Line of maximal elevation along a groundwater ridge where the water table slopes downward in opposite directions or streamline that borders a catchment area.

Groundwater model. Model of a groundwater system that is manipulated in order to obtain information on the response of the natural groundwater system. The most widely used groundwater models are numerical models.

Half-life. Time required to reduce the concentration of a given radioactive isotope or solute to half its initial value by radioactive decay or by first-order biodegradation.

Head. Sum of datum level, elevation head, and pressure head (in unconfined aquifers equal to the groundwater elevation).

Head-dependent boundary (or condition). *See* semipermeable boundary.

Hele-Shaw model. *See* viscous fluid model.

Henry's law constant. Coefficient that represents the equilibrium partitioning factor between water and vapor phases.

Horizontal-plane model. Two-dimensional, depth-integrated model.

Hydraulic approach. *See* depth-integration.

Hydraulic model. Porous media or viscous fluid model.

Hydrologic budget. *See* mass-balance model.

Hydrolysis. Reaction of the solute in water, usually with the introduction of a hydroxyl (OH) group into the solute's chemical structure, and commonly with the loss of a leaving group.

Immiscible liquids. Liquids that do not mix with water.

Immobile water. Crystalline water contained in minerals, adsorbed water referred to as hygroscopic water bound by electrostatic and van der Waal's forces on soil grains or water enclosed in disconnected voids.

Impermeable boundary. *See* no-flow boundary.

Implicit model. Model that makes use of the unknown hydraulic heads or concentrations at the end of the time step in the numerical solution procedure.

Initial conditions. Head or concentration within the model area at some initial time, usually denoted as $t = 0$.

Initial guess. Initial approximation of the solution.

Instability. Numerical solution leads to calculation of unrealistic or unbalanced heads or concentrations.

Internal boundary. Boundaries such as rivers, lakes, or faults that are within the model area.

Internodal velocity. Velocity along the direct connection line between two nodes in FD models.

Inverse model. Model to determine characteristics of the groundwater system from known heads or concentrations.

Irregular grid. Grid where blocks or elements differ from each other.

Iteration. Process of repeatedly cycling through the solutions for the unknowns in a matrix equation, replacing old values of an unknown by the most recently computed values.

Iteration residual error. Remaining error in iterative solution of matrix equations when iterations are terminated by a specified error criterion.

Kinetic reaction. Reaction for which local equilibrium cannot be assumed.

Kriging. Statistical interpolation method that accounts for the spatial structure of the variable in question.

Leakage. Groundwater flow through a semipermeable layer.

Leakage factor. Measure of the amount of leakage.

Leakage principle. Introducing sources and sinks of water in the model as flux through a semipermeable layer.

Leaky aquifer. *See* semiconfined aquifer.

Linear isotherm. Equilibrium adsorption isotherm in which the adsorbed concentration is assumed to be proportional to concentration in the water phase.

Line source. Any source including ditches, channels, rivers, from which water and/or contaminants are discharged over a long section.

Longitudinal dispersion coefficient. Dispersion coefficient in flow direction.

Longitudinal dispersivity. Dispersivity in flow direction.

Lumped parameter model. *See* empirical model.

Macrodispersion. Dispersion caused by irregularities of groundwater flow at field scale.

Mass-balance model. Models that balance mass fluxes, either groundwater or solutes, over large volumes of the groundwater system.

Mathematical model. Formulation of governing flow and transport processes in mathematical terms.

Mechanical dispersion. Dispersion caused by irregularities of groundwater flow at pore-grain scale.

Membrane model. Groundwater model that makes use of the analogy of the deformation of the water table due to groundwater extraction and the deflection of thin membranes due to vertical displacement at discrete points.

Mesh. *See* grid.

Mesh-centered model. Type of FD model in which solutions are calculated at nodes that coincide with model block faces.

Method of characteristics (MOC). Solving for transport by tracing equally distributed particles through the flow field.

Method of fragments. Solution method for groundwater flow in which the flow system is subdivided into individual sections using straight equipotential lines.

Method of images. Combined use of the superposition principle and symmetry to introduce various boundary conditions to the model area.

Miscible liquids. Liquids that can be dissolved in water.

Mixed boundary condition. *See* semipermeable boundary for flow. In transport modeling, mixed boundaries combine advective and dispersive transport.

Mobile water. Water filling interconnected pore space and being transmitted freely through the pores of rock or soil or, alternatively, through rock fractures.

Model accuracy. Ability of the model to approximate the real system.

Model area. That part of the groundwater system for which an approximated solution is derived using a model.

Model audit. Comparison of model predictions and field data subject to changed future conditions.

Model errors. Difference between model predictions and actual conditions due to errors inherent in the numerical approximation, in the limited accuracy of the computer, and from simplifications of the natural system introduced in the model.

Modeling report. Presentation of modeling activities and predicted results.

Model input data. Information on the investigated groundwater systems that is required by the numerical model in order to obtain an approximate solution.

Model output data. Numerical or graphical results of the model calculations.

Model quality assurance. Process that is used to ensure that an adequate modeling analysis has been conducted.

Model selection. Choice of a model that fits the investigated groundwater problem.

Monitoring well. Well used to measure water levels or to obtain water samples for water-quality analysis.

Moving boundary. Moving free surface (water table) in vertical-plane models.

Multi-layer model. Quasi-three-dimensional model that consists of a sequence of two-dimensional, depth-integrated models are linked together by sources and sinks.

Multiphase flow. Movement of more than one fluid in the pore space. Simultaneous flow of more than two fluids exists in the unsaturated zone (water and air), in oil reservoirs (oil, gas, and water), and near to groundwater contamination sources of nonmiscible pollutants.

Multi-well flow. Groundwater flow caused by the operation of more than one well.

Neumann's condition. See prescribed flux boundary.

Neumann criterion. Stability criterion that relates dispersive flux to space discretization.

Node. Point in FD models that represents the center of the cell.

No-flux boundary. Special type of prescribed flux boundary where flux is zero.

Nonlinearity. Mathematical relationship with terms that depend on the unknown solution.

Nonunique solution. Different combinations of model input data allowing the calculation of identical head or concentration distributions.

Numerical dispersion. Numerical errors in the numerical model that appear in the simulations as dispersive transport.

Numerical error. Errors in the numerical solution such as truncation errors, roundoff errors, numerical dispersion or oscillation.

Numerical model. Computer program based on the numerical approximation of the flow and/or transport equation.

Octanol-water partition coefficient. Coefficient that represents the equilibrium of an organic compound between an organic phase (octanol) and the aqueous phase.

One-dimensional model. Model that accounts for one space dimension.

Organic. Being, containing, or relating to carbon compounds, especially in which hydrogen is attached to carbon, whether derived from living organisms or not; usually distinguished from inorganic or mineral.

Organic carbon partition coefficient. Ratio of concentration of solute in octanol phase to concentration of solute in aqueous phase. As a consequence measure of the preference of a hydrophobic organic to leave the aqueous phase in preference to the solid organic phase present in the soil.

Organic carbon weight fraction. Coefficient that represents the weight fraction of organic carbon in the soil material defined as the percentage of organic carbon.

Oscillations. *See* overshoot and undershoot.

Overrelaxation. Modifying the most recently computed value in an iterative solution technique by extrapolation.

Overshoot. Heads or concentrations that are calculated to be larger than the actual ones.

Parallel plate model. *See* viscous fluid model.

Parameter estimation. *See* model calibration.

Particle tracking. Numerical method to simulate transport by placing a particle into the flow field and numerically integrating the flow path.

Postprocessor. Numerical tool such as plot software package to process and to present model output data.

Peclet criterion. Limit of Peclet number to control numerical dispersion.

Peclet number. Measure of relative importance of convection versus dispersion processes. Can be expressed as ratio of convective to dispersive terms in the transport equation.

Phased modeling. Modeling approach in which a sequence of successively more complex models are used.

Phreatic aquifer. *See* unconfined aquifer.

Plane symmetry. System with symmetry with respect to a vertical plane.

Plume. Zone of contaminated groundwater containing dissolved contaminants.

Point source. Any source including tanks, wells, spills, from which water and/or contaminants are discharged at discrete points.

Pollutant. Solute that causes groundwater pollution.

Pollution. Adverse groundwater quality changes that render the groundwater unfit for domestic consumption.

Porous media model. Porous media or bench-scale model representing a small scale replica of the investigated groundwater system.

Preprocessor. Numerical tools to process and to prepare model input data.

Prescribed concentration boundary. Boundary on which the concentration is known.

Prescribed flux boundary. Boundary on which water flux or solute flux is known.

Prescribed head boundary. Boundary on which the head is known.

Profile model. *See* vertical-plane model.

Program. *See* code.

Quality assurance label. Labels attached to model results, printouts, or plots to demonstrate checking of key model items such as mass balances.

Radial converging flow. Flow system with symmetry with respect to a vertical axis.

Random walk (RW) model. Transport model where transport is calculated by following the random path of many particles through the flow field.

Recharge boundary. *See* prescribed flux boundary.

Recharge node. Node in a FD model that represents a recharge well or a recharge boundary.

Reflective boundary. *See* no-flux boundary.

Relaxation factor. Factor that modifies the most recently computed values for the unknowns in an iterative solution technique before using these values to replace old values.

Representative element volume (REV). Representative volume of the aquifer that allows the simulation of the actual aquifer with a representative continuum in which macroscopic laws can be used to describe flow or transport at macroscopic scale.

Retardation factor. Ratio of transport velocity of a nonreactive tracer to transport velocity of solute.

River node. Node that represents a river in the model.

Roundoff error. Errors due to the limit in significant figures stored in a computer.

Second-kind (or type) condition. *See* prescribed flux boundary.

Semianalytical model. Transport model in which velocities are determined analytically and streamlines are calculated numerically.

Semiconfined aquifer. Aquifer that receives or loses water to adjacent aquifers through an aquitard.

Semipermeable boundary. Model boundaries that represent leakage from surface water bodies where the exchange of water between the aquifer and the surface water depends on the head difference, and hydraulic properties of intervening layers, between surface water and groundwater.

Sensitivity analysis. Model analysis to demonstrate the model response to variations in model input data.

Single-cell model. *See* mass-balance model.

Single-phase flow. Movement of groundwater and dissolved contaminants.

Smearing. *See* numerical dispersion.

Solubility. Equilibrium partitioning between the pure liquid phase and water.

Solute balance. Accounting of all solute inputs and outputs within a system.

Sorption. Adhesion or release of molecules or ions on the soil or rock surface.

Space discretization. Subdivisions of the model area.

Specific discharge. Flux per unit area per unit time.

Stability. Behavior of numerical errors during computation. A solution in which errors decrease over time is considered to be stable.

Statistical model. Models that describe random movements of solutes using statistical theory.

Steady state. Flow or transport conditions that do not change over time.

Streamline. *See* flow line.

Stochastic modeling. Addressing uncertainty in model study by introducing model input data as random variables.

Superposition principle. Principle that individual linear solutions can be combined to give a net solution.

Third-kind (or type) condition. *See* semipermeable boundary or mixed boundary.

Thomas algorithm. Algorithm used to solve tridiagonal matrices.

Three-dimensional model. Model that accounts for three space dimensions.

Time discretization. Subdivision of time domain in time steps.

Time step. Time interval used to discretize continuous events by step functions.

Time stepping. Discretizing time into time levels separated by time intervals and solving flow and transport for each time level, starting with the initial condition.

Tracer. Conservative solutes that do not influence the viscosity and density of groundwater and that are employed to track the movement of water.

Transient. Flow or transport conditions that change over time.

Transport modeling. Describing solute transport using numerical models.

Transport processes. Physical processes such as advection or dispersion and chemical/biological processes such as sorption or biodegradation that influence the transport of a solute in the subsurface.

Transverse dispersion coefficient. Dispersion coefficient transverse to the flow direction.

Transverse dispersivity. Dispersivity transverse to the flow direction.

Trial-and-error method. Adjusting model input data manually by trial and error to match observed responses of the groundwater system.

Truncation error. Error in FD models due to neglecting terms of higher order in the Taylor series expansion.

Two-dimensional model. Model that accounts for two space dimensions.

Unconfined aquifer. Aquifer that has a free water table.

Underrelaxation. Modifying the most recently computed value in an iterative solution technique by weighting it with the value from the previous iteration.

Undershoot. Heads or concentrations that are calculated to be smaller than the actual ones.

Uniform groundwater flow. Groundwater flow in which, at every point, the specific discharge has the same magnitude and direction.

Unstable solution. *See* instability.

Upwind difference. Finite-difference approximation of advective transport in which upgradient values dominate the solution.

User friendly. A model is called user friendly, if the handling of model input and output data is relatively easy or facilitated by pre- and postprocessors.

Validation. Comparison of numerical model results with a data set not used in model calibration.

Verification. *See* validation. (Sometimes verification is used to describe comparison of numerical model results with analytical or laboratory results.)

Vertical-plane model. Two-dimensional vertical model.

Viscous fluid model. Groundwater model that is based on the analogy of the movement of a viscous fluid such as glycerine between two closely spaced parallel plates and two-dimensional groundwater flow.

Volatilization. Evaporation of contaminants.

Water balance. Accounting for all water inputs and all water outputs within a system.

Water vapor. Gaseous water distributed in the unsaturated groundwater zone.

Zero-dimension model. *See* mass-balance model.

REFERENCES

Abramowitz, M., and I. A. Stegun, 1970. *Handbook of Mathematical Functions*, 9th ed., New York: Dover, p. 1046.

Abriola, L. M., 1988. Multiphase flow and transport models for organic chemicals: A review and assessment. EA-5976, Electric Power Res. Inst., Palo Alto, CA.

Anderson, M. P., and W. W. Woessner, 1992. *Applied Groundwater Modeling—Simulation of Flow and Advective Transport*, San Diego: Academic Press.

Aris, R., 1956. On the dispersion of a solute in a fluid flowing through a tube. *Proc. Roy. Soc.* Ser. A 235, 67.

ASTM, 1984. Standard practice for evaluating environmental fate models of chemicals. Am. Soc. Test, Mater., Philadelphia, E978-84.

ASTM, 1993. *Standard Guide for Application of a Ground-Water Flow Model to a Site-Specific Problem*. ASTM D 5447-93.

Atkinson, R., T. P. Cherrill, A. W. Herbert, D. P. Hodgkinson, C. P. Jackson, J. Rae, and P. C. Robinson, 1984. Review of the groundwater flow and radionuclide transport modeling in KBS-3. Harwell Report AERE-R. 11140, HMSO, London.

Baca, R. G., J. C. Walton, A. S. Rood, and M. D. Otis, 1988. Organic contaminant release from a mixed waste disposal site: A computer simulation study of transport through the vadose zone and site remediation. Paper presented at Tenth Annual Low-Level Waste Management Forum DOE, Denver, CO, August.

Bachmat, Y., and J. Bear, 1964. The general equations of hydrodynamic dispersion in homogeneous, isotropic porous mediums. *J. Geophys. Res.* 69.

Baetsle, L. H., 1969. Migration of radionuclides in porous media. *Progress in Nuclear Energy*. Series XII, Health Physics, ed. A. M. F. Duhamel. Elmsford, NY: Pergamon Press, pp. 707–730.

Ball, J. W., E. A. Jenne, and M. W. Cantrell, 1981. WATEQ3—A geochemical model with uranium added. USGS Open File Report, 81-1183, p. 84.

Ball, J. W., D. K. Nordstrom, and D. W. Zachmann, 1987. WATEQ4F—A personal computer FORTRAN translation of the geochemical model WATEQ2 with revised database. USGS Open File Report, 87–50.

Barker, J. F., G. C. Patrick, and D. Major, 1987. Natural attenuation of aromatic hydrocarbons in a shallow sand aquifer. *Groundwater Monitoring Review*. Winter.

Baveye, P., and A. Valocchi, 1989. An evaluation of mathematical models of the transport of biologically reacting solutes in saturated soils. *Water Resources Res.* 25(6), 1413–1421.

Bear, J., 1961a. Some experiments in dispersion. *J. Geophys. Res.* 66(8), 2427–2455.

Bear, J., 1961b. On the tensor form of dispersion in porous media. *J. Geophys. Res.* 66(4), 1185–1198.

Bear, J., 1972. *Dynamics of Fluids in Porous Media.* New York: American Elsevier, p. 764.

Bear, J., 1979. *Hydraulics of Groundwater.* McGraw-Hill Series in Water Resources and Environmental Engineering, New York: McGraw-Hill, p. 569.

Bear, J., and Verruijt, A., 1987. Modeling groundwater flow and pollution. Reidel, Boston.

Beljian, M. S., 1988. Representation of individual wells in two-dimensional ground water modeling, *Proc., Solving Ground Water Problems with Models*, February 10–12, Denver.

Birdsell, K. H., P. G. Stringer, L. F. Brown, G. A. Cederberg, B. J. Travis, and A. E. Norris, 1988. Modeling tracer diffusion in fractured and unfractured, unsaturated, porous media, *J. Contam. Hydrol.* 3(2–4), 145–170.

Bond, F., and S. Hwang, 1988. Selection criteria for mathematical models used in exposure assessments: Ground-Water Models. EPA/600/8-88/075, U.S. EPA.

Boonstra, J., 1989. *SATEM: Selected Aquifer Test Evaluation Methods. A Microcomputer Program.* ILRI publication 48. The Netherlands: ILRI, p. 80.

Boonstra, J., and N. A. de Ridder, 1981. *Numerical Modeling of Groundwater Basins: A User-Oriented Manual.* ILRI publication 29. The Netherlands: ILRI.

Boulton, N. S., 1954. The drawdown of the water table under non-steady conditions near a pumped well in an unconfined formation. *Proc. Inst. Civil Engrs.* 3, 564–579.

Boulton, N. S., 1963. Analysis of data from nonequilibrium pumping tests allowing for delayed yield from storage. *Proc. Inst. Civil Engrs.* 26, 469–482.

Bradley, M. D., 1983. *The Scientist and Engineer in Court.* Washington, DC: American Geophysical Union.

Bredehoeft, J. D., and W. L. F. Konikow, 1993. Groundwater models: Validate or invalidate. Editorial. *Groundwater* 31(2), 178–179.

Carslaw, H. S., and J. C. Jaeger, 1959. *Conduction of Heat in Solids*, 2d ed. Oxford: Oxford University Press, p. 510.

CH2M Hill, 1990. VLEACH, A one-dimensional finite difference vadose zone leaching model, version 1.02. Prepared for the U.S. Environmental Protection Agency, Region 9, San Francisco, CA.

Chan, T., V. Guvanasen, and J. A. K. Reid, 1987. Numerical modeling of coupled fluid, heat and solute transport in deformable fractured rock. *Coupled Processes Associated with Nuclear Waste Repositories. Proc. Int. Symp.*, Berkeley, CA ed. C. F. Tsang. San Diego: Academic Press, 605–625.

Chan, T., and N. W. Scheier, 1987. Finite element simulation of groundwater flow and heat and radionuclide transport in a plutonic rock mass. *Proc. 6th Int. Congress on Rock Mechanics*, Montreal, August 30–September 3, pp. 41–46. Balkema Press.

Chan, T., 1988. An overview of groundwater flow and radionuclide transport modeling in the Canadian nuclear fuel waste management program. *Proc. Conference on Geostatistical, Sensitivity and Uncertainty Methods for Groundwater Flow and Radionuclide Transport Modeling*, September 15–17, San Francisco. pp. 39–61. Battelle Press.

Chandrasekhar, S., 1943. Stochastic problems in physics and astronomy. *Rev. of Modern Physics*, ISCIS, pp. 1–90.

Chemical Rubber Company, 1985. *CRC Handbook of Chemistry and Physics.* 66th ed., Boca Raton, FL: CRC Press.

Chemical Rubber Company, 1986. Handbook of Chemistry and Physics. Boca Raton, FL: CRC Press.

Cherry, J. A., 1990. Short course: Dense, immiscible phase liquid contaminants (DNAPLs) in porous and fractured media. Waterloo Centre for Groundwater Research, Kitchener, Ontario, November 26–29.

Chiou, C. T., L. J. Peters, and V. H. Freed, 1979. A physical concept of soil-water equilibrium for nonionic organic compounds. *Science* 206, 831–832.

Chow, V. T., 1952. On the determination of transmissivity and storage coefficients from pumping test data. *Am. Geophys. Union Trans.* 33, 397–404.

Chu, W., E. W. Strecker, and D. P. Lettenmaier, 1987. An evaluaton of data requirements for groundwater contaminant transport modeling. *Water Resources Res.* 23(13), 408–424.

Cleary, R. W., and M. J. Ungs, 1994. PRINCE user guide version 3.0. Princeton analytical models of flow and mass transport. Waterloo Hydrogeologic Software, Waterloo, Ontario.

Coats, K. H., and B. D. Smith, 1964. Dead-end pore volume and dispersion in porous media. Pet. Trans AIME, 231, SPEJ 73.

Cohen, R. M., and W. J. Miller, 1983. Use of analytical models for evaluating corrective actions at hazardous waste disposal facilities. *Proc., Third National Symposium on Aquifer Restoration and Groundwater Monitoring.* National Water Well Association, Worthington, OH, pp. 85–97.

Cooper, H. H., and C. E. Jacob, 1946. A generalised graphical method for evaluation formation constants and summarizing well field history. *Am. Geophys. Union Trans.* 27, 526–534.

Crank, J., 1956. *Mathematics of Diffusion.* New York: Oxford University Press, p. 347.

Custodio, E., A. Gurgui, J. Ferreira, and P. Lobo (eds.), 1988. Groundwater flow and quality modeling, NATO ASI Ser. C. 224.

Dacosta, J. A., and R. R. Bennett, 1960. The pattern of flow in the vicinity of a recharging and discharging pair of wells in an aquifer having areal parallel flow. *Internationaler Kongreß für wissenschaftliche Hydrologie*, Helsinki, vols. 51–52, 1–2.

Dagan, G. 1982. Stochastic modeling of groundwater flow by unconditional and conditional probabilities. The solute transport. *Water Resources Res.* 18, 835–848.

Dale, T., and P. A. Domenico, 1990. The Aggie Solution Technique to the Contaminant Transport Problem. *User's Manual.* College Station: Texas A&M University.

Dames & Moore, 1995. TARGET: Mathematical model for ground-water flow and solute transport. Code documentation.

Darcy, H., 1856. *Les Fontaines Publiques de la Ville de Dijon.* Paris: V. Dalmont, p. 647.

Davis, L. A., and G. Segol, 1985. Documentation and user's guide; GS2 and GS3—Variably saturated flow and mass transport models. Report NUREG/CR-3901 and WWL/TM-1791-2, U.S. Nucl. Reg. Comm., Washington, DC.

Davis, S.N., and R. J. M. De Wiest, 1966. Hydrogeology. New York: Wiley, p. 463.

De Glee, G. J., 1930. Over grondwaterstromingen bij wateronttrekking door middel van putten. Thesis. Delft, The Netherlands: J. Waltman, p. 175.

De Glee, G. J., 1951. Berekeningsmethoden voor de winning van grondwater. *Drinkwatervoorziening, 3e Vacantie cursus.* The Hague, The Netherlands. Moorman's periodieke pers, pp. 38–80.

De Josselin de Jong, G., 1958. Longitudinal and transverse diffusion in granular deposits. *Am. Geophys. Union Trans.* 39, 67–74.

Desaulniers, D., J. A. Cherry, and P. Fritz, 1981. Origin, age and movement of pore water in argillaceous quaternary deposits at four sites in southwestern Ontario. *J. of Hydrol.* 150, 231–257.

Domenico, P. A., and G. A. Robbins, 1985. A new method of contaminant plume and analysis. *Groundwater* 23(4), 476–485.

Driscoll, F. G., 1986. *Groundwater and Wells.* 2d ed., Johnson Division, St. Paul, Minnesota, p. 1089.

Duguid, J. O., and M. Reeves, 1976. Material transport in porous media: A finite element Galerkin model. Report ORNL-4928, Oak Ridge Natl. Lab., Oak Ridge, TN.

Dupuit, J. 1863. Études théoriques et pratiques sur le mouvement des eaux dans les canaux découverts et à travers les terrains perméables, 2d ed. Paris: Dunot, p. 304.

Evans, D. D., and T. J. Nicholson, 1987. *Flow and Transport Through Unsaturated Fractured Rock*, eds. D. D. Evans and T. J. Nicholson. Geophysical Monograph 42. Washington, DC: American Geophysical Union.

Faust, C. R., 1985. Transport of immiscible fluids within and below the unsaturated zone: A numerical model. *Water Resources Res.* 21, 587–596.

Faust, C. R., J. H. Guswa, and J. W. Mercer, 1989. Simulation of three-dimensional flow of immiscible fluids within and below the unsaturated zone: *Water Resources Res.* 25(12), 2449–2464.

Faust, C. R., P. N. Sims, C. P. Spalding, P. F. Andersen, and D. E. Stephenson, 1990. FTWORK: A three-dimensional groundwater flow and solute transport code. Westinghouse Savannah River Company Report WSRC-RP- 89-1085, Aiken, SC.

Felmy, A. R., D. Girvin, and E. A. Jenne, 1983. MINTEQ: A computer program for calculating aqueous geochemical equilibria. Report, U.S. Environmental Protection Agency, Washington, DC.

Franke, O. L., T. E. Reilly, and G. D. Bennett, 1987. Definition of boundary and initial conditions in the analysis of saturated groundwater flow systems—An introduction. USGS Techniques of Water Resources Investigations, bk. 3, ch. B5. USGS, Reston, VA.

FAZ, 1994. *Frankfurter Allgemeine Zeitung*, June 1, 1994.

Franz, T., and N. Guiger, 1994. FLOWPATH version 4. Steady state two-dimensional horizontal aquifer simulation model. Waterloo Hydrogeologic Software, Waterloo, Ontario.

Freeze, R. A., and J. A. Cherry, 1979. *Groundwater.* Englewood Cliffs, NJ: Prentice-Hall, p. 604.

Freeze, R. A., J. Massmann, L. Smith, T. Sperling, and B. James, 1990. Hydrogeologic decision analysis: 1. A framework. *Groundwater* 28(5), 738–766.

Freyberg, D. L., 1986. A natural gradient experiment on solute transport in a sand aquifer: 2. Spatial movements and the advection and dispersion of nonreactive tracers. *Water Resources Res.* 22(13), p. 2031–2047.

Freyburg, D. L., 1988. An exercise in ground-water model calibration and prediction. *Groundwater* 26(3), 350–360.

Frind, E. O., and G. E. Hokkanen, 1987. Simulation of the borden plume using the alternating direction Galerkin technique. *Water Resources Res.* 23(5), 918–930.

Frind, E. O., and G. B. Montanga, 1985. The dual formulation of flow for contaminant transport modeling: 1. Review of theory and accuracy aspects. *Water Resources Res.* 21, 159–169.

Frind, E. O., G. B. Matanga, and J. A. Cherry, 1985. The dual formulation of flow for contaminant transport modeling: 2. The Borden aquifer. *Water Resources Res.* 21, p. 170.

Frind, E. O., and G. F. Pinder, 1982. The principal direction technique for solution of the advection-dispersion equation. *Proc. Tenth World Congress of the International Association for Mathematics and Computers in Simulation* (IMACS), Concordia University, Montreal.

Frind, E. O., E. A. Sudicky, and S. L. Schellenberg, 1987. Micro-scale modeling in the study of plume evolution in heterogenous media. *Stochastic Hydrol. Hydraul.* 1, p. 263–279.

Garven, G., and R. A. Freeze, 1984a. Theoretical analysis of the role of groundwater flow in the genesis of stratabound ore deposits: 1. Mathematical and numerical model. *Am. J. Sci.* 284(10), 1085–1124.

Garven, G., and R. A. Freeze, 1984b. Theoretical analysis of the role of groundwater flow in the genesis of stratabound ore deposits: 2. Quantitative results. *Am. J. Sci.* 284(10), 1125–1174.

Gelhar, L. W., and C. L. Axness, 1983. Three-dimensional stochastic analysis of macrodispersion in aquifers. *Water Resources Res.*, 19(1), 161–180.

Gelhar, L. W., C. Welty, and K. R. Rehfeldt, 1992. A critical review of data on field-scale dispersion in aquifers. *Water Resources Res.* 28(7), 1955–1974.

Germain, D. M., 1988. Contaminant migration in fractured porous media: Modeling and analysis of advective-diffusive interaction. Ph.D. thesis, Univ. of Waterloo, Ontario, p. 166.

Gomez-Hernandez, J. J., and S. M. Gorelick, 1989. Effective ground water parameter values: Influence of spatial variability of hydraulic conductivity, leakage and recharge. *Water Resources Res.* 25(3), 405–419.

Grove, D. B., 1977. The use of Galerkin finite element methods to solve mass transport equations. USGS Water Resource Investigations 77–49, p. 55. (Available as PB-277 532 from Natl. Tech. Inf. Serv., Springfield, VA).

Guiger, N., J. Molson, T. Franz, and E. Frind, 1994. FLOTRANS user guide version 2.2, two-dimensional steady-state flownet and advective-dispersive contaminant transport model. Waterloo Hydrogeologic Software/Waterloo Centre for Groundwater Research, Waterloo, Ontario.

Gupta, S. K., C. T. Kincaid, P. R. Meyer, C. A. Newbill, and C. R. Cole, 1982. A multidimensional finite element code for the analysis of coupled fluid, energy and solute transport (CFEST). Report PNL-4260. Pac. Northwest Lab., Richland, WA.

Guvanasen, V., J. A. K. Reid, and B. W. Nakka, 1985. Predictions of hydrogeological perturbations due to the construction of the underground research laboratory. Atomic Energy of Canada Limited Technical Record, TR-344 (5000, AECL, Chalk River, Ontario K0J1J0).

Hall, R. C., M. A. Jones, and C. S. Mawbey, 1990. A review of computer programs for modeling the effects of saline intrusion in groundwater flow systems. U.K. DOE Report DOE/RW/90.013.

Hansen, V. E., 1949. Evaluation of unconfined flow to multiple wells by membrane analysis. Thesis, Iowa State Univ., Ames.

Hantush, M. S., and C. E. Jacob, 1955. Non-steady radial flow in an infinite leaky aquifer. *Am. Geophys. Union Trans.* 36, 95–100.

Harleman, D. R. F., and R. R. Rumer, 1963. The analytical solution for the injection of a tracer slug in a plane. *Fluid Mechanics*, 16.

Hem, J. D., 1970. Study and interpretation of the chemical characteristics of natural water, 2d ed. USGS Water-Supply Paper 1473, p. 363.

Herbert, A. W., C. P. Jackson, and D. A. Lever, 1987. Coupled groundwater flow and solute transport with fluid density strongly dependent upon concentration. Harwell Report TP. 1207, Harwell Laboratory.

Herr, M., 1985. Grundlagen der hydraulischen Sanierung verunreinigter Porengrundwasserleiter. Mitteilungen des Institut für Wasserbau, Universität Stuttgart. No. 63, p. 174.

Hill, M. C., 1993. MODFLOWP. U.S. Geological Survey code documentation.

Hornbeck, R. W., 1975. *Numerical Methods.* Englewood Cliffs, NJ: Prentice-Hall, p. 310.

Hostetler, C. J., and R. L. Erikson, 1989. FASTCHEM Package, vol. 5. User's guide to the EICM coupled geohydrochemical transport code. Report EA-5870-CCM, Elec. Power Res. Inst., Palo Alto, CA.

Huyakorn, P.S., J. W. Mercer, and P. F. Andersen, 1984. SWICHA: A three-dimensional finite-element code for analyzing seawater intrusion in coastal aquifers. GeoTrans code documentation.

Intera Environmental Consultants, 1983a. PHREEQE: A geochemical speciation and mass transfer code suitable for nuclear waste performance assessment. Report ONWI-435. Prepared for Off. of Nucl. Waste Isol., Battelle Mem. Inst., Columbus, OH.

Intera Environmental Consultants, 1983b. SWENT: A three-dimensional finite-different code for the simulation of fluid, energy, and solute radionuclide transport. Report ONWI-457. Prepared for Off. of Nucl. Waste Isol., Battelle Mem, Inst., Columbus, OH.

Intera Environmental Consultants, 1983d. Geochemical models suitable for performance assessment of nuclear waste storage: Comparison of PHREEQE and EQ3/EQ6. Report ONWI-473. Prepared for Off. of Nucl. Waste Isol., Battelle Mem. Inst., Columbus, OH.

Isherwood, D. J., 1984. Application of the ruthenium and technetium thermodynamic data bases used in the EQ3/6 geochemical codes. Report UCRL-53594. Laurence Livermore Natl. Lab., Livermore, CA.

Ito, K., 1951. *On Stochastic Differential Equations.* New York: Am. Math. Soc.

Jacob, C. E., 1950. Flow of groundwater. *Engineering Hydraulics*, ed. H. Rouse. New York: Wiley, pp. 321–386.

Javandel, I., C. Doughty, and C. Tsang, 1984. *Groundwater Transport: Handbook of Mathematical Models.* Water Resources Monograph Series 10. Washington, DC: American Geophysical Union, p. 288.

Karickhoff, S. W., D. S. Brown, and T. A. Scott, 1979. Sorption of hydrophobic pollutants on natural sediments. *Water Res.*, 13, 241–248.

Kauch, E. P., 1982. Zur Situierung von Brunnen im Grundwasserstrom. Österreichische Wasserwirtschaft. No. 7/8.

Kerrisk, J. 1984. Reaction-path calculations of groundwater chemistry and mineral formation of Rainier mesa, Nevada. Report LANL-9912-MS. Los Alamos Natl. Lab., Los Alamos, NM.

Kincaid, C. T., and J. R. Morrey, 1984. Geochemical models for solute migration: Vol. 2. Preliminary evaluation of selected computer codes. Report EA-3417. Elec. Power Res. Inst., Palo Alto, CA.

Kincaid, C. T., J. R. Morrey, and J. E. Rogers. 1984. Geohydrochemical models for solute migration: 1. Process description and computer code selection. Report EA-3417, Power Res. Inst., Palo Alto, CA.

Kinzelbach, W., 1985. Numerische Modellierung des Transports von Schadstoffen im Grundwasser. Report of the Institut für Wasserbau. Universität Stuttgart.

Kinzelbach, W., and P. Ackerer, 1986. Modélisation du transport de d'écoulement nonpermanent. *Hydrogéologie* 2, 192–206.

Kinzelbach, W., 1986. *Groundwater Modeling: An Introduction with Sample Programs in BASIC.* New York: Elsevier Science.

Kipp, K. L., Jr., 1986. Adaptation of the Carter-Tracy water influx calculation to groundwater flow simulation: *Water Resources Res.* 22 (3), 423–428.

Kipp, K. L. Jr., 1987. HST3D; A computer code for simulation of heat and solute transport in three dimensional groundwater systems. USGS Water Resources Investigations, 86-4095.

Konikow, L. F., and J. D. Bredehoeft, 1978. Computer model of two-dimensional solute transport and dispersion in groundwater. *USGS Techniques of Water-Resources Investigations*, bk. 7, ch. C2. Washington, DC: GPO, p. 90.

Konikow, L. F., 1986. Predictive accuracy of a groundwater model—Lessons from a post-audit. *Groundwater* 24(2), 677–690.

Krupka, K. M., and E. A. Jenne, 1982. WATEQ3 geochemical model: Thermodynamic data for several additional solids. Report PNL-4276, p. 58. Pac. Northwest Lab., Richland, WA.

Krupka, K. M., and J. K. Morrey, 1985. MINTEQ geochemical reaction code: Status and applications. Proc. Conference on the Application of Geochemical Models to High-Level Nuclear Waste Repository Assessment, Oak Ridge, TN, October 2–5, 1984, eds. G. K. Jacobs and S. K. Whatley. Report NUREG/CP-0062 and ORNL/TM-9585, pp. 46–53. U.S. Nucl. Regul. Comm., Washington, DC.

Kruseman, G. P., and N. A. de Ridder, 1970. Analysis and evaluation of pumping test data. *Intern. Inst. Land Reclamation and Improvement Bull.* (Wageningen, The Netherlands), (11), 200.

Kueper, B. H., and E. O. Frind, 1991a. Two-phase flow in heterogenous porous media: 1. Model development. *Water Resources Res.* 27(7), 1049–1057.

Kueper, B. H., and E. O. Frind, 1991b. Two-phase flow in heterogenous porous media, 2. Model application. *Water Resources Res.* 27(6), 1059–1070.

Lee, D. R., and J. A. Cherry, 1978. A field exercise on groundwater flow using seepage meters and mini-piezometers. *J. Geolog. Educ.* 27, 6–10.

Lehr, J. H., 1990. The scientific process, part II: Can we learn it? Editorial. *Groundwater*, 28(6), 850–855.

Leppert, S. C., and T. O. Bengsston, 1990. Comparison of density-coupled transport codes for a saltwater intrusion study in Coastal West-Central Florida. *EOS* 71(71).

List, E. J., 1965. The stability and mixing in a density-stratified horizontal flow in a saturated porous medium. W. Keck Laboratory of Hydraulics and Water Resources Dept. kH-R-11, Calif. Inst. of Tech., Pasadena.

Loague, K., and R. E. Green, 1991. Statistical and graphical methods for evaluating solute transport models: overview and application, *J. Contam. Hydrol.* 7, 51–73.

Mackay, D. M., P. V. Roberts, and J. A. Cherry, 1985. Transport of organic contaminants in groundwater. *Environ. Sci. Technol.* 19(5), 15–23.

MacQuarrie, K. T. B., and E. A. Sudicky, 1990. Simulation of biodegradable organic contaminants in groundwater: 2. Plume behaviour in uniform and random flow fields. *Water Resources Res.*, 26 (2), 223–239.

Mabey, W. R., and T. Mill, 1978. Critical review of hydrolysis of organic compounds in water under environmental conditions. *J. Phys. Chem. Ref. Data* 7, 383.

Mambert, W. A., 1985. *Effective Presentation.* 2d ed. New York: Wiley.

Mangold, D. C., and C. F. Tsang, 1983. A study of nonisothermal chemical transport in geothermal systems by a three-dimensional coupled thermal and hydrologic parcel model. *Trans. Geotherm. Resour. Counc.* 7, 455–459.

Mangold, D. C., and C. F. Tsang, 1991. A summary of subsurface hydrological and hydrochemcial models. *Rev. of Geophysics* 29(1), 51–79.

Marlon-Lambert, J., 1978. Computer programs for groundwater flow and solute transport analysis. Report N25090, Golder Assoc., Vancouver, BC.

Massmann, J., and R. A. Freeze, 1989. Updating random hydraulic conductivity fields: A two-step procedure. *Water Resources Res.* 25(7), 1763–1765.

Massmann, J., R. A. Freeze, L. Smith, T. Sperling, and B. James, 1991. Hydrogeologic decision analysis: 2. Applications to groundwater contamination. *Groundwater* 29(4), 536–548.

McDonald, G., and A. W. Harbaugh, 1989. A modular three-dimensional finite difference groundwater flow model. *USGS Techniques of Water Resource Investigations*, bk. 6, ch. A1. Washington, DC:GPO.

Mehlhorn, H., K. Spitz, and H. Kobus, 1981. Kurzschlußströmung zwischen Schluck- und Entnahmebrunnen—Kritischer Abstand und Rückströmrate. *Wasser und Boden* 4.

Mehlhorn, H., and D. Flinsbach, 1983. Der Großpumpversuch des Zweckverbandes Landesversorgung im Erolzheimer Feld-Illertal. *Wasserwirtschaft* 73(12).

Mercer, J. W., and R. M. Cohen, 1990. A review of immiscible fluids in the subsurface: Properties, models, characterization and remediation. *J. Contam. Hydrol.* 6, 107–163.

Melchior, D. C., and R. L. Bassett, (eds.), 1990. *Chemical Modeling of Aqueous Systems II.* Washington, DC: American Chemical Society.

Molson, J., 1988. Three-dimensional numerical simulation of groundwater flow and contaminant transport at the Borden Landfill. M.Sc. thesis. Univ. of Waterloo, Ontario.

Montgomery. J. H., 1991. *Groundwater Chemicals Desk Reference*, vol. 2. Chelsea, MI: Lewis Publishers.

Montgomery, J. H., and L. M. Welkom, 1989. *Groundwater Chemicals Desk Reference*, vol. 1. Chelsea, MI: Lewis Publishers.

Moreno, J. L., 1989. Three-dimensional simulation of the migration and cleanup of trichloroethylene. Proc. Fourth International Conference on Solving Groundwater Problems with Models, Indianapolis, IN.

Moreno, J. L., and P.O. Sinton, 1994. Are density effects of dissolved contaminants important? Proc. Ground Water Modeling Conference, Colorado State Univ., Fort Collins.

Morrey, J. R., C. R. Kincaid, and C. J. Hostetler, 1986. Geohydrochemical models for solute migration, vol. 3. Evaluation of selected computer codes. Report EA-3417. Elec. Power Res. Inst. Palo Alto, CA.

Munson, L. S., 1984. *How to Conduct Training Seminars.* New York: McGraw-Hill.

Muskat, M., 1949. *Physical Principles of Oil Production.* New York: McGraw-Hill.

Naff, R., 1984. Definition of global dispersion coefficients. Bericht PSE–No. 83/4B. Projekt Sicherheitstudien Entsorgung. TU Berlin. Hahn-Meitner-Institut.

Narasimhan, T.N., and M. Alavi, 1986. A technique for handling tensorial quantities in the integral finite difference method. Earth Sciences Division Annual Report 1985, pp. 111–113. Lawrence Berkeley Lab., Berkeley, CA.

Narasimhan, T. N., A. F. White, and T. Tokunaga, 1985. Hydrology and geochemistry of the uranium mill tailings pile at Riverton, Wyoming: II. History matching. Report LBL-18526. Lawrence Berkeley Lab., Berkeley, CA.

Narasimhan, T. N., A. F. White, and T. Tokunaga, 1986. Groundwater contamination from an inactive uranium mill tailings pile: 2. Application of a dynamic mixing model. *Water Resources Res.* 22, 1820–1834.

National Research Council, 1990. *Ground Water Models: Scientific and Regulatory Applications.* Washington, DC: National Academy Press.

Nea/Ski, 1988. The International HYDROCOIN Project: Level 1 code verification, organization for economic cooperation and development. Paris: France.

Niederer, U., 1988. Perception of safety in waste disposal: The review of the Swiss project GEWAHR 1985. Proc. GEOVAL 1987 Symposium in Stockholm, April 7–9, 1987, pp. 11–26. The Swedish Nuclear Power Inspectorate, Stockholm.

Nielsen, D. R., M. Th. Genuchten, and J. W. Biggar, 1986. Water flow and solute transport processes in the unsaturated zone. *Water Resources Res.* 22(9), 89S–108S.

Nofzinger, D. L., K. Rajender, S. K. Nagudu, and P. Su, 1989. CHEMFLO: One-dimensional water and chemical movement in saturated soils. EPA/600/8-89/076. Robert S. Kerr Environmental Research Laboratory, Ada, OK.

Nordstrom, D. K., L. N. Plummer, D. Langmuir, E. Busenberg, H. M. May, B. F. Jones, and D. L. Parkhurst, 1990. Revised chemical equilibrium data for major water-mineral reactions and their limitations. *Chemical Modeling of Aqueous Systems II,* eds. D. C. Melchior and R. L. Bassett. Washington, DC: American Chemical Society, pp. 398–413.

Ogata, A., and R. B. Banks, 1961. A solution of the differential equation of longitudinal dispersion in porous media. USGS, Professional Paper Nr. 411-A.

Parker, J. C., 1989. Multiphase flow and transport in porous media. *Rev. Geophys.* 27, 311–328.

Parkhurst, D. L., D. C. Thorstenson, and L. N. Plummer, 1980. PHREEQE—A computer program for geochemical calculations. Report USGS/WRI-80-96. USGS, Water Resources Division, Reston, VA.

Parkhurst, D. L., L. M. Plummer, and D. C. Thorstenson, 1982. BALANCE—A computer program for calculating mass transfer for geochemical reactions in groundwater. USGS, Water Resources Division, Reston, VA.

Paulousky, N. N., 1956. *Collected Works.* Leningrad: Akad, Nauk USSR.

Peterson, S.R., A. R. Felmy, R. J. Serne, and G. W. Gee, 1983. Predictive geochemical modeling of interactions between uranium mill tailings solutions and sediments in a flow-through system: Model formulations and preliminary results. Report NUREG/CR-4782 (PNL-4782). Pac. Northwest Lab., Richland, WA.

Pinder, G. F., and W. G. Gray, 1977. *Finite Element Simulation in Surface and Subsurface Hydrology.* New York: Academic Press, p. 295.

Pinder, G. F., 1990. *Princeton Transport Code Class.* Princeton, NJ: Princeton Univ.

Plummer, L. N., B. F. Jones, and A. H. Truesdell, 1976. WATEQF—A FORTRAN IV version of WATEQ, a computer program for calculating chemical equilibrium of natural waters. USGS, Reston, VA. Water Resources Investigations 76-13. (Revised January 1984).

Plummer, L. N., and D. L. Parkhurst, 1985. PHREEQE: Status and applications in Proceedings of the conference on the application of geochemical models to high-level nuclear waste repository assessment. Oak Ridge, TN, October 2–5, 1984, eds. G. K. Jacobs and S. K. Whatley. Reports NUREG/CP-0062 and ORNL/TM-9585, pp. 37–45, U.S. Nucl. Regul. Comm., Washington, DC.

Plummer, L. N. and D. L. Parkhurst, 1990. Application of the Pitzer equations to the PHREEQE geochemical model. *Chemical Modeling of Aqueous Systems II*, eds., D. C. Melchior and R. L. Bassett, Washington, DC: American Chemical Society, pp. 128–137.

Plummer, L. N., D. L. Parkhurst, and D. C. Thorstenson, 1983. Development of reaction models for groundwater systems. *Geochim. Cosmochim. Acta* 47, 665–686.

Pollock, D. W., 1989. Documentation of computer programs to compute and display pathlines using results from the U.S. geological survey modular three-dimensional finite-difference groundwater flow model. USGS Open File Report 89-381.

Polubarinova-Kochina, P. Ya., 1952. *Theory of Groundwater Movement* (in Russian). Moscow: Gostekhizdat. English transl. R. J. M. de Wiest. Princeton, NJ: Princeton Univ. Press, 1962.

Prickett, T. A., and G. Lonnquist, 1971. Selected digital computer techniques for groundwater resource evaluation. Illinois State Water Survey, Bull. 55, Dept. of Registration and Education, p. 62.

Prickett, T. A., T. G. Naymik, and C. G. Lonnquist, 1981. A "random walk" solute transport model for selected groundwater quality evaluations. Illinois State Water Survey, Bull. 65, Dept. of Registration and Education, p. 103.

Pruess, K., 1983. Development of the general purpose simulator MULKOM. Earth Sciences Division Annual Report 1982, Report LBL-15500, pp. 133–134. Lawrence Berkeley Lab., Berkeley, CA.

Pruess, K., 1986. TOUGH User's Guide, Rep. NUREG/CR-4645 and SAND 86-7104. Prepared for Div. of Waste Manage., Off. of Nucl. Mater. Safety and Safeguards, U.S. Nucl., Regul. Comm., Washington, DC. (Also available as Report LBL-20700, Lawrence Berkeley Lab., Berkeley, CA.)

Pruess, K., and T. N. Narasimhan, 1985. A practical method for modeling fluid and heat flow in fractured porous media. *SPEJ.*, 41(2), 14–21.

Pruess, K., and J. S. Y. Wang, 1984. TOUGH—A numerical model for nonisothermal unsaturated flow to study waste canister heating effects. Scientific Basis for Nuclear Waste Management VII, Mater. Res. Soc. Symp. Proc., 26, 1031–1038.

Pruess, K., Y. W. Tsang, and J. S. Y. Wang, 1985. Modeling of strongly heat driven flow in partially saturated fractured porous media. Proc., IAH 17th International Congress on the Hydrogeology of Rocks of Low Permeability, Tucson: Univ. of Arizona Press, p. 486–497.

Pruess, K., and J. S. Y. Wang, 1987. Numerical modeling of isothermal and nonisothermal flow in unsaturated fractured rock—A review. In *Flow and Transport through Unsaturated Fractured Rock.* Washington, DC: American Geophysical Union.

Reeves, M., and R. M. Cranwell, 1981. User's manual for the Sandia waste-isolation flow and transport model. Report SAND81-2516 and NUREG/CR-2324, Sandia Natl. Lab., Albuquerque, NM.

Reeves, M., D. S. Ward, P. A. Davis, and E. J. Bonano, 1986. Theory and implementation for SWIFT II. Release 4.84, Report NUREG/CR-3328 and SAND83-1159, Sandia Natl. Lab., Albuquerque, NM.

Remson, I., G. M. Hornberger, and F. J. Malz, 1971. *Numerical Methods in Subsurface Hydrology.* New York: Wiley-Interscience.

Reid, J. A. K., and T. Chan, 1987. Sensitivity of Whiteshell geosphere modeling results to dimensionality of the computer simulations employed. Proc. Conference on Geostatistical Sensitivity and Uncertainty Methods for Groundwater Flow and Radionuclide Transport Modeling, San Francisco, CA, September 15–17.

Rifau, H. S., P. B. Bedient, R. C. Borden, and J. F. Haasbeek, 1989. Bioplume II—Computer model of two-dimensional contaminant transport under the influence of oxygen limited biodegradation in groundwater (User's Manual Version 1.0, Preprocessor Service Code Version 1.0, Source Code Version 1.0) EPA/600/8-88/093, NTIS PB 89-151 120/AS.

Roberts, P. V., M. N. Goltz, and D. A. MacKay, 1986. A natural gradient experiment on solute transport in a sand aquifer: 3. Retardation estimates and mass balances for organic solutes. *Water Resources Res.* 22(13), 2047–2059.

Robinson, P. C., C. P. Jackson, A. W. Herbert, and R. Atkinson, 1986. Review of the groundwater flow modeling of the Swiss Project Gewahr. Harwell Report AERE-R.11929, HMSO, London.

Rulon, J., G. S. Bodvarsson, and P. Montazer, 1986. Preliminary numerical simulations of groundwater flow in the unsaturated zone. Yucca Mountain, NV. Report LBL-20553, Lawrence Berkeley Lab., Berkeley, CA.

Runchal, A. K., 1985. PORFLOW: A general purpose model for fluid flow, heat transfer and mass transport in anisotropic, inhomogeneous, equivalent porous media. Tech. Note TN-011, Anal. and Comput. Res., Inc., Los Angeles, CA.

Runchal, A. K., 1987. Theory and application of the PORFLOW model for analysis of coupled flow, heat and radionuclide transport in porous media. *Coupled Processes Associated with Nuclear Waste Repositories*, ed. C. F. Tsang, Berkeley, CA: Lawrence Berkeley Laboratory, pp. 495–516.

Runchal, A. K., J. Treger, and G. Segal, 1979. Program EP21 (GWTHERM): Two-dimensional fluid flow, heat and mass transport in porous media. Tech. Note TN-LA-34, Adv. Technol. Group, Dames & Moore, Los Angeles, CA.

Saffman, P. G., 1960. Dispersion due to molecular diffusion and macroscopic mixing in flow through a network of capillaries. *Fluid Mechanics*, 2(7), 194–208.

Sanford, W. E., and L. F. Konikow, 1989. A two-constituent solute-transport model for groundwater having variable density. Holcomb Research Institute. TNO-QGV Institute. Golden, CO: IGWMC-FOS36.

Sauty, J., 1978. Identification des parameters du transport hydrodispersif dans les aquiferes par interpretation de tracages en ecoulement cyclindrique convergent ou divergent. *J. of Hydrol.* 39(3/4), 69–103.

Sauty, J., 1980. An analysis of hydrodispersive transfer in aquifers. *Water Resources Res.* 24(1), 145–158.

Savinskii, I. D., 1965. *Probability Tables for Locating Elliptical Underground Masses with Rectilinear Grid.* Consultants Bureau. New York: Plenum Press, p. 100.

Sayre, W. W., 1968. Dispersion of mass in open channel flow. Hydraulics Paper, Nr. 3, 73, Colorado State Univ.

Sayre, W. W., 1973. Natural mixing processes in rivers. *Environmental Impact on Rivers*, ed. H. W. Shen. Fort Collins, CO, ch. 6.

Scheiddegger, A. E., 1954. Statistical hydrodynamics in porous media. *J. Appl. Phys.* 25(8), 994–1001.

Schroeder, P. R., C. M. Lloyd and P. A. Zappi, 1994. Users guide for HELP Version 3 for experienced users. EPA/600/R-94/168a. Washington, DC: U.S. Environmental Protection Agency, Office of Research and Development.

Schwartz, F. W., 1977. Macroscopic dispersion in porous media: The controlling factors. *Water Resources Res.*, 13(4), 743–752.

Schwarzenbach, R. P., and J. Westfall, 1981. Transport of non polar organic compounds from surface water to groundwater. *Env. Sci. and Techn.* 15(11).

Schwille, F., 1988. Dense chlorinated solvents in porous and fractured media: Model experiments. Transl. from German by J. F. Pankow. Chelsea, MI: Lewis Publishers, p. 146.

Segol, G. A., 1976. A three-dimensional Galerkin finite element model for the analysis of contaminant transport in variably saturated porous media: User's guide. Report, Dept. of Earth Sci., Univ. of Waterloo, Ontario.

Sharma, D., 1981. Some applications of a novel computational procedure for solving the Richards equation. Proc., AGU Fall Meeting, San Francisco, CA.

Sharma, D., M. I. Asgian, W. R. Highland, and J. L. Moreno, 1983. Analysis of complex seepage problems with the disposal of uranium tailings: Selected case studies. *Mineral and Energy Resources* (Colorado School of Mines), 26(1).

Smith, L. T., T. Clemo, and M. D. Robertson, 1990. New approaches to the simulation of field-scale solute transport in fractured rocks. Paper presented at 5th Canadian/American Conference on Hydrogeology, National Groundwater Assoc., Calgary, Canada.

Spitz, K., 1985. Dispersion in porösen Medien: Einfluß von Inhomogenitäten und Dichteunterschieden. Report of the Institut für Wasserbau no. 60, p. 131, Universität Stuttgart.

Spitz, K., 1989. Groundwater flow modeling—Course manual for the Directorate of Environmental Geology, Indonesia (unpublished).

Strack, O., 1989. *Groundwater Mechanics*. Englewood Cliffs, NJ: Prentice Hall, p. 732.

Sudicky, E. A., J. A. Cherry, and E. O. Frind, 1983. Migration of contaminants in groundwater at a landfill: A case study, 4, a natural-gradient dispersion test. *J. Hydrol.* 63, 81–08.

Sudicky, E. A., 1986. A natural gradient experiment on solute transport in a sand aquifer: Spatial variability of hydraulic conductivity and its role in the dispersion process. *Water Resources Res.* 22, p. 2069–2082.

Sykes, J. F., R. B. Lantz, S. B. Pahwa, and D. S. Ward, 1982a. Numerical simulation of thermal energy storage experiment conducted by Auburn University. *Groundwater* 20, 569–576.

Sykes, J. F., S. B. Pahwa, R. B. Lantz, and D. S. Ward, 1982b. Numerical simulation of flow and contaminant migration at an extensively monitored landfill. *Water Resources Res.* 18, 1687–1704.

Taylor, G., 1954. Conditions under which dispersion of a solute in a stream of solvent can be used to measure molecular diffusion. *Proc. Roy. Soc.* Ser. A 225, 473–477.

Theis, C. V., 1935. The relation between the lowering of the piezometric surface and the rate and duration of discharge of a well using groundwater storage. *Am. Geophys. Union Trans.* 16, 519–524.

Thiem, G. 1906. *Hydrologische Methoden.* Leipzig: Gebhardt, p. 56.

Thomas, S. D., B. Ross, and J. W. Mercer, 1982. A summary of repository siting models. Report NUREG/CR-2782, U.S. Nucl. Regul. Comm., Washington, DC.

Thomas, T. R., 1988. Modeling hypothetical groundwater transport of nitrates, chromium, and cadmium at the Idaho chemical processing plant. Report WINCO-1060, Westinghouse Idaho Nucl. Co., Idaho Falls.

Travis, B. J., 1984. TRACR3D: A model of flow and transport in porous/fractured media. Report LA-9667-MS, Los Alamos Natl. Lab., Los Alamos, NM.

Travis, B. J., and H. E. Nuttall, 1987. Analysis of colloid transport. *Coupled Processes Associated with Nuclear Waste Repositories*, ed. C. F. Tsang. San Diego: Academic Press, pp. 517–528.

Tsang, C. F., and C. Doughty, 1985. Detailed validation of a liquid and heat flow code against field performance. Proc. Eighth SPE Symposium on Reservoir Simulation, Dallas, February 10–13, 1985. Report SPE-13503, Soc. of Pet. Eng., Richardson, TX. 1985. (Also available as Rep. LBL-18833, Lawrence Berkeley Laboratory, Berkeley, CA).

Uffink, G. J. M., 1985. Macrodispersie in gelaagde pakketten—Deel 1: Een rekenmodel. Laboratorium voor Bodem- en Grondwateronderzoek, RIVM, Leidschendam, The Netherlands.

U.S. Department of the Interior, Bureau of Land Management and Bureau of Indian Affairs, 1986. Final Jackpile-Paguate Uranium Mine Reclamation Project, Environmental Impact Statement, vol. 1, BLM-NM-ES-86-018-4134.

USEPA, 1989. Drinking water regulations and health advisories. Washington, DC: U.S. Environmental Protection Agency.

Van den Berg, R., and J. M. Roels, 1991. Beurteilung der Gefährdung des Menschen und der Umwelt durch Exposition gegenüber Bodenverunreinigungen— Integration der Teilaspekte. Leitraad Bodemsanering (revised draft). Reichsinstitut fur Volksgesundheit und Umwelthygiene (Bilthoven): Bericht Nr. 725201007.

van der Heijde, P. K. M., 1987. Quality assurance in computer simulations of groundwater contamination. *Environmental Software* 2(1) 19–28.

van der Heijde, P. K. M., and R. A. Park, 1986. U.S. EPA Groundwater Modeling Policy Study Group. Report of findings and discussion of selected groundwater modeling issues. International Ground Water Modeling Center, Holcomb Res. Inst., Butler Univ., Indianapolis.

Vassilios, Kaleris, 1986. Erfassung des Austausches von Oberflächen-und Grundwasser in horizontalebenen Grundwassermodellen. Report of the Institut für Wasserbau, no. 62, p. 137, Universität Stuttgart.

Verschueren, K., 1983. *Handbook of Environmentai Data on Organic Chemicals*, 2d ed., New York: Van Nostrand Reinhold.

Voss, C. I., 1984. Saturated-unsaturated transport (SUTRA), USGS Water Resource Investigations 84-4369.

Walton, W. C., 1970. *Groundwater Resource Evaluation*. New York: McGraw-Hill, p. 664.

Ward, D. S., M. Reeves, and L. E. Duda, 1984. Verification and field comparison of the Sandia waste-isolation flow and transport model (SWIFT), Report SAND83-1154 and NUREG/CR-3316. Sandia Natl. Lab., Albuquerque, NM.

Ward, D. S., D. R. Buss, J. W. Mercer, and S. S. Hughes, 1987a. Evaluation of a groundwater corrective action at the Chem-Dyne hazardous waste site using a telescopic mesh refinement modeling approach. *Water Resources Res.* 23, 603–617.

Ward, D. S., D. R. Buss, D. W. Morganwalp, and T. D. Wadsworth, 1987b. Waste confinement performance of deep injection wells. Paper presented at Solving Ground Water Problems with Models, Natl. Water Well Assoc., February 10–12, Denver, CO.

Wolery, T. J., 1979. Calculation of chemical equilibrium between aqueous solution and minerals: The EQ3/6 software package. Report UCRL-52658, Lawrence Livermore Natl. Lab., Livermore, CA.

Yeh, G. T., and D. S. Ward, 1979. FEMWATER: A finite-element model of water flow through saturated-unsaturated porous media. Report ORNL-5567. Oak Ridge Natl. Lab. Oak Ridge, TN.

Yeh, G. T., and D. S. Ward, 1981. FEMWASTE: A finite-element model of waste transport through saturated-unsaturated porous media. Report ORNL-5601, Oak Ridge Natl. Lab., Oak Ridge, TN.

Zheng, C., 1992. MT3D: A modular three-dimensional transport model, version 1.5. S.S. Papadopoulos & Associates, Inc., Bethesda, MD. Code Documentation.

INDEX

A page number shown in *italic* indicates an illustration.

www.ingramcontent.com/pod-product-compliance
Lightning Source LLC
LaVergne TN
LVHW011900200926
848484LV00002B/490